Lecture Notes in Mathematics 2004

Editors:
J.-M. Morel, Cachan
F. Takens, Groningen
B. Teissier, Paris

Kai Diethelm

The Analysis of Fractional Differential Equations

An Application-Oriented Exposition Using Differential Operators of Caputo Type

 Springer

Kai Diethelm
GNS Gesellschaft
für Numerische Simulation mbH
Am Gaußberg 2
38114 Braunschweig
Germany
diethelm@gns-mbh.com

ISBN: 978-3-642-14573-5   e-ISBN: 978-3-642-14574-2
DOI: 10.1007/978-3-642-14574-2
Springer Heidelberg Dordrecht London New York

Lecture Notes in Mathematics ISSN print edition: 0075-8434
                          ISSN electronic edition: 1617-9692

Library of Congress Control Number: 2010933969

Mathematics Subject Classification (2010): 34A08, 34A12, 34-02, 34-01, 26A33, 33E12

Cover design: SPi Publisher Services

Printed on acid-free paper

springer.com

Preface

> *There is a universe of mathematics lying in between*
> *the complete differentiations and integrations.*

— O. Heaviside

This book is devoted to some questions in Fractional Calculus, that is, the theory of differential and integral operators of non-integer order, and in particular to differential equations containing such operators. Even though the first steps of the theory itself date back to the first half of the nineteenth century, the subject only really came to life over the last few decades. A particular feature is that engineers and scientists have developed new models that involve fractional differential equations. These models have been applied successfully, e.g., in mechanics (theory of viscoelasticity and viscoplasticity), (bio-)chemistry (modelling of polymers and proteins), electrical engineering (transmission of ultrasound waves), medicine (modelling of human tissue under mechanical loads), etc. The mathematical theory seems to be lagging behind the needs of those applications but the wealth of applications indeed indicates the truth of the above quote from Heaviside [93, §437]. There are some books dealing with the aspects that can be summarized as the "pure mathematical" side of the problems without taking into consideration those questions that arise in the applications mentioned above, and some that the engineer's point of view without a rigorous mathematical justification of the ideas. This book attempts to fill the gap between these two approaches: We try to establish a mathematically sound theory of the differential equations that have been shown to be relevant in practice and provide a thorough mathematical analysis. In order to be self-contained, we repeat the fundamentals of fractional calculus before coming to the main topic. A particular goal of this book is to provide a solid foundation that may later be used for the construction of efficient and reliable numerical methods for fractional differential equations. The author strongly believes that a successful development and a thorough understanding of such numerical schemes is not possible without such a stable analytical background.

The reader is assumed to be familiar with classical calculus (differential and integral calculus and the elementary theory of differential equations). A working knowledge of Lebesgue integration theory is helpful now and then, but not absolutely essential.

It is my pleasure to thank a number of people for their constant support, interest, encouragement, and many useful discussions, namely Heinz-Wilhelm Alten and Klaus-Jürgen Förster (Universität Hildesheim), Helmut Braß and Marc Weilbeer (Technische Universität Braunschweig), Thomas Hennecke (Universität Kassel) Neville J. Ford (University of Chester), Alan D. Freed (formerly at NASA's John H. Glenn Research Center in Cleveland, now at Saginaw Valley State University), Paul L. Butzer (RWTH Aachen), Rudolf Gorenflo (Freie Universität Berlin), Francesco Mainardi (Universitá di Bologna), and André Schmidt (Universität Stuttgart). Moreover I would like to thank Mrs Ute McCrory at Springer-Verlag for her support during the manuscript preparation process.

Parts of the book have been used as a text for a graduate course on fractional differential equations that I taught to students of mathematics, physics and engineering at Technische Universität Braunschweig.

Braunschweig, June 2010   *Kai Diethelm*

Contents

Part I   Fundamentals of Fractional Calculus

1   Introduction 3
   1.1   Motivation 3
   1.2   The Basic Idea 7
   1.3   An Example Application of Fractional Calculus...................... 10
   Exercises... 12

2   Riemann–Liouville Differential and Integral Operators 13
   2.1   Riemann–Liouville Integrals........................... 13
   2.2   Riemann–Liouville Derivatives........................... 26
   2.3   Relations Between Riemann–Liouville Integrals and Derivatives 39
   2.4   Grünwald–Letnikov Operators........................... 41
   Exercises... 46

3   Caputo's Approach 49
   3.1   Definition and Basic Properties 49
   3.2   Nonclassical Representations of Caputo Operators.................... 60
   Exercises... 65

4   Mittag-Leffler Functions 67
   Exercises... 72

Part II   Theory of Fractional Differential Equations

**5   Existence and Uniqueness Results
for Riemann–Liouville Fractional Differential Equations**................ 77
   Exercises... 82

**6   Single-Term Caputo Fractional Differential Equations:
Basic Theory and Fundamental Results**................................. 85
   6.1   Existence of Solutions................................... 85
   6.2   Uniqueness of Solutions 93
   6.3   Influence of Perturbed Data...............................109

6.4   Smoothness of the Solutions 116
6.5   Boundary Value Problems 127
Exercises 132

**7   Single-Term Caputo Fractional Differential Equations:
       Advanced Results for Special Cases** 133
7.1   Initial Value Problems for Linear Equations 133
7.2   Boundary Value Problems for Linear Equations 154
7.3   Stability of Fractional Differential Equations 157
7.4   Singular Equations 163
Exercises 166

8   Multi-Term Caputo Fractional Differential Equations 167
Exercises 185

Appendix

A   List of Symbols 189

B   A Table of Caputo Derivatives 193

C   Numerical Solution of Fractional Differential Equations 195
C.1   An Algorithm for Single-Term Equations 195
C.2   Numerical Schemes for Multi-Term Equations 211
Exercise 225

D   Useful Results from Analysis 227
D.1   Euler's Gamma Function 227
D.2   Fixed Point Theorems 229
D.3   The Laplace Transform 230
D.4   Hadamard's Finite-Part Integral 233
D.5   Approximation Theory 234
Exercises 235

References 237

Index 245

Part I
Fundamentals of Fractional Calculus

Chapter 1
Introduction

1.1 Motivation

This book is about problems arising in the area of *fractional calculus* – a branch
of mathematics that is, in a certain sense, as old as classical calculus as we know it
today: The origins can be traced back [162] to the end of the seventeenth century,
the time when Newton and Leibniz developed the foundations of differential and
integral calculus. In particular, Leibniz introduced the symbol

$$\frac{d^n}{dx^n} f(x)$$

to denote the nth derivative of a function f. When he reported this in a letter to
de l'Hospital (apparently with the implicit assumption that $n \in \mathbb{N}$), de l'Hospital
replied: "What does $\frac{d^n}{dx^n} f(x)$ mean if $n = 1/2$?" This letter from de l'Hospital, written in 1695, is nowadays commonly accepted as the first occurrence of what we
today call a *fractional derivative*, and the fact that de l'Hospital specifically asked
for $n = 1/2$, i.e. a *fraction* (rational number), actually gave rise to the name of this
part of mathematics. This name has remained in use ever since, even though it is
well known by now that there is no reason to restrict n to the set of rational numbers.
Indeed, as we shall see in this book, any real number – rational or irrational – will
do just as well, at least for the analytical considerations that we shall concentrate
on. (Certain, but not all, numerical methods for the solution of some types of differential equations may encounter problems when arbitrary real numbers are admitted;
cf. [39].) As a matter of fact, even complex numbers may be allowed, but this is well
beyond the scope of this book.

What is the scope of this book, then? Well, the writing of this book has essentially
been motivated by the enormous numbers of very interesting and novel applications
of fractional differential equations in physics, chemistry, engineering, finance, and
other sciences that have been developed in the last few decades. Some early examples are given in the book of Oldham and Spanier [146] (diffusion processes)
and the classical papers of Bagley and Torvik [184], Caputo [23], and Caputo and
Mainardi [24, 25] (these four papers dealing with the modelling of the mechanical

K. Diethelm, *The Analysis of Fractional Differential Equations*,
Lecture Notes in Mathematics 2004, DOI 10.1007/978-3-642-14574-2_1,
© Springer-Verlag Berlin Heidelberg 2010

properties of materials) as well as in the publications of Marks and Hall [130] (signal processing) and Olmstead and Handelsman [147] (also dealing with diffusion problems); more recent results are described, e.g., in the work of Benson [17] (advection and dispersion of solutes in natural porous or fractured media), Bai and Feng [12] and Cuesta and Finat Codes [31] (image processing), Chern [28], Diethelm and Freed [50, 51] and Diethelm, Freed and Luchko [69] (modelling of the behaviour of viscoelastic and viscoplastic materials under external influences), Dokoumetzidis et al. [57], Popović et al. [156] and Verotta [187] (pharmacokinetics), Freed and Diethelm [68] and Magin [124] (bioengineering), Gaul, Klein, and Kempfle [71] (description of mechanical systems subject to damping), Glöckle and Nonnenmacher [75] (relaxation and reaction kinetics of polymers), Gorenflo and Rutman [82] (so-called ultraslow processes), Gorenflo, Mainardi et al. [78, 83, 128, 169, 170] (connections to the theory of random walks, the latter two papers especially with respect to applications to mathematical models in finance), Joulin [99] and Roquejoffre et al. [8, 110] (modelling of combustion), Metzler et al. [134] (relaxation in filled polymer networks), Podlubny [152] and Caponetto et al. [22] (control theory; the latter publication also with details on the hardware implementation of fractional order controllers), Podlubny et al. [155] (heat propagation), and Shaw, Warby and Whiteman [176] (modelling of viscoelastic materials). A completely different and very novel application field is the area of mathematical psychology where fractional-order systems may be used to model the behaviour of human beings [5, 178]. Specifically, the way in which a person reacts to external influences depends on the experience he or she has made in the past. In other words, humans have memories, and we shall see later in this book (e.g., in Remark 6.4) that fractional operators are a very natural tool to model memory-dependent phenomena. Surveys or collections of such applications can also be found in Baleanu et al. [13], Gorenflo and Mainardi [81], Hilfer [94], Klages et al. [105], Le Mehauté et al. [111], Mainardi [125,126], Matignon and Montseny [133], Nonnenmacher and Metzler [141], Podlubny [153], Sabatier et al. [166], Taş et al. [180] and Uchaikin [186]. In addition there are some applications of fractional calculus within various fields of mathematics itself, e.g. in the analytical investigation of various types of special functions [104]. Finally we refer to the work of Woon [193] that essentially mentions mathematical applications that, in turn, have important implications in other sciences like physics.

It turned out that many of these applications gave rise to a type of equations that has not been covered in the standard mathematical literature. This is connected to the fact that, in a certain sense, the answer to de l'Hospital's question (which Leibniz was not able to find, except for the special case $f(x) = x$) is not unique. There are very many possible generalizations of $\frac{d^n}{dx^n} f(x)$ to the case $n \notin \mathbb{N}$. We shall only discuss two of them, the *Riemann–Liouville derivative* (cf. Chap. 2) and the *Caputo derivative* (Chap. 3). The former concept is historically the first (developed in works of Abel, Riemann and Liouville in the first half of the nineteenth century) and the one for which the mathematical theory has been established quite well by now, but it has certain features that lead to difficulties when applying it to "real-world" problems. As a consequence, the latter concept was developed. It is closely related to the Riemann–Liouville idea, but certain modifications were introduced in order to avoid the above-mentioned difficulties. The mathematical implications of these modifications have not been investigated fully so far. In this book we intend to give as much information on this topic as presently possible.

The structure of this book is arranged in the following way. We begin by recalling some classical facts from calculus that form the basis of our intended generalization and by looking at a simple example application of fractional calculus in mechanics. Then, in Chaps. 2 and 3, we introduce the fundamental concepts and definitions of fractional calculus. This includes, in particular, some basic results concerning Riemann–Liouville differential and integral operators. As mentioned above, the main goal of this book is to present a comprehensive overview over the properties of the operators of Caputo's type and over the theory of differential equations involving such operators, but nevertheless we have decided to include these statements on the Riemann–Liouville operators for a number of reasons:

- This decision allows us to provide a comparison of the two approaches in a self-contained way. Such a comparison should be useful both for the reader who is familiar with the classical theory of the Riemann–Liouville operators and wants to learn about Caputo's version, and for the novice in fractional calculus who can then decide whether he or she wants to move on to other works in order to gain a deeper understanding of Riemann–Liouville derivatives and integrals.
- Even though Caputo's version of the fractional calculus requires a modification of the Riemann–Liouville type *differential* operators of fractional order, the Riemann–Liouville *integral* operators of fractional order do not need to undergo any changes. They can be used in Caputo's fractional calculus in their original form.
- Finally it actually turns out that a number of the proofs of our results on Caputo operators can be given in a relatively simple way by using related properties of Riemann–Liouville operators and the precise knowledge of how the two types of operators are interrelated.

Next, in Chap. 4, we introduce a class of functions that is of fundamental importance in the theory of fractional differential equations, the Mittag-Leffler functions. We shall meet these functions again in various places in the later chapters of our text, and Chap. 4 will provide some basic knowledge about them that we will make use of then.

The core of the book is the second part that is devoted to the analytical study of fractional differential equations. Once again, for reasons similar to those stated above, we first take a brief look at the theory of such equations with Riemann–Liouville operators in the short Chap. 5 before turning our attention to equations with Caputo operators in Chaps. 6–8. We address questions of existence and uniqueness of solutions, and we investigate the properties of the solutions. A proper knowledge of these properties is not only of interest in its own right but also plays a major role in the successful construction of numerical methods. We believe that the results of the second part will be very useful in such contexts. To demonstrate this very briefly, we have provided an appendix where one specific numerical method is introduced and analyzed. This method can also help overcome the problems caused by the shortage of analytical methods for the computation of solutions to fractional differential equations. Of course, a few properly understood analytical approaches exist, and we shall present them in this text (e.g., the Picard iteration scheme discussed in Sect. 6.2 and the formulas for linear equations mentioned in Sect. 7.1), but – as can be expected from the corresponding observation in the classical theory of first-order differential equations – there is no generally applicable method to find an analytic solution to an arbitrarily given fractional differential equation. We need to mention however that, apart from the approaches that we shall treat explicitly, several other numerical and analytical methods for solving differential equations of fractional order have been developed. This includes the decomposition method usually attributed to Adomian [3] that can however actually be traced back to a series of much older papers by Perron [148–151] and that is known to have rather poor convergence properties in general [60,64,159], the variational iteration method [91,92], the homotopy analysis method [114], the homotopy perturbation method [139], the generalized differential transform method [145] and a few others for all of which convergence proofs are available only under rather restrictive conditions (see, e.g., [181]) and many of which are known not to converge in a satisfactory way anyway [113]. Therefore we shall refrain from dealing with these approaches in detail in this book.

Rather, we shall complete our treatment of the theory of fractional differential equations by providing additional appendices devoted to the collection of various other types of useful information: A list of all the symbols used in the book and a brief table of the Caputo derivatives of certain important functions (corresponding tables for Riemann–Liouville and other fractional derivatives already exist in the literature, cf., e.g., [153, 167]). Finally we shall make use of some classical results from analysis about topics like the Gamma function and Laplace transforms. For the sake of completeness, we give an account of those results in an appendix too.

Of course it is possible to set up a theory (and, based on this theory, to discuss numerical methods) for *partial* fractional differential equations, i.e. differential equations in more than one variable, where at least one of the partial derivatives involved is not an integer-order operator. Many of the results presented in the following chapters can be shown to be useful building blocks in such a context, but a thorough treatment of these problems is not what we are aiming at in this book. We will rather try to give a comprehensive treatment of univariate problems – in other words, *ordinary* differential equations. (Note that some authors prefer to reserve the term *ordinary* for equations of integer order and use the expression *extraordinary* for fractional differential equations in one variable. We shall not follow their nomenclature because it is the author's opinion that these equations can and should be used as an absolutely normal, and by no means extraordinary, mathematical tool that is useful for plenty of applications.)

During the course of the text we will occasionally state theorems from classical analysis for purposes of comparison with their fractional counterparts or in order to illustrate the ideas behind the generalizations. These classical theorems are usually well known and no proofs will be provided for them. To make the distinction clear, they will not follow the standard numbering scheme of the other theorems; instead they will be assigned a label consisting of the number of the chapter where they will be found, followed by a capital letter such as, e.g., Theorem 1.A.

1.2 The Basic Idea

The basic idea behind fractional calculus is intimately related to a classical standard result from (classical) differential and integral calculus, the fundamental theorem [165, Theorem 6.18]:

Theorem 1.A (Fundamental Theorem of Classical Calculus). *Let $f : [a,b] \rightarrow \mathbb{R}$ be a continuous function, and let $F : [a,b] \rightarrow \mathbb{R}$ be defined by*

$$F(x) := \int_a^x f(t)\,\mathrm{d}t.$$

Then, F is differentiable and

$$F' = f.$$

Therefore we have a very close relation between differential operators and integral operators. It is one of the goals of fractional calculus to retain this relation in a suitably generalized sense. Hence there is also a need to deal with fractional integral operators, and actually it turns out to be useful to discuss these first before coming to fractional differential operators (and thus to an answer of de l'Hospital's question).

It has proven to be convenient to use the notational conventions introduced in the following definition.

Definition 1.1. (a) By D, we denote the operator that maps a differentiable function onto its derivative, i.e.

$$Df(x) := f'(x).$$

(b) By J_a, we denote the operator that maps a function f, assumed to be (Riemann) integrable on the compact interval $[a,b]$, onto its primitive centered at a, i.e.

$$J_a f(x) := \int_a^x f(t)\,\mathrm{d}t$$

for $a \leq x \leq b$.

(c) For $n \in \mathbb{N}$ we use the symbols D^n and J_a^n to denote the n-fold iterates of D and J_a, respectively, i.e. we set $D^1 := D$, $J_a^1 := J_a$, and $D^n := DD^{n-1}$ and $J_a^n := J_a J_a^{n-1}$ for $n \geq 2$.

The key question now is: How can we extend the concepts of Definition 1.1 (c) to $n \notin \mathbb{N}$? Once we will have provided such an extension, we then need to ask for the mapping properties of the resulting operators, and in particular this includes the question for their domains and ranges.

Note that Theorem 1.A reads, in our notation,

$$DJ_a f = f$$

which implies that

$$D^n J_a^n f = f \tag{1.1}$$

for $n \in \mathbb{N}$, i.e. D^n is the left inverse of J_a^n in a suitable space of functions. We wish to retain this property. However, as we shall see, it is by no means straightforward to generalize the conditions of Theorem 1.A to the fractional case $n \notin \mathbb{N}$ in such a way that everything can be kept intact easily. It is a classical error made very often that known properties from standard calculus are generalized to the fractional setting too directly and without sufficient caution.

In Chap. 2 we want to give a first generalization of the concepts of Definition 1.1 (c) to $n \notin \mathbb{N}$. As already mentioned, there are various possible generalizations. We only discuss those that have got major significance for practical applications. A thorough investigation of many other possibilities (that however excludes the practically very important case of the Caputo operator which is to be introduced in Chap. 3 below) is contained in the encyclopaedic monograph of Samko, Kilbas and Marichev [167] and in the more recent book of Kilbas, Srivastava and Trujillo [100].

Following the outline given above, we begin with the integral operator J_a^n. In the case $n \in \mathbb{N}$, it is well known (and easily proved by induction) [167, eq. (2.16)] that we can replace the recursive definition of Definition 1.1 (c) by the following explicit formula.

Lemma 1.1. *Let f be Riemann integrable on $[a,b]$. Then, for $a \le x \le b$ and $n \in \mathbb{N}$, we have*

$$J_a^n f(x) = \frac{1}{(n-1)!} \int_a^x (x-t)^{n-1} f(t) \, dt.$$

Moreover, it is an immediate consequence of (1.1) (and therefore a consequence of the fundamental theorem) that the following relation holds for the operators D and J_a:

Lemma 1.2. *Let $m, n \in \mathbb{N}$ such that $m > n$, and let f be a function having a continuous nth derivative on the interval $[a,b]$. Then,*

$$D^n f = D^m J_a^{m-n} f.$$

Proof. By (1.1), we have $f = D^{m-n} J_a^{m-n} f$. Applying the operator D^n to both sides of this relation and using the fact that $D^n D^{m-n} = D^m$, the statement follows. $\qquad \square$

These two Lemmata are fundamental for the generalizations coming up in Chaps. 2 and 3. A look at Lemma 1.1 reveals that it will be useful to generalize the factorial to non-integer arguments. Such a generalization exists and is well-known: *Euler's Gamma function*. The definition is as follows.

Definition 1.2. The function $\Gamma : (0, \infty) \to \mathbb{R}$, defined by

$$\Gamma(x) := \int_0^\infty t^{x-1} e^{-t} \, dt,$$

is called *Euler's Gamma function* (or *Euler's integral of the second kind*).

In order to give a reasonably self-contained treatment of the topic, we state some of the key properties of the Gamma function in Appendix D.1. The most important one, for our purposes, is the following the proof of which is also given in Appendix D.1.

Theorem 1.3. *For $n \in \mathbb{N}$, we have $(n-1)! = \Gamma(n)$.*

Before we start the main work, we shall introduce some function spaces in which we are going to discuss matters. Since these are classical spaces, we can be rather brief here.

Definition 1.3. Let $0 < \mu \leq 1$, $k \in \mathbb{N}_0$ and $1 \leq p$.

$$L_p[a,b] := \left\{ f : [a,b] \to \mathbb{R}; f \text{ is measurable on } [a,b] \text{ and } \int_a^b |f(x)|^p \, dx < \infty \right\},$$

$$L_\infty[a,b] := \{ f : [a,b] \to \mathbb{R}; f \text{ is measurable and essentially bounded on } [a,b] \},$$

$$H_\mu[a,b] := \{ f : [a,b] \to \mathbb{R}; \exists c > 0 \, \forall x, y \in [a,b] : |f(x) - f(y)| \leq c|x-y|^\mu \},$$

$$C^k[a,b] := \{ f : [a,b] \to \mathbb{R}; f \text{ has a continuous } k\text{th derivative} \},$$

$$C[a,b] := C^0[a,b],$$

$$H_0[a,b] := C[a,b].$$

In other words, $L_p[a,b]$ is (for $1 \leq p \leq \infty$) the usual *Lebesgue space*, whereas $H_\mu[a,b]$ is a *Hölder space* or *Lipschitz space* of order μ. When the interval $[a,b]$ in question is clear from the context, we will often choose not to mention it explicitly and use the simpler notation L_p instead of $L_p[a,b]$, etc.

Now and then we shall also use a slightly less standard function space:

Definition 1.4. By H^* or $H^*[a,b]$ we denote the set of functions $f : [a,b] \to \mathbb{R}$ with the property that there exists some constant $L > 0$ such that

$$|f(x+h) - f(x)| \leq L|h| |\ln|h||^{-1}$$

whenever $|h| < 1/2$ and $x, x+h \in [a,b]$.

Obviously, this set is slightly larger than H_1.

When working in a Lebesgue space rather than in a space of continuous functions, we can still retain the main part of the statement of the fundamental theorem:

Theorem 1.B (Fundamental Theorem in Lebesgue Spaces). *Let $f \in L_1[a,b]$. Then, $J_a f$ is differentiable almost everywhere in $[a,b]$, and $DJ_a f = f$ also holds almost everywhere on $[a,b]$.*

A proof of this theorem can be found in [160, §23].

We shall occasionally also use the following set of functions.

Definition 1.5. By A^n or $A^n[a,b]$ we denote the set of functions with an absolutely continuous $(n-1)$st derivative, i.e. the functions f for which there exists (almost everywhere) a function $g \in L_1[a,b]$ such that

$$f^{(n-1)}(x) = f^{(n-1)}(a) + \int_a^x g(t)\,dt.$$

In this case we call g the *(generalized) nth derivative* of f, and we simply write $g = f^{(n)}$.

A note of caution is in order here. It is clear from this definition that a function $f \in A^1$ possesses (almost everywhere) a derivative $f' \in L_1$. However this implication cannot be reversed in general. There exist, for example (cf. [160, §24]), non-constant functions f that are differentiable almost everywhere with $f' \equiv 0$, which surely is an L_1 function. Obviously, in such a case f cannot be represented as the primitive of f' as required in Definition 1.5, and hence such a function f is not in A^1.

1.3 An Example Application of Fractional Calculus

Before we come to a detailed study of the mathematical properties of fractional differential operators and fractional differential equations, let us take a brief look at a simple but not unrealistic example of a model arising in mechanics where fractional derivatives can be used successfully. The model has been originally proposed as a theory by Nutting [142, 143]; the works of Scott Blair et al. [172] were among the first to confirm its value in practice.

Specifically, we want to describe the behaviour of certain materials under the influence of external forces. The traditional way to deal with such questions in mechanics uses the laws of Hooke and Newton. The relation that we are interested in is the relation between stress $\sigma(t)$ and strain $\varepsilon(t)$, both of which are taken as functions of time t. If we are dealing with viscous liquids, then Newton's law

$$\sigma(t) = \eta D^1 \varepsilon(t) \tag{1.2}$$

is the tool of our choice. Here the material constant η is the so-called viscosity of the material. Hooke's law

$$\sigma(t) = E D^0 \varepsilon(t) \tag{1.3}$$

on the other hand is the correct way of modelling the stress–strain relationship for elastic solids. The constant E is known as the modulus of elasticity of the material. Of course it is common practice not to mention the operator D^0 (i.e. the identity operator) in (1.3) explicitly, but we have deviated from this path to stress the formal similarity between the two laws.

Now consider an experiment where the strain is manipulated in a controlled fashion such that, say, $\varepsilon(t) = t$ for $t \in [0, T]$ with some $T > 0$. It then follows that the stress behaves as

$$\sigma(t) = Et$$

in the case of an elastic solid and

$$\sigma(t) = \eta = \text{const}$$

for a viscous liquid. We may summarize these equations in the form

$$\psi_k = \frac{\sigma(t)}{\varepsilon(t)} t^k \tag{1.4}$$

where $\psi_0 = E$ and $\psi_1 = \eta$. Evidently the case $k = 0$ corresponds to Hooke's law for solids and $k = 1$ refers to Newton's law for liquids.

In practice it is not uncommon to find so-called viscoelastic materials that exhibit a behaviour somewhere between the pure viscous liquid and the pure elastic solid, i.e. where one would observe a relationship of the form (1.4) with $0 < k < 1$. In this case it is appropriate to interpret k as a second material constant in addition to ψ_k. Classical examples are polymers, but some types of biological tissue may also share this property as well as a number of metals (aluminium, for example) at least under certain temperature and pressure conditions. It should be noted that for the case of a constant strain ε, the stress in such a material would develop according to the formula

$$\sigma(t) = \text{const} \cdot t^{-k}$$

and thus converges to zero for very long observation times. In this respect it once again lies between a viscous liquid for which σ vanishes identically and an elastic solid whose stress σ is a nonzero constant.

In view of all these "interpolation" properties it is natural to assume that it is also possible to model the relation between stress and strain for such a viscoelastic material via an equation of the form

$$\sigma(t) = \nu D^k \varepsilon(t) \tag{1.5}$$

where v is a material constant and $k \in (0,1)$ is the parameter introduced above. This equation "interpolates" between (1.2) and (1.3) in a similar spirit. In view of the above mentioned theoretical foundations of (1.5) laid by Nutting, this relation is frequently called *Nutting's law*.

We shall see in the following chapters that it is indeed justified to argue in this way if we define the differential operator D^k properly. In particular it will turn out that all the relations mentioned above can be kept intact.

Exercises

Exercise 1.1. Give a proof for Lemma 1.1.

Exercise 1.2. Show that, for $0 < \mu_1 < \mu_2 < 1$,

$$H_0 \supset H_{\mu_1} \supset H_{\mu_2} \supset H_1 \supset C^1$$

and

$$H_{\mu_2} \supset H^* \supset H_1.$$

Provide some examples showing that all the inclusions are strict.

Exercise 1.3. Show that $H_1[a,b] \subset A^1[a,b]$. Is the function f with $f(x) = (x-a)^\alpha$ for some $0 < \alpha < 1$ an element of these two sets?

Chapter 2
Riemann–Liouville Differential and Integral Operators

We are now in a position to give a first definition for fractional integral and differential operators J_a^n and D^n, $n \notin \mathbb{N}$. As indicated above, we begin with the integral operator.

2.1 Riemann–Liouville Integrals

In view of the considerations of the previous chapter, the following concept seems rather natural.

Definition 2.1. Let $n \in \mathbb{R}_+$. The operator J_a^n, defined on $L_1[a,b]$ by

$$J_a^n f(x) := \frac{1}{\Gamma(n)} \int_a^x (x-t)^{n-1} f(t) \, dt$$

for $a \leq x \leq b$, is called the *Riemann–Liouville fractional integral operator of order n*.

For $n = 0$, we set $J_a^0 := I$, the identity operator.

The definition for $n = 0$ is quite convenient for future manipulations. It is evident that the Riemann–Liouville fractional integral coincides with the classical definition of J_a^n in the case $n \in \mathbb{N}$, except for the fact that we have extended the domain from Riemann integrable functions to Lebesgue integrable functions (which will not lead to any problems in our development). Moreover, in the case $n \geq 1$ it is obvious that the integral $J_a^n f(x)$ exists for every $x \in [a,b]$ because the integrand is the product of an integrable function f and the continuous function $(x - \cdot)^{n-1}$. In the case $0 < n < 1$ though, the situation is less clear at first sight. However, the following result asserts that this definition is justified.

Theorem 2.1. *Let $f \in L_1[a,b]$ and $n > 0$. Then, the integral $J_a^n f(x)$ exists for almost every $x \in [a,b]$. Moreover, the function $J_a^n f$ itself is also an element of $L_1[a,b]$.*

K. Diethelm, *The Analysis of Fractional Differential Equations*,
Lecture Notes in Mathematics 2004, DOI 10.1007/978-3-642-14574-2_2,
© Springer-Verlag Berlin Heidelberg 2010

Proof. We write the integral in question as

$$\int_a^x (x-t)^{n-1} f(t)\,dt = \int_{-\infty}^{\infty} \phi_1(x-t)\phi_2(t)\,dt$$

where

$$\phi_1(u) = \begin{cases} u^{n-1} & \text{for } 0 < u \le b-a, \\ 0 & \text{else,} \end{cases}$$

and

$$\phi_2(u) = \begin{cases} f(u) & \text{for } a \le u \le b, \\ 0 & \text{else.} \end{cases}$$

By construction, $\phi_j \in L_1(\mathbb{R})$ for $j \in \{1,2\}$, and thus by a classical result on Lebesgue integration [190, Theorem 4.2d] the desired result follows. $\qquad\square$

One important property of integer-order integral operators is preserved by our generalization:

Theorem 2.2. *Let $m, n \ge 0$ and $\phi \in L_1[a,b]$. Then,*

$$J_a^m J_a^n \phi = J_a^{m+n} \phi$$

holds almost everywhere on $[a,b]$. If additionally $\phi \in C[a,b]$ or $m+n \ge 1$, then the identity holds everywhere on $[a,b]$.

Corollary 2.3. *Under the assumptions of Theorem 2.2,*

$$J_a^m J_a^n \phi = J_a^n J_a^m \phi.$$

There is an algebraic way to state this result.

Theorem 2.4. *The operators $\{J_a^n : L_1[a,b] \to L_1[a,b]; n \ge 0\}$ form a commutative semigroup with respect to concatenation. The identity operator J_a^0 is the neutral element of this semigroup.*

Proof (of Theorem 2.2). We have

$$J_a^m J_a^n \phi(x) = \frac{1}{\Gamma(m)\Gamma(n)} \int_a^x (x-t)^{m-1} \int_a^t (t-\tau)^{n-1} \phi(\tau)\,d\tau\,dt.$$

In view of Theorem 2.1, the integrals exist, and by Fubini's theorem we may interchange the order of integration, obtaining

$$J_a^m J_a^n \phi(x) = \frac{1}{\Gamma(m)\Gamma(n)} \int_a^x \int_\tau^x (x-t)^{m-1}(t-\tau)^{n-1} \phi(\tau)\,dt\,d\tau$$

$$= \frac{1}{\Gamma(m)\Gamma(n)} \int_a^x \phi(\tau) \int_\tau^x (x-t)^{m-1}(t-\tau)^{n-1}\,dt\,d\tau.$$

The substitution $t = \tau + s(x - \tau)$ yields

$$J_a^m J_a^n \phi(x) = \frac{1}{\Gamma(m)\Gamma(n)} \int_a^x \phi(\tau) \int_0^1 [(x - \tau)(1 - s)]^{m-1}$$

$$\times [s(x - \tau)]^{n-1}(x - \tau) \, ds \, d\tau$$

$$= \frac{1}{\Gamma(m)\Gamma(n)} \int_a^x \phi(\tau)(x - \tau)^{m+n-1} \int_0^1 (1 - s)^{m-1} s^{n-1} \, ds \, d\tau.$$

In view of Theorem D.6, $\int_0^1 (1 - s)^{m-1} s^{n-1} \, ds = \Gamma(m)\Gamma(n)/\Gamma(n+m)$, and thus

$$J_a^m J_a^n \phi(x) = \frac{1}{\Gamma(m+n)} \int_a^x \phi(\tau)(x - \tau)^{m+n-1} \, d\tau = J_a^{m+n} \phi(x)$$

almost everywhere on $[a, b]$.

Moreover, by the classical theorems on parameter integrals, if $\phi \in C[a, b]$ then also $J_a^n \phi \in C[a, b]$, and therefore $J_a^m J_a^n \phi \in C[a, b]$, and $J_a^{m+n} \phi \in C[a, b]$ too. Thus, since these two continuous functions coincide almost everywhere, they must coincide everywhere.

Finally, if $\phi \in L_1[a, b]$ and $m + n \geq 1$ we have, by the result above

$$J_a^m J_a^n \phi = J_a^{m+n} \phi = J_a^{m+n-1} J_a^1 \phi$$

almost everywhere. Since $J_a^1 \phi$ is continuous, we also have that $J_a^{m+n} \phi = J_a^{m+n-1} J_a^1 \phi$ is continuous, and once again we may conclude that the two functions on either side of the equality almost everywhere are continuous; thus they must be identical everywhere. □

We now consider some mapping properties of the operator J_a^n. Roughly speaking, we shall see that fractional integration improves the smoothness properties of functions. To be a bit more precise, we can say that $J_a^n \phi$ is the sum of two expressions one of which (denoted by Φ in the theorem below) is usually better behaved than ϕ itself, whereas the other one may be non-smooth at the point a (with a precisely known behaviour there) and is a C^∞ function elsewhere.

Theorem 2.5. *Let $\phi \in H_\mu[a, b]$ for some $\mu \in [0, 1]$, and let $0 < n < 1$. Then*

$$J_a^n \phi(x) = \frac{\phi(a)}{\Gamma(n+1)}(x - a)^n + \Phi(x)$$

with some function Φ. This function Φ satisfies

$$\Phi(x) = O\left((x - a)^{\mu+n}\right)$$

as $x \to a$. *Moreover,*

$$\Phi \in \begin{cases} H_{\mu+n}[a,b] & \text{if } \mu+n < 1, \\ H^*[a,b] & \text{if } \mu+n = 1, \\ H_1[a,b] & \text{if } \mu+n > 1. \end{cases}$$

Remark 2.1. It is possible to show an even stronger statement in the case $\mu+n > 1$: Under this assumption we have that $\Phi \in C^1[a,b]$ and $\Phi' \in H_{\mu+n-1}[a,b]$, cf. [167, Theorem 3.1].

Proof (of Theorem 2.5). We have

$$J_a^n \phi(x) = \frac{\phi(a)}{\Gamma(n)} \int_a^x (x-t)^{n-1}\, dt + \frac{1}{\Gamma(n)} \int_a^x \frac{\phi(t)-\phi(a)}{(x-t)^{1-n}}\, dt.$$

This yields the desired representation with

$$\Phi(x) = \frac{1}{\Gamma(n)} \int_a^x \frac{\phi(t)-\phi(a)}{(x-t)^{1-n}}\, dt.$$

In view of $\phi \in H_\mu$,

$$|\Phi(x)| \leq \frac{1}{\Gamma(n)} \int_a^x \frac{L|t-a|^\mu}{(x-t)^{1-n}}\, dt = \frac{L}{\Gamma(n)} \int_a^x (t-a)^\mu (x-t)^{n-1}\, dt$$

$$= \frac{L}{\Gamma(n)} (x-a)^{\mu+n} \int_0^1 s^\mu (1-s)^{n-1}\, ds = O\left((x-a)^{\mu+n}\right).$$

Now we set $g(x) := (\phi(x) - \phi(a))/\Gamma(n)$. Moreover let $h > 0$ and $x, x+h \in [a,b]$. Then,

$$\Phi(x+h) - \Phi(x) = \int_a^{x+h} g(t)(x+h-t)^{n-1}\, dt - \int_a^x g(t)(x-t)^{n-1}\, dt$$

$$= \int_a^x g(t) \left[(x+h-t)^{n-1} - (x-t)^{n-1}\right]\, dt$$

$$+ \int_x^{x+h} g(t)(x+h-t)^{n-1}\, dt$$

$$= \underbrace{\int_a^x (g(t) - g(x)) \left[(x+h-t)^{n-1} - (x-t)^{n-1}\right]\, dt}_{=:K_1}$$

$$+ \underbrace{\int_x^{x+h} (g(t) - g(x))(x+h-t)^{n-1}\, dt}_{=:K_2} + K_3$$

where K_3 contains the remaining terms. An explicit calculation shows that

$$K_3 = g(x) \left(\int_a^x \left[(x+h-t)^{n-1} - (x-t)^{n-1} \right] dt + \int_x^{x+h} (x+h-t)^{n-1} dt \right).$$

We estimate the terms K_1, K_2, and K_3 separately. In view of our assumption $\phi \in H_\mu$, it is clear that $g \in H_\mu$ too, and hence

$$|K_1| = \left| \int_0^{x-a} (g(x-u) - g(x)) \left[(u+h)^{n-1} - u^{n-1} \right] du \right|$$

$$\leq L \int_0^{x-a} u^\mu \left[u^{n-1} - (u+h)^{n-1} \right] du$$

$$= Lh \int_0^{(x-a)/h} (ht)^\mu \left[(th)^{n-1} - (th+h)^{n-1} \right] dt$$

$$= Lh^{\mu+n} \int_0^{(x-a)/h} t^\mu \left[t^{n-1} - (t+1)^{n-1} \right] dt.$$

At $t \to 0$, there is no problem with the convergence of the integral since the integrand behaves as $t^{\mu+n-1}$ there (the exponent is strictly greater than -1). In the case $x - a < h$, the integral is bounded by $\int_0^1 t^{\mu+n-1} dt = 1/(\mu+n)$, and thus $K_1 = O(h^{\mu+n})$ in this case. If $x - a \geq h$, we find

$$\int_0^{(x-a)/h} t^\mu \left[t^{n-1} - (t+1)^{n-1} \right] dt$$

$$= \int_0^1 t^\mu \left[t^{n-1} - (t+1)^{n-1} \right] dt + \int_1^{(x-a)/h} t^\mu \left[t^{n-1} - (t+1)^{n-1} \right] dt$$

$$< \frac{1}{\mu+n} + (1-n) \int_1^{(x-a)/h} t^{\mu+n-2} dt$$

in view of the mean value theorem of differential calculus. The remaining integral can easily be calculated explicitly. We find for $\mu + n < 1$ that

$$|K_1| \leq O(h^{\mu+n}) \left(\frac{1}{\mu+n} + (1-n) \int_1^{(x-a)/h} t^{\mu+n-2} dt \right)$$

$$\leq O(h^{\mu+n}) \left(\frac{1}{\mu+n} + (1-n) \int_1^\infty t^{\mu+n-2} dt \right) \leq O(h^{\mu+n})$$

because of $\mu + n < 1$ and $x - a \geq h$. For $\mu + n = 1$ a similar calculation gives

$$|K_1| \leq O(h^{\mu+n}) \left(1 + \int_1^{(x-a)/h} t^{-1} dt \right) = O(h \ln h^{-1}).$$

Finally for $\mu + n > 1$ we have, since $x \le b$,

$$|K_1| \le O(h^{\mu+n}) \int_1^{(b-a)/h} t^{\mu+n-2} \, dt = O(h^{\mu+n}) \left(\frac{b-a}{h} \right)^{\mu+n-1} = O(h).$$

Thus

$$K_1 = \begin{cases} O(h^{\mu+n}) & \text{for } \mu+n < 1, \\ O(h \ln h^{-1}) & \text{for } \mu+n = 1, \\ O(h) & \text{for } \mu+n > 1. \end{cases}$$

Next we estimate K_2. We again use the fact that $g \in H_\mu$ to derive (using the substitution $s = (t-x)/h$)

$$|K_2| \le L \int_x^{x+h} (t-x)^\mu (x+h-t)^{n-1} \, dt = L h^{\mu+n} \int_0^1 s^\mu (1-s)^{n-1} \, ds = O(h^{\mu+n})$$

irrespective of μ and n. Obviously this bound is stronger than the above bound for K_1.

Finally for K_3 we use the Hölder assumption on ϕ which implies that $|g(x)| \le L(x-a)^\mu$ for some constant L. Evaluating the integrals analytically, we find

$$|K_3| \le \frac{L}{n} (x-a)^\mu \left[(x-a+h)^n - (x-a)^n \right].$$

In the case $x - a \le h$, this is bounded from above by $O(h^{\mu+n})$. If $x - a > h$, we estimate the term in brackets by the mean value theorem of differential calculus and find (taking into account that $n < 1$)

$$K_3 = O(1)(x-a)^\mu h (x-a)^{n-1} = O(h)(x-a)^{\mu+n-1}.$$

Once again we look at the three cases separately and find $K_3 = O(h)$ for $\mu + n = 1$, $|K_3| \le O(h)h^{\mu+n-1} = O(h^{\mu+n})$ for $\mu + n < 1$, and $|K_3| \le O(h)(b-a)^{\mu+n-1} = O(h)$ for $\mu + n > 1$. Again, these estimates are stronger than those we obtained for K_1.

Combining all the estimates, we derive

$$\Phi(x+h) - \Phi(x) = \begin{cases} O(h^{\mu+n}) & \text{for } \mu+n < 1, \\ O(h \ln h^{-1}) & \text{for } \mu+n = 1, \\ O(h) & \text{for } \mu+n > 1. \end{cases} \qquad \square$$

A similar result may be obtained when we assume the integrand ϕ to be in a suitable Lebesgue class.

Theorem 2.6. *Let $n > 0$, $p > \max\{1, 1/n\}$, and $\phi \in L_p[a,b]$. Then*

$$J_a^n \phi(x) = o\left((x-a)^{n-1/p} \right)$$

as $x \to a+$. If additionally $n - 1/p \notin \mathbb{N}$, then $J_a^n \phi \in C^{\lfloor n-1/p \rfloor}[a,b]$, and $D^{\lfloor n-1/p \rfloor} J_a^n \phi \in H_{n-1/p-\lfloor n-1/p \rfloor}[a,b]$.

Remark 2.2. In the case $n - 1/p \in \mathbb{N}$, a slightly modified statement may be shown. The Hölder condition on the derivative of $J_a^n \phi$ must be replaced by a condition similar to the one defining the set H^*. We refer to [167, Theorem 3.6] for details.

Proof (of Theorem 2.6). For the given p, we introduce the conjugate exponent $q \in [1, \infty)$ by the relation $p^{-1} + q^{-1} = 1$. Then, by definition of the Riemann–Liouville integral operator and Hölder's inequality,

$$|J_a^n \phi(x)| \leq \frac{1}{\Gamma(n)} \left(\int_a^x |\phi(t)|^p \, dt \right)^{1/p} \left(\int_a^x |x-t|^{(n-1)q} \, dt \right)^{1/q}$$

$$\leq \frac{(x-a)^{n-1/p}}{\Gamma(n) ((n-1)q+1)^{1/q}} \left(\int_a^x |\phi(t)|^p \, dt \right)^{1/p} = o\left((x-a)^{n-1/p} \right).$$

For the proof of the smoothness result, we discuss the case $n - 1/p < 1$ first. Here, we find that

$$J_a^n \phi(x+h) - J_a^n \phi(x) = \underbrace{\frac{1}{\Gamma(n)} \int_x^{x+h} (x+h-t)^{n-1} \phi(t) \, dt}_{=:K_1}$$

$$+ \underbrace{\frac{1}{\Gamma(n)} \int_a^x \left[(x+h-t)^{n-1} - (x-t)^{n-1} \right] \phi(t) \, dt}_{=:K_2}.$$

As above we use Hölder's inequality to derive that

$$|K_1| \leq \left(\int_x^{x+h} |\phi(t)|^p \, dt \right)^{1/p} \left(\int_x^{x+h} (x+h-t)^{(n-1)q} \, dt \right)^{1/q} \leq ch^{n-1/p}$$

with some constant c. Moreover, also by Hölder's inequality,

$$|K_2| \leq \|\phi\|_{L_p} \left(\int_a^x |(x+h-t)^{n-1} - (x-t)^{n-1}|^q \, dt \right)^{1/q}$$

$$\leq \|\phi\|_{L_p} h^{n-1/p} \left(\int_0^{(x-a)/h} |s^{n-1} - (s+1)^{n-1}|^q \, ds \right)^{1/q}.$$

In the case $x - a \leq h$ the latter integral is bounded by a constant. In the complementary case $x - a > h$, we use (as in the proof of the previous theorem) the mean value theorem of differential calculus and find

$$\int_0^{(x-a)/h} |s^{n-1} - (s+1)^{n-1}|^q \, ds$$

$$= \int_0^1 |s^{n-1} - (s+1)^{n-1}|^q \, ds + \int_1^{(x-a)/h} |s^{n-1} - (s+1)^{n-1}|^q \, ds$$

$$\leq c + |n-1| \int_1^{(x-a)/h} s^{(n-2)q} \, ds$$

$$= c + \frac{|n-1|}{q(n-2)+1} \left[s^{q(n-2)+1} \right]_1^{(x-a)/h}$$

with some constant c. We look at the denominator in the last expression and find that

$$q(n-2)+1 = q\left(n-1-1+\frac{1}{q}\right) = q\left(n-1-\frac{1}{p}\right) < 0$$

in view of $q > 0$ and our assumption that $n - 1/p < 1$. Thus we may continue the estimation for the integral by

$$\int_0^{(x-a)/h} \left| s^{n-1} - (s+1)^{n-1} \right|^q \, ds$$

$$\leq c + \left| \frac{n-1}{q(n-2)+1} \right| \left(1 - \left(\frac{x-a}{h} \right)^{q(n-2)+1} \right) = O(1).$$

Hence

$$K_2 = O(h^{n-1/p})$$

too, and therefore $J_a^n \phi \in H_{n-1/p}$ in this case.

Now we discuss the remaining case $n - 1/p > 1$. Then, in particular, $n > 1$, and in view of the semigroup property of fractional integration we may write

$$J_a^n \phi = J_a^{\lfloor n-1/p \rfloor} J_a^{n-\lfloor n-1/p \rfloor} \phi.$$

Therefore, by the fundamental theorem of calculus,

$$D^{\lfloor n-1/p \rfloor} J_a^n \phi = J_a^{n-\lfloor n-1/p \rfloor} \phi = J_a^{\tilde{n}} \phi.$$

Here we have that $\tilde{n} = n - \lfloor n - 1/p \rfloor \geq n - \lfloor n \rfloor \geq 0$ and $\tilde{n} - 1/p = n - 1/p - \lfloor n - 1/p \rfloor < 1$ because, by assumption, $n - 1/p \notin \mathbb{N}$. Thus we may apply the result that we just proved, with n replaced by $\tilde{n}$, and find that $D^{\lfloor n-1/p \rfloor} J_a^n \phi = J_a^{\tilde{n}} \phi \in H_{\tilde{n}-1/p-\lfloor \tilde{n}-1/p \rfloor} = H_{n-1/p-\lfloor n-1/p \rfloor}$. (The indices of the Hölder spaces are identical here because the difference $\tilde{n} - n$ is an integer.)   □

Example 2.1. Let $f(x) = (x-a)^\beta$ for some $\beta > -1$ and $n > 0$. Then,

$$J_a^n f(x) = \frac{\Gamma(\beta+1)}{\Gamma(n+\beta+1)} (x-a)^{n+\beta}.$$

In view of the well known corresponding result in the case $n \in \mathbb{N}$, this result is precisely what one would expect from a sensible generalization of the integral operator. In view of Theorem D.6, the derivation is direct:

$$J_a^n f(x) = \frac{1}{\Gamma(n)} \int_a^x (t-a)^\beta (x-t)^{n-1} \, dt$$

$$= \frac{1}{\Gamma(n)} (x-a)^{n+\beta} \int_0^1 s^\beta (1-s)^{n-1} \, ds = \frac{\Gamma(\beta+1)}{\Gamma(n+\beta+1)} (x-a)^{n+\beta}.$$

Next we discuss the interchange of limit operation and fractional integration.

Theorem 2.7. *Let $n > 0$. Assume that $(f_k)_{k=1}^\infty$ is a uniformly convergent sequence of continuous functions on $[a,b]$. Then we may interchange the fractional integral operator and the limit process, i.e.*

$$\left(J_a^n \lim_{k \to \infty} f_k \right)(x) = \left(\lim_{k \to \infty} J_a^n f_k \right)(x).$$

In particular, the sequence of functions $(J_a^n f_k)_{k=1}^\infty$ is uniformly convergent.

Proof. We denote the limit of the sequence (f_k) by f. It is well known that f is continuous. We then find

$$|J_a^n f_k(x) - J_a^n f(x)| \leq \frac{1}{\Gamma(n)} \int_a^x |f_k(t) - f(t)|(x-t)^{n-1} \, dt$$

$$\leq \frac{1}{\Gamma(n)} \|f_k - f\|_\infty \int_a^x (x-t)^{n-1} \, dt$$

$$= \frac{1}{\Gamma(n+1)} \|f_k - f\|_\infty (x-a)^n$$

$$\leq \frac{1}{\Gamma(n+1)} \|f_k - f\|_\infty (b-a)^n$$

which converges to zero as $k \to \infty$ uniformly for all $x \in [a,b]$.   □

Corollary 2.8. *Let f be analytic in $(a-h, a+h)$ for some $h > 0$, and let $n > 0$. Then*

$$J_a^n f(x) = \sum_{k=0}^\infty \frac{(-1)^k (x-a)^{k+n}}{k!(n+k)\Gamma(n)} D^k f(x)$$

for $a \leq x < a + h/2$, and

$$J_a^n f(x) = \sum_{k=0}^\infty \frac{(x-a)^{k+n}}{\Gamma(k+1+n)} D^k f(a)$$

for $a \leq x < a+h$. In particular, $J_a^n f$ is analytic in $(a, a+h)$.

Proof. For the first statement, we use the definition of the Riemann–Liouville integral operator J_a^n, viz.

$$J_a^n f(x) = \frac{1}{\Gamma(n)} \int_a^x f(t)(x-t)^{n-1} \, dt,$$

and expand $f(t)$ into a power series about x. Since $x \in [a, a+h/2)$, the power series converges in the entire interval of integration. Thus, by Theorem 2.7, it is legal to exchange summation and integration. Then we use the explicit representation for the fractional integral of the power function that we had derived in Example 2.1 to find the final result.

For the second statement, we proceed in a similar way, but we now expand the power series at a and not at x. This allows us again to conclude the convergence of the series in the required interval.

The analyticity of $J_a^n f$ follows immediately from the second statement. □

The statements of these last two results allow us to look at another instructive example.

Example 2.2. Let $f(x) = \exp(\lambda x)$ with some $\lambda > 0$. Compute $J_0^n f(x)$ for $n > 0$.

In the case $n \in \mathbb{N}$ we obviously have $J_0^n f(x) = \lambda^{-n} \exp(\lambda x)$. However, this result does not generalize in a straightforward way to the case $n \notin \mathbb{N}$. Rather, in view of the well known series expansion of the exponential function, Theorem 2.7 and Example 2.1, we find

$$J_0^n f(x) = J_0^n \left[\sum_{k=0}^{\infty} \frac{(\lambda \cdot)^k}{k!} \right] (x) = \sum_{k=0}^{\infty} \frac{\lambda^k}{k!} J_0^n [(\cdot)^k](x)$$

$$= \sum_{k=0}^{\infty} \frac{\lambda^k}{\Gamma(k+n+1)} x^{k+n} = \lambda^{-n} \sum_{k=0}^{\infty} \frac{(\lambda x)^{k+n}}{\Gamma(k+n+1)},$$

and here the series on the right-hand side is not $\exp(\lambda x)$. In Chap. 4 we shall encounter a class of functions that we can conveniently use to express the series, but for the moment we only note that fractional integral operators of Riemann–Liouville type do not reproduce exponential functions in the same way as integrals of integer order do.

Incidentally the same problem arises when we compute fractional integrals of other non-polynomial functions that have very simple integer-order integrals such as the sine or cosine function. We encourage the reader to work out the details as an exercise.

In the last two theorems of this section we discuss another important property of fractional integral operators, namely the continuity with respect to the order of the operator. We first look at the case that we work in Lebesgue spaces. Under this assumption the situation is very simple:

Theorem 2.9. *Let* $1 \leq p < \infty$ *and let* $(m_k)_{k=1}^{\infty}$ *be a convergent sequence of nonnegative numbers with limit* m. *Then, for every* $f \in L_p[a,b]$,

$$\lim_{k \to \infty} J_a^{m_k} f = J_a^m f$$

where the convergence is in the sense of the $L_p[a,b]$ *norm.*

A proof of this result may be found in [167, Theorem 2.6].

However, when we deal with the space $C[a,b]$ equipped with the Chebyshev norm (which is much more interesting and important for the applications that we have in mind), then the situation is more complicated. In order to illustrate this, we provide a very simple example before coming to the theorem itself that will describe the general case.

Example 2.3. Let $f(x) = 1$ and consider a strictly monotonic and convergent sequence $(m_k)_{k=1}^{\infty}$ of nonnegative numbers with limit $m \geq 0$. By Example 2.1 we find that

$$J_a^{m_k} f(x) = \frac{1}{\Gamma(m_k + 1)} (x - a)^{m_k}.$$

Then we have to introduce a distinction of two cases that exhibit very different types of behaviour:

- If $m > 0$ then

$$\|J_a^{m_k} f - J_a^m f\|_{\infty} = \sup_{x \in [a,b]} \left| \frac{(x-a)^{m_k}}{\Gamma(m_k + 1)} - \frac{(x-a)^m}{\Gamma(m+1)} \right| \to 0 \qquad (2.1)$$

  as $m_k \to m$, which can be shown with a lengthy but simple estimation. We leave the details of this special case to the reader (the result will of course follow from Theorem 2.10 below) and only note that we have convergence in the Chebyshev norm in this case.
- If $m = 0$ then the sequence (m_k) must be decreasing. Moreover we have that $J_a^{m_k} f(a) = 0$ for all k, whereas $J_a^m f(a) = J_a^0 f(a) = f(a) = 1$, i.e. we do not even have pointwise convergence, let alone convergence in the Chebyshev norm (uniform convergence).

A complete description of the situation looks as follows.

Theorem 2.10. *Let $f \in C[a,b]$ and $m \geq 0$. Moreover assume that (m_k) is a sequence of positive numbers such that $\lim_{k \to \infty} m_k = m$. Then, for every $\varepsilon > 0$,*

$$\lim_{k \to \infty} \sup_{x \in [a+\varepsilon,b]} |J_a^{m_k} f(x) - J_a^m f(x)| = 0.$$

If additionally $m > 0$ or $f(x) = O((x-a)^{\delta})$ as $x \to a$ for some $\delta > 0$ then

$$\lim_{k \to \infty} \|J_a^{m_k} f - J_a^m f\|_{\infty} = 0.$$

Proof. We begin with the case $m > 0$. Without loss of generality we may assume that $m_k > m/2$ for all k. It then follows for arbitrary $x \in [a,b]$ that

$$|J_a^{m_k} f(x) - J_a^m f(x)| \leq \int_a^x |f(t)| \left| \frac{(x-t)^{m_k-1}}{\Gamma(m_k)} - \frac{(x-t)^{m-1}}{\Gamma(m)} \right| dt$$

$$\leq \|f\|_{\infty} \int_a^x (x-t)^{m/4} \left| \frac{(x-t)^{m_k-1-m/4}}{\Gamma(m_k)} - \frac{(x-t)^{-1+3m/4}}{\Gamma(m)} \right| dt$$

$$= \|f\|_\infty \int_0^{x-a} u^{m/4} \left| \frac{u^{m_k-1-m/4}}{\Gamma(m_k)} - \frac{u^{-1+3m/4}}{\Gamma(m)} \right| du$$

$$\leq \|f\|_\infty \int_0^{b-a} u^{m/4} \left| \frac{u^{m_k-1-m/4}}{\Gamma(m_k)} - \frac{u^{-1+3m/4}}{\Gamma(m)} \right| du$$

$$\leq \|f\|_\infty (b-a)^{m/4} \int_0^{b-a} \left| \frac{u^{m_k-1-m/4}}{\Gamma(m_k)} - \frac{u^{-1+3m/4}}{\Gamma(m)} \right| du.$$

It is easy to see that the integrand only changes its sign for

$$u = c_k := \left(\frac{\Gamma(m_k)}{\Gamma(m)} \right)^{1/(m_k-m)},$$

and an explicit calculation using de l'Hospital's rule reveals that $\lim_{k\to\infty} c_k = \exp(-\Psi(m))$ where $\Psi = \Gamma'/\Gamma$ denotes the Digamma function. Thus we may continue our estimation above according to

$$|J_a^{m_k} f(x) - J_a^m f(x)| \leq \|f\|_\infty (b-a)^{m/4} \int_0^{b-a} \left| \frac{u^{m_k-1-m/4}}{\Gamma(m_k)} - \frac{u^{-1+3m/4}}{\Gamma(m)} \right| du$$

$$\leq \|f\|_\infty (b-a)^{m/4} \left| \int_0^{c_k} \left(\frac{u^{m_k-1-m/4}}{\Gamma(m_k)} - \frac{u^{-1+3m/4}}{\Gamma(m)} \right) du \right.$$

$$\left. - \int_{c_k}^{b-a} \left(\frac{u^{m_k-1-m/4}}{\Gamma(m_k)} - \frac{u^{-1+3m/4}}{\Gamma(m)} \right) du \right|$$

$$= \|f\|_\infty (b-a)^{m/4} \left| \frac{2 c_k^{m_k-m/4}}{\Gamma(m_k)\left(m_k - \frac{m}{4}\right)} - \frac{8 c_k^{3m/4}}{3m\Gamma(m)} \right.$$

$$\left. - \frac{(b-a)^{m_k-m/4}}{\Gamma(m_k)\left(m_k - \frac{m}{4}\right)} + \frac{4(b-a)^{3m/4}}{3m\Gamma(m)} \right|$$

and evidently the term inside the absolute value operation converges to 0 as $m_k \to m$, which completes the proof in the case $m > 0$.

In the case $m = 0$ this argument is not applicable. Instead, we proceed as follows. For some $\varepsilon_* \in [0, x-a]$ that will be specified more precisely later, we write

$$|J_a^{m_k} f(x) - f(x)| = \left| \int_a^x f(t) \frac{(x-t)^{m_k-1}}{\Gamma(m_k)} dt - f(x) \right|$$

$$\leq \left| \int_a^{x-\varepsilon_*} f(t) \frac{(x-t)^{m_k-1}}{\Gamma(m_k)} dt \right| + \left| \int_{x-\varepsilon_*}^x f(t) \frac{(x-t)^{m_k-1}}{\Gamma(m_k)} dt - f(x) \right|$$

$$\leq \frac{\varepsilon_*^{m_k-1} \|f\|_\infty (x-\varepsilon_*-a)}{\Gamma(m_k)} + \left| \int_0^{\varepsilon_*} f(x-u) \frac{u^{m_k-1}}{\Gamma(m_k)} du - f(x) \right|.$$

Using the generalized mean value theorem, we find that the integral can be represented in the form

$$\int_0^{\varepsilon_*} f(x-u)\frac{u^{m_k-1}}{\Gamma(m_k)}\,du = f(\xi)\int_0^{\varepsilon_*}\frac{u^{m_k-1}}{\Gamma(m_k)}\,du = f(\xi)\frac{\varepsilon_*^{m_k}}{\Gamma(m_k+1)}$$

with some $\xi \in [x-\varepsilon_*, x]$. Thus,

$$|J_a^{m_k}f(x) - f(x)| \le \frac{\varepsilon_*^{m_k-1}\|f\|_\infty (b-\varepsilon_*-a)}{\Gamma(m_k)} + \left|\frac{f(\xi)\varepsilon_*^{m_k}}{\Gamma(m_k+1)} - f(x)\right| \qquad (2.2)$$

for all $x \in [a+\varepsilon_*, b]$, where $\xi \in [x-\varepsilon_*, x]$. Now we set $\varepsilon_* := m_k^{1/2}$. For the value ε mentioned in the claim (that has been fixed in advance), we know that (since $m_k \to 0$) there exists some $k_0 \in \mathbb{N}$ such that for all $k \ge k_0$ we have $\varepsilon_* = m_k^{1/2} < \varepsilon$. Thus, for these k the inequality (2.2) is valid for all $x \in [a+\varepsilon, b]$. With our special choice of ε_* it takes the form

$$|J_a^{m_k}f(x) - f(x)|$$

$$\le \frac{m_k^{(m_k-1)/2}}{\Gamma(m_k)}(b-a)\|f\|_\infty + \left|\frac{f(\xi)m_k^{m_k/2}}{\Gamma(m_k+1)} - f(\xi) + f(\xi) - f(x)\right|$$

$$\le \frac{m_k^{(m_k-1)/2}}{\Gamma(m_k)}(b-a)\|f\|_\infty + \|f\|_\infty \left(\frac{m_k^{m_k/2}}{\Gamma(m_k+1)} - 1\right) + \omega(f; x-\xi) \qquad (2.3)$$

where

$$\omega(g;h) := \sup\{|g(y_1) - g(y_2)| : y_1, y_2 \in [a,b], |y_1 - y_2| \le h\}$$

denotes the classical modulus of continuity of the function $g : [a,b] \to \mathbb{R}$. An explicit calculation, using the rule of de l'Hospital, yields

$$\frac{m_k^{(m_k-1)/2}}{\Gamma(m_k)} \to 0 \qquad \text{and} \qquad \frac{m_k^{m_k/2}}{\Gamma(m_k+1)} - 1 \to 0$$

as $k \to \infty$. Moreover, our results above imply that $\xi \to x$ as $k \to \infty$, and since f is continuous this implies that $\omega(f; x-\xi) \to 0$ as well. Thus we have the uniform convergence to 0 on $[a+\varepsilon, b]$.

Finally we have to prove the uniform convergence on all of $[a,b]$ in the case $m = 0$ if $f(x) = O((x-a)^\delta)$. To this end we proceed much as in the previous case, and in particular we still find the bound (2.3) for $x \geq a + \varepsilon_* = a + m_k^{1/2}$. In the case $x = a$ we trivially have

$$|J_a^{m_k} f(x) - f(x)| = 0 - 0 = 0,$$

and hence it remains to prove the uniform convergence for $x \in [a, a + m_k^{1/2}]$. Since

$$|f(x)| \leq C(x-a)^\delta \leq C m_k^{\delta/2}$$

for this range of x (where C is a constant independent of k) and

$$
\begin{aligned}
|J_a^{m_k} f(x)| &= \frac{1}{\Gamma(m_k)} \left| \int_a^x (x-t)^{m_k-1} f(t)\, dt \right| \\
&\leq \frac{C}{\Gamma(m_k)} \int_a^x (x-t)^{m_k-1} (t-a)^\delta\, dt \\
&= \frac{C}{\Gamma(m_k)} (x-a)^{m_k+\delta} \int_0^1 (1-s)^{m_k-1} s^\delta\, ds \\
&= C \frac{\Gamma(\delta+1)}{\Gamma(m_k+\delta+1)} (x-a)^{m_k+\delta} \\
&\leq C \frac{\Gamma(\delta+1)}{\Gamma(m_k+\delta+1)} m_k^{(m_k+\delta)/2}
\end{aligned}
$$

by the substitution $s = (t-a)/(x-a)$ and the fundamental properties of the Beta function, we find for $x \in [a, a + m_k^{1/2}]$

$$
\begin{aligned}
|J_a^{m_k} f(x) - f(x)| &\leq |J_a^{m_k} f(x)| + |f(x)| \\
&\leq C \frac{\Gamma(\delta+1)}{\Gamma(m_k+\delta+1)} m_k^{(m_k+\delta)/2} + C m_k^{\delta/2}.
\end{aligned}
$$

Recalling that, by de l'Hospital's rule, $m_k^{m_k/2} \to 1$ and hence $m_k^{(m_k+\delta)/2} \to 0$ as $k \to \infty$, we indeed obtain the required uniform convergence result. □

An alternative proof of the second part of this theorem will be given in Remark 6.11.

2.2 Riemann–Liouville Derivatives

Having established these fundamental properties of Riemann–Liouville integral operators, we now come to the corresponding differential operators. To motivate the definition coming up, we recall Lemma 1.2 that (under certain conditions) states the identity

$$D^n f = D^m J_a^{m-n} f$$

where m and n were integers such that $m > n$. Now assume that n is not an integer. Then we may still choose an integer m such that $m > n$. In view of the theory developed in the previous section, the right-hand side of the identity remains meaningful. However, there is one major difference to the classical case where both m and n are integers: Now we find that the operator obtained in this way depends on the choice of the point a. We hence come to the following definition if we choose the value of the integer m to be as small as possible:

Definition 2.2. Let $n \in \mathbb{R}_+$ and $m = \lceil n \rceil$. The operator D_a^n, defined by

$$D_a^n f := D^m J_a^{m-n} f$$

is called the *Riemann–Liouville fractional differential operator of order n*.
  For $n = 0$, we set $D_a^0 := I$, the identity operator.

  Once again we see that, as a consequence of Lemma 1.2, the newly defined operator D_a^n coincides with the classical differential operator D^n whenever $n \in \mathbb{N}$.
  In Lemma 1.2 we had not required m to be as small as possible; indeed arbitrary natural numbers for m were allowed as long as the inequality $m > n$ was satisfied. A similar statement holds here.

Lemma 2.11. *Let $n \in \mathbb{R}_+$ and let $m \in \mathbb{N}$ such that $m > n$. Then,*

$$D_a^n = D^m J_a^{m-n}.$$

Proof. Our assumptions on m imply that $m \geq \lceil n \rceil$. Thus,

$$D^m J_a^{m-n} = D^{\lceil n \rceil} D^{m-\lceil n \rceil} J_a^{m-\lceil n \rceil} J_a^{\lceil n \rceil - n} = D^{\lceil n \rceil} J_a^{\lceil n \rceil - n} = D_a^n$$

in view of the semigroup property of fractional integration and (1.1). $\qquad\square$

  The next result contains a very simple sufficient condition for the existence of $D_a^n f$.

Lemma 2.12. *Let $f \in A^1[a,b]$ and $0 < n < 1$. Then $D_a^n f$ exists almost everywhere in $[a,b]$. Moreover $D_a^n f \in L_p[a,b]$ for $1 \leq p < 1/n$ and*

$$D_a^n f(x) = \frac{1}{\Gamma(1-n)} \left(\frac{f(a)}{(x-a)^n} + \int_a^x f'(t)(x-t)^{-n} \, dt \right).$$

Proof. We use the definition of the Riemann–Liouville differential operator and the fact that $f \in A^1$. This yields

$$D_a^n f(x) = \frac{1}{\Gamma(1-n)} \frac{d}{dx} \int_a^x f(t)(x-t)^{-n} \, dt$$

$$= \frac{1}{\Gamma(1-n)} \frac{d}{dx} \int_a^x \left(f(a) + \int_a^t f'(u) \, du \right) (x-t)^{-n} \, dt$$

$$= \frac{1}{\Gamma(1-n)} \frac{d}{dx} \left(f(a) \int_a^x \frac{dt}{(x-t)^n} + \int_a^x \int_a^t f'(u)(x-t)^{-n} \, du \, dt \right)$$

$$= \frac{1}{\Gamma(1-n)} \left(\frac{f(a)}{(x-a)^n} + \frac{d}{dx} \int_a^x \int_a^t f'(u)(x-t)^{-n} \, du \, dt \right).$$

By Fubini's Theorem we may interchange the order of integration in the double integral. This yields

$$D_a^n f(x) = \frac{1}{\Gamma(1-n)} \left(\frac{f(a)}{(x-a)^n} + \frac{d}{dx} \int_a^x f'(u) \frac{(x-u)^{1-n}}{1-n} \, du \right).$$

The standard rules on the differentiation of parameter integrals then give the desired representation. The integrability statement is an immediate consequence of this representation using classical results from Lebesgue integration theory.   □

As an immediate consequence of this definition, we shall state the fractional derivatives of some elementary functions.

Example 2.4. Let $f(x) = (x-a)^\beta$ for some $\beta > -1$ and $n > 0$. Then, in view of Example 2.1,

$$D_a^n f(x) = D^{\lceil n \rceil} J_a^{\lceil n \rceil - n} f(x) = \frac{\Gamma(\beta+1)}{\Gamma(\lceil n \rceil - n + \beta + 1)} D^{\lceil n \rceil} [(\cdot - a)^{\lceil n \rceil - n + \beta}](x).$$

Specifically, if $n - \beta \in \mathbb{N}$, the right-hand side is the $\lceil n \rceil$-th derivative of a classical polynomial of degree $\lceil n \rceil - (n - \beta) \in \{0, 1, \ldots, \lceil n \rceil - 1\}$, and so the expression vanishes, i.e.

$$D_a^n [(\cdot - a)^{n-m}](x) = 0 \text{ for all } n > 0, \, m \in \{1, 2, \ldots, \lceil n \rceil\}.$$

On the other hand, if $n - \beta \notin \mathbb{N}$, we find

$$D_a^n [(\cdot - a)^\beta](x) = \frac{\Gamma(\beta+1)}{\Gamma(\beta+1-n)} (x-a)^{\beta-n}.$$

Both these relations are straightforward generalizations of what we know for integer-order derivatives. Note that in the last expression the argument of the Gamma function may be negative. We discuss the interpretation of this in Appendix D.1.

The following example shows, however, that not every relation can be carried over in a direct fashion from the classical setting to fractional derivatives.

Example 2.5. Let $f(x) = \exp(\lambda x)$ for some $\lambda > 0$, and let $n > 0$, $n \notin \mathbb{N}$. Then

$$D_a^n f(x) = \frac{\exp(\lambda a)}{\Gamma(1-n)} (x-a)^{-n} {}_1F_1(1; 1-n; \lambda(x-a)).$$

Here $_1F_1$ denotes Kummer's confluent hypergeometric function. This expression looks somewhat different from the familiar

$$D^n f(x) = \lambda^n \exp(\lambda x) \tag{2.4}$$

that holds for $n \in \mathbb{N}$. This strange behaviour is in a certain sense related to the non-uniqueness of fractional integral (and differential) operators. If we had chosen $a = -\infty$, then we could have obtained (2.4). The corresponding operators are known as *Liouville fractional integrals* and have been investigated; cf. e.g. [167]. For our purposes however they have no major role to play because they would naturally lead us to an analysis on unbounded intervals, and that would not be a natural setting for the differential equations to be considered later on. So we shall not use them any further.

We have seen in Theorem 2.2 that the Riemann–Liouville integral operators form a semigroup. It is an immediate consequence of their definition that the classical differential operators $\{D^n : n \in \mathbb{N}_0\}$ also have a semigroup property. Therefore it is natural to ask when the Riemann–Liouville differential operators have got such a structure. We begin our investigations in this direction with a positive result.

Theorem 2.13. *Assume that $n_1, n_2 \geq 0$. Moreover let $\phi \in L_1[a,b]$ and $f = J_a^{n_1+n_2} \phi$. Then,*

$$D_a^{n_1} D_a^{n_2} f = D_a^{n_1+n_2} f.$$

Note that in order to apply this identity we do not need to know the function ϕ explicitly; it is sufficient to know that such a function exists. In view of Theorem 2.6, the condition on f implies not only a certain degree of smoothness but also the fact that, as $x \to a$, $f(x) \to 0$ sufficiently fast.

Proof. By our assumption on f and the definition of the Riemann–Liouville differential operator,

$$D_a^{n_1} D_a^{n_2} f = D_a^{n_1} D_a^{n_2} J_a^{n_1+n_2} \phi = D^{\lceil n_1 \rceil} J_a^{\lceil n_1 \rceil - n_1} D^{\lceil n_2 \rceil} J_a^{\lceil n_2 \rceil - n_2} J_a^{n_1+n_2} \phi.$$

The semigroup property of the integral operators allows us to rewrite this expression as

$$D_a^{n_1} D_a^{n_2} f = D^{\lceil n_1 \rceil} J_a^{\lceil n_1 \rceil - n_1} D^{\lceil n_2 \rceil} J_a^{\lceil n_2 \rceil + n_1} \phi$$
$$= D^{\lceil n_1 \rceil} J_a^{\lceil n_1 \rceil - n_1} D^{\lceil n_2 \rceil} J_a^{\lceil n_2 \rceil} J_a^{n_1} \phi.$$

Because of (1.1) and the fact that the orders of the integral and differential operators involved are natural numbers, we find that this is equivalent to

$$D_a^{n_1} D_a^{n_2} f = D^{\lceil n_1 \rceil} J_a^{\lceil n_1 \rceil - n_1} J_a^{n_1} \phi = D^{\lceil n_1 \rceil} J_a^{\lceil n_1 \rceil} \phi$$

where we have once again used the semigroup property of fractional integration. We may now use (1.1) one more time and find that

$$D_a^{n_1} D_a^{n_2} f = \phi.$$

The proof that $D_a^{n_1+n_2} f = \phi$ goes along similar lines.   □

The smoothness and zero condition in this theorem is not just a technicality. The following examples show some cases where the condition is not satisfied. They prove that an unconditional semigroup property of fractional differentiation in the Riemann–Liouville sense does not hold.

Example 2.6. Let $f(x) = x^{-1/2}$ and $n_1 = n_2 = 1/2$. Then, as shown in Example 2.4, $D_0^{n_1} f(x) = D_0^{n_2} f(x) = 0$, and hence also $D_0^{n_1} D_0^{n_2} f(x) = 0$, but $D_0^{n_1+n_2} f(x) = D^1 f(x) = -(2x^{3/2})^{-1}$.

Example 2.7. Let $f(x) = x^{1/2}$, $n_1 = 1/2$ and $n_2 = 3/2$. Then, again using Example 2.4, $D_0^{n_1} f(x) = \sqrt{\pi}/2$ and $D_0^{n_2} f(x) = 0$. This implies $D_0^{n_1} D_0^{n_2} f(x) = 0$ but $D_0^{n_2} D_0^{n_1} f(x) = -x^{-3/2}/4 = D^2 f(x) = D_0^{n_1+n_2} f(x)$.

In other words, the first of these two examples shows that it is possible to have

$$D_a^{n_1} D_a^{n_2} f = D_a^{n_2} D_a^{n_1} f \neq D_a^{n_1+n_2} f,$$

whereas the second one exemplifies the case where

$$D_a^{n_1} D_a^{n_2} f \neq D_a^{n_2} D_a^{n_1} f = D_a^{n_1+n_2} f$$

holds.

Recall that one of the key features that we wanted to obtain was (1.1). It turns out that the Riemann–Liouville definitions indeed have this property.

Theorem 2.14. *Let $n \geq 0$. Then, for every $f \in L_1[a,b]$,*

$$D_a^n J_a^n f = f$$

almost everywhere.

Proof. The case $n = 0$ is trivial for then D_a^n and J_a^n are both the identity operator.

For $n > 0$ we proceed as in the proof of Theorem 2.13: Let $m = \lceil n \rceil$. Then, by the definition of D_a^n, the semigroup property of fractional integration and (1.1) (which may be applied here since $m \in \mathbb{N}$),

$$D_a^n J_a^n f(x) = D^m J_a^{m-n} J_a^n f(x) = D^m J_a^m f(x) = f(x).$$ □

Essentially this result and its proof have already been known to Abel [1] even though he has not denoted the operators involved as integrals and derivatives of fractional order, respectively.

Now we come to an analogue of Theorem 2.7.

Theorem 2.15. *Let $n > 0$. Assume that $(f_k)_{k=1}^\infty$ is a uniformly convergent sequence of continuous functions on $[a,b]$, and that $D_a^n f_k$ exists for every k. Moreover assume that $(D_a^n f_k)_{k=1}^\infty$ converges uniformly on $[a+\varepsilon, b]$ for every $\varepsilon > 0$. Then, for every $x \in (a,b]$, we have*

$$\left(\lim_{k \to \infty} D_a^n f_k \right)(x) = \left(D_a^n \lim_{k \to \infty} f_k \right)(x).$$

Proof. We recall that $D_a^n = D^{\lceil n \rceil} J_a^{\lceil n \rceil - n}$. By Theorem 2.7, the sequence $(J_a^{\lceil n \rceil - n} f_k)_k$ is uniformly convergent, and we may interchange the limit operation and the fractional integral. By assumption, the $\lceil n \rceil$th derivative of this series converges uniformly on every compact subinterval of $(a,b]$. Thus, by a standard theorem from analysis, we may also interchange the limit operator and the differential operator whenever $a < x \le b$. $\square$

We can immediately deduce an analogue of Corollary 2.8.

Corollary 2.16. *Let f be analytic in $(a-h, a+h)$ for some $h > 0$, and let $n > 0$, $n \notin \mathbb{N}$. Then*

$$D_a^n f(x) = \sum_{k=0}^\infty \binom{n}{k} \frac{(x-a)^{k-n}}{\Gamma(k+1-n)} D^k f(x)$$

for $a < x < a + h/2$, and

$$D_a^n f(x) = \sum_{k=0}^\infty \frac{(x-a)^{k-n}}{\Gamma(k+1-n)} D^k f(a)$$

for $a < x < a + h$. In particular, $D_a^n f$ is analytic in $(a, a+h)$.

In this result, the binomial coefficients $\binom{n}{k}$ for $n \in \mathbb{R}$ and $k \in \mathbb{N}_0$ are defined by

$$\binom{n}{k} := \frac{n(n-1)(n-2)\cdots(n-k+1)}{k!}. \tag{2.5}$$

Proof. We use Corollary 2.8 and the definition of the operator D_a^n,

$$D_a^n f(x) = D^{\lceil n \rceil} J_a^{\lceil n \rceil - n} f(x).$$

This immediately yields the last two statements. For the first claim, we proceed in a similar way, using the fact that $k! \Gamma(n)(n+k)\binom{-n}{k} = (-1)^k \Gamma(k+1+n)$ (cf. Exercise 2.2). This allows us to rewrite the first statement of Corollary 2.8 as

$$J_a^{\lceil n \rceil - n} f(x) = \sum_{k=0}^\infty \binom{n - \lceil n \rceil}{k} \frac{(x-a)^{k+\lceil n \rceil - n}}{\Gamma(k+1+\lceil n \rceil - n)} D^k f(x).$$

Differentiating $\lceil n \rceil$ times with respect to x, we find

$$D_a^n f(x) = \sum_{k=0}^\infty \binom{n - \lceil n \rceil}{k} \frac{1}{\Gamma(k+1+\lceil n \rceil - n)} D^{\lceil n \rceil} \left[(\cdot - a)^{k+\lceil n \rceil - n} D^k f \right](x).$$

The classical version of Leibniz' formula (cf. Theorem 2.A below) then yields

$$D_a^n f(x) = \sum_{k=0}^{\infty} \binom{n - \lceil n \rceil}{k} \frac{1}{\Gamma(k+1+\lceil n \rceil - n)}$$

$$\times \sum_{j=0}^{\lceil n \rceil} \binom{\lceil n \rceil}{j} D^{\lceil n \rceil - j} \left[(\cdot - a)^{k+\lceil n \rceil - n} \right] (x) D^{k+j} f(x)$$

$$= \sum_{k=0}^{\infty} \binom{n - \lceil n \rceil}{k} \sum_{j=0}^{\lceil n \rceil} \binom{\lceil n \rceil}{j} \frac{(x-a)^{k+j-n}}{\Gamma(k+1+j-n)} D^{k+j} f(x).$$

By definition, $\binom{\mu}{j} = 0$ if $\mu \in \mathbb{N}$ and $\mu < j$. Thus we may replace the upper limit in the inner sum by ∞ without changing the expression. The substitution $j = \ell - k$ gives

$$D_a^n f(x) = \sum_{k=0}^{\infty} \sum_{\ell=k}^{\infty} \binom{n - \lceil n \rceil}{k} \binom{\lceil n \rceil}{\ell - k} \frac{(x-a)^{\ell-n}}{\Gamma(\ell+1-n)} D^\ell f(x)$$

$$= \sum_{\ell=0}^{\infty} \sum_{k=0}^{\ell} \binom{n - \lceil n \rceil}{k} \binom{\lceil n \rceil}{\ell - k} \frac{(x-a)^{\ell-n}}{\Gamma(\ell+1-n)} D^\ell f(x).$$

An explicit calculation yields

$$\sum_{k=0}^{\ell} \binom{n - \lceil n \rceil}{k} \binom{\lceil n \rceil}{\ell - k} = \binom{n}{\ell},$$

(see also Exercise 2.2) and thus the first claim follows.   □

Another interesting peculiarity of fractional differentiation comes up when we look at the generalizations of the classical rules for differentiating functions that are composed from other functions in a certain way. The first result in this connection is trivial.

Theorem 2.17. *Let f_1 and f_2 be two functions defined on $[a,b]$ such that $D_a^n f_1$ and $D_a^n f_2$ exist almost everywhere. Moreover, let $c_1, c_2 \in \mathbb{R}$. Then, $D_a^n(c_1 f_1 + c_2 f_2)$ exists almost everywhere, and*

$$D_a^n(c_1 f_1 + c_2 f_2) = c_1 D_a^n f_1 + c_2 D_a^n f_2.$$

Proof. This linearity property of the fractional differential operator is an immediate consequence of the definition of D_a^n.   □

When it comes to products of functions, the situation is completely different. In the classical case, we have the following well known result (that we have already used in the proof of Corollary 2.16):

Theorem 2.A (Leibniz' formula). *Let $n \in \mathbb{N}$, and let $f, g \in C^n[a,b]$. Then,*

$$D^n[fg] = \sum_{k=0}^{n} \binom{n}{k} (D^k f)(D^{n-k} g).$$

We point out two special properties of this result: The formula is symmetric, i.e. we may interchange f and g on both sides of the equation without altering the expression, and in order to evaluate the nth derivative of the product fg, we only need derivatives up to the order n of both factors. In particular, none of the factors needs to have an $(n+1)$st derivative. The following theorem transfers Leibniz' formula to the fractional setting, and it is immediately evident that both these properties are lost.

Theorem 2.18 (Leibniz' formula for Riemann–Liouville operators). *Let $n > 0$, and assume that f and g are analytic on $(a - h, a + h)$ with some $h > 0$. Then,*

$$D_a^n[fg](x) = \sum_{k=0}^{\lfloor n \rfloor} \binom{n}{k}(D_a^k f)(x)(D_a^{n-k}g)(x) + \sum_{k=\lfloor n \rfloor+1}^{\infty} \binom{n}{k}(D_a^k f)(x)(J_a^{k-n}g)(x)$$

for $a < x < a + h/2$.

Note in the theorem that k runs through the nonnegative integers; therefore we could have written $D^k f$ instead of $D_a^k f$ on the right-hand side. Moreover, k runs through *all* the nonnegative integers, and thus we need to have $f \in C^\infty[a,b]$ in order to have a right-hand side that makes sense. The smoothness requirements of g seem to be much less restrictive (derivatives of g are only required up to the order n), but for our proof we need analyticity of the product fg, and this is generally only assured if g is analytic too. This shows that the two main properties stated above are indeed lost. Finally we mention that Riemann–Liouville *integral* operators arise on the right-hand side. No such expressions were present in the classical formulation. In spite of all these differences we recover the classical result from the fractional result by using an integer value for n because then the binomial coefficients $\binom{n}{k}$ are zero for $k > n$, so that the second sum (the one that causes all the differences) vanishes.

Proof. In view of Corollary 2.16 we have

$$D_a^n[fg](x) = \sum_{k=0}^{\infty} \binom{n}{k}\frac{(x-a)^{k-n}}{\Gamma(k+1-n)}D^k[fg](x).$$

Now we apply the standard Leibniz formula to $D^k[fg]$ and interchange the order of summation. This yields

$$D_a^n[fg](x) = \sum_{k=0}^{\infty} \binom{n}{k}\frac{(x-a)^{k-n}}{\Gamma(k+1-n)} \sum_{j=0}^{k} \binom{k}{j}D^j f(x)D^{k-j}g(x)$$

$$= \sum_{j=0}^{\infty}\sum_{k=j}^{\infty} \binom{n}{k}\frac{(x-a)^{k-n}}{\Gamma(k+1-n)} \binom{k}{j}D^j f(x)D^{k-j}g(x)$$

$$= \sum_{j=0}^{\infty} D^j f(x) \sum_{\ell=0}^{\infty} \binom{n}{\ell+j}\frac{(x-a)^{\ell+j-n}}{\Gamma(\ell+j+1-n)} \binom{\ell+j}{j}D^\ell g(x).$$

The observation

$$\binom{n}{\ell+j}\binom{\ell+j}{j} = \binom{n}{j}\binom{n-j}{\ell}$$

gives us

$$
\begin{aligned}
D_a^n[fg](x) &= \sum_{j=0}^{\infty} D^j f(x) \binom{n}{j} \sum_{\ell=0}^{\infty} \binom{n-j}{\ell} \frac{(x-a)^{\ell+j-n}}{\Gamma(\ell+j+1-n)} D^\ell g(x) \\
&= \sum_{j=0}^{\lfloor n \rfloor} \binom{n}{j} D^j f(x) \sum_{\ell=0}^{\infty} \binom{n-j}{\ell} \frac{(x-a)^{\ell+j-n}}{\Gamma(\ell+j+1-n)} D^\ell g(x) \\
&\quad + \sum_{j=\lfloor n \rfloor+1}^{\infty} \binom{n}{j} D^j f(x) \sum_{\ell=0}^{\infty} \binom{n-j}{\ell} \frac{(x-a)^{\ell+j-n}}{\Gamma(\ell+j+1-n)} D^\ell g(x).
\end{aligned}
$$

By the first parts of Corollaries 2.16 and 2.8, respectively, we may replace the inner sums, and the desired result follows.   □

The other important rule for the evaluation is the chain rule,

$$D[g(f(\cdot))](x) = (Dg)(f(x))Df(x).$$

In the same sense as Leibniz' formula is the generalization of the product rule for first derivatives to derivatives of arbitrary order $n \in \mathbb{N}$, the chain rule can also be generalized to the case of nth derivatives with $n \in \mathbb{N}$. The result is known as Faà di Bruno's formula. It can be written in various forms the best known of which is probably the so-called *set partition version* that we now recall.

Theorem 2.B (Faà di Bruno's formula). *If g and f are functions with a sufficient number of derivatives and $n \in \mathbb{N}$, then*

$$D^n[g(f(\cdot))](x) = \sum (D^k g)(f(x)) \prod_{\mu=1}^{n} (D^\mu f(x))^{b_\mu}$$

where the sum is over all partitions of $\{1,2,\dots,n\}$ and for each partition, k is its number of blocks and b_j is the number of blocks with exactly j elements.

A nicely readable account of the history of this formula, including a proof and a discussion of many related aspects, is given in [98]. For our purposes, we only illustrate the formula by an example:

Example 2.8. The fourth derivative of $g(f(\cdot))$ is given by

$$
\begin{aligned}
\frac{d^4}{dx^4} g(f(x)) &= g'(f(x))f^{(4)}(x) + 4g''(f(x))f'(x)f'''(x) + 3g''(f(x))[f''(x)]^2 \\
&\quad + 6g'''(f(x))[f'(x)]^2 f''(x) + g^{(4)}(f(x))[f'(x)]^4.
\end{aligned}
$$

In order to confirm this statement via Faà di Bruno's formula, we need to write up all partitions of $\{1,2,3,4\}$ and count their blocks and the elements in each block. The possible partitions are

$\{1,2,3,4\}$;

$\{1\}, \{2,3,4\}$; $\{2\}, \{1,3,4\}$; $\{3\}, \{1,2,4\}$; $\{4\}, \{1,2,3\}$;

$\{1,2\}, \{3,4\}$; $\{1,3\}, \{2,4\}$; $\{1,4\}, \{2,3\}$;

$\{1\},\{2\}, \{3,4\}$; $\{1\},\{3\}, \{2,4\}$; $\{1\},\{4\}, \{2,3\}$;

$\{2\},\{3\}, \{1,4\}$; $\{2\},\{4\}, \{1,3\}$; $\{3\},\{4\}, \{1,2\}$;

$\{1\},\{2\}, \{3\},\{4\}$.

Hence we have one partition consisting of only one block with four elements (which corresponds to $b_4 = 1$, $b_1 = b_2 = b_3 = 0$), four partitions consisting of two blocks with one and three elements, respectively (i.e. with $b_1 = b_3 = 1$, $b_2 = b_4 = 0$), three partitions consisting of three blocks with 1, 1 and 2 elements, respectively ($b_1 = 2$, $b_2 = 1$, $b_3 = b_4 = 0$), six partitions consisting of two blocks with two elements each ($b_2 = 2$, $b_1 = b_3 = b_4 = 0$), and finally one partition consisting of four blocks with one element each ($b_1 = 4$, $b_2 = b_3 = b_4 = 0$). Taking the sum over all partitions we thus get the required formula.

For this formula, a replacement in the fractional setting is known too [153, §2.7.3]. We state it here without proof for the sake of completeness.

Theorem 2.19 (Faà di Bruno's formula for Riemann–Liouville operators).
Under suitable assumptions on the functions f and g we have

$$D_a^n[f(g(\cdot))](x)$$

$$= \sum_{k=1}^{\infty} \binom{n}{k} \frac{k!(x-a)^{k-n}}{\Gamma(k-n+1)} \sum_{\ell=1}^{k} (D^\ell f)(g(x)) \sum_{(a_1,\ldots,a_k)\in A_{k,\ell}} \prod_{r=1}^{k} \frac{1}{a_r!} \left(\frac{D^r g(x)}{r!} \right)^{a_r}$$

$$+ \frac{(x-a)^{-n}}{\Gamma(1-n)} f(g(x)) \tag{2.6}$$

where $(a_1,\ldots,a_k) \in A_{k,\ell}$ means that

$$a_1,\ldots,a_k \in \mathbb{N}_0, \qquad \sum_{r=1}^{k} r a_r = k \quad and \quad \sum_{r=1}^{k} a_r = \ell.$$

The structure of the right-hand side of (2.6) is so complex that it hardly seems to be of any practical use. We shall therefore not elaborate on this rule any further and instead turn to a completely different type of problems related to fractional differential operators.

Specifically, a natural question to ask is: What can be said about $D_a^n f$ as $n \to m \in \mathbb{N}$? We start by looking at a special case.

Example 2.9. Let $f(x) = (x-a)^k$ for some $k > 0$. Then we have, because of Example 2.4,

$$D^m f(x) = \frac{\Gamma(k+1)}{\Gamma(k+1-m)}(x-a)^{k-m} \quad \text{and} \quad D_a^n f(x) = \frac{\Gamma(k+1)}{\Gamma(k+1-n)}(x-a)^{k-n}.$$

Comparing these two expressions, we find in the limit $n \to m$:

$$D_a^n f(x) \to D^m f(x) \text{ uniformly on } [a,b] \text{ if } m < k.$$

For $m = k$ we obtain that $D^m f(x) = \Gamma(m+1)$, which is a non-zero constant, whereas

$$D_a^n f(a) = \begin{cases} 0 & \text{if } n < m, \\ \infty & \text{if } n > m. \end{cases}$$

So we have pointwise convergence on $(a,b]$, but not at the point a, and therefore uniform convergence on the complete interval $[a,b]$ is not possible. In the remaining case $m > k$ we only obtain pointwise convergence on $(a,b]$ because the difference of the two expressions is always unbounded at a.

In the case where f is of a very general form and not as simple as in this example, we can state a similar, albeit slightly weaker, observation.

Theorem 2.20. *Let $f \in C^m[a,b]$ for some $m \in \mathbb{N}$. Then,*

$$\lim_{n \to m-} D_a^n f = D^m f$$

in a pointwise sense on $(a,b]$. The convergence is uniform on $[a,b]$ if additionally $f(x) = O((x-a)^{m+\delta})$ for some $\delta > 0$ as $x \to a+$.

Proof. In view of the smoothness assumption on f, we may perform a Taylor expansion of f centered at a and find $f(x) = T_{m-1}[f;a](x) + R_{m-1}[f;a](x)$ where

$$T_{m-1}[f;a](x) = \sum_{k=0}^{m-1} \frac{f^{(k)}(a)}{k!}(x-a)^k$$

and R_{m-1} denotes the remainder term. We have that

$$D_a^n f(x) - D^m f(x) = D_a^n T_{m-1}[f;a](x) - D^m T_{m-1}[f;a](x)$$
$$+ D_a^n R_{m-1}[f;a](x) - D^m R_{m-1}[f;a](x).$$

Since T_{m-1} is a polynomial of degree $m-1$, we find that $D^m T_{m-1}[f;a] \equiv 0$. Moreover, by Example 2.4,

$$D_a^n T_{m-1}[f;a](x) = \sum_{k=0}^{m-1} \frac{f^{(k)}(a)}{\Gamma(k+1-n)}(x-a)^{k-n}.$$

Thus

$$D_a^n f(x) - D^m f(x) = D^m J_a^{m-n} R_{m-1}[f;a](x) + \sum_{k=0}^{m-1} \frac{f^{(k)}(a)}{\Gamma(k+1-n)}(x-a)^{k-n}$$
$$- D^m R_{m-1}[f;a](x).$$

The sum is identically zero under the assumption that $f(x) = O((x-a)^m)$ because the relevant derivatives of f vanish at a. Without that additional assumption, it vanishes in the limit as $n \to m$ in a pointwise sense on $(a,b]$ because the arguments of the Gamma functions converge to a nonpositive integer. Since the Gamma function has poles at the nonpositive integers, the limit is zero.

It remains to prove that the difference $D_a^n R_{m-1}[f;a] - D^m R_{m-1}[f;a]$ has the required convergence properties. To do this, we use the integral representation of R_{m-1} which, in our notation, can be written in the form

$$R_{m-1}[f;a](x) = \frac{1}{\Gamma(m)} \int_a^x f^{(m)}(u)(x-u)^{m-1}\, \mathrm{d}u = J_a^m D^m f(x).$$

Thus $D^m R_{m-1}[f;a] = D^m J_a^m D^m f = D^m f$. For the other term we can write

$$D_a^n R_{m-1}[f;a] = D^m J_a^{m-n} R_{m-1}[f;a] = D^m J_a^{m-n} J_a^m D^m f$$
$$= D^m J_a^m J_a^{m-n} D^m f = J_a^{m-n} D^m f$$

in view of the semigroup property of fractional integration and Theorem 2.14. By assumption, $D^m f$ is continuous on $[a,b]$. Thus, pointwise convergence of $J_a^{m-n} D^m f$ against $J_a^0 D^m f = D^m f$ on $(a,b]$ follows from Theorem 2.10.

Moreover, if $f(x) = O((x-a)^{m+\delta})$ then $D^m f(x) = O((x-a)^\delta)$ as $x \to a$, and thus we have uniform convergence on the full interval $[a,b]$ by the last statement of Theorem 2.10. This completes the proof. □

Remark 2.3. There is one more fundamental difference between differential operators of integer order and the Riemann–Liouville fractional derivatives: The former are local operators, the latter are not. The meaning of the word *local* here is as follows. In order to calculate $D^n f(x)$ for $n \in \mathbb{N}$, it is sufficient to know f in an arbitrarily small neighbourhood of x. This follows from the classical representation of D^n as a limit of a difference quotient. However, to calculate $D_a^n f(x)$ for $n \notin \mathbb{N}$, the definition tells us that we need to know f throughout the entire interval $[a,x]$.

This property is exhibited in an even more evident way in the next Lemma that provides an alternative representation of the Riemann–Liouville derivative. This new representation has proven to be rather useful in the development and presentation of certain numerical algorithms (see, e.g., [34]).

Lemma 2.21. *Let $n > 0$, $n \notin \mathbb{N}$, and $m = \lceil n \rceil$. Assume that $f \in C^m[a,b]$ and $x \in [a,b]$. Then,*

$$D_a^n f(x) = \frac{1}{\Gamma(-n)} \int_a^x (x-t)^{-n-1} f(t)\, dt.$$

In this statement, the integral needs some further explanation. The integrand exhibits a singularity of order $n+1$ which is strictly greater than one, and thus the integral will in general exist neither in the proper nor in the improper sense. Therefore we define such an integral according to Hadamard's finite-part integral concept as explained in Appendix D.4.

A very short and simple proof of the Lemma can be given using methods from the theory of generalized functions (distributions) [72]. However, we do not want to introduce this machinery here. We therefore give a more elementary proof that essentially follows the ideas of Elliott [61]. Note that, according to that reference, certain parts of the proof can already be found in a paper of Marchaud [129] dating back to 1927.

Proof. By definition, $D_a^n f(x) = D^m J_a^{m-n} f(x)$ and

$$J_a^{m-n} f(x) = \frac{1}{\Gamma(m-n)} \int_a^x (x-t)^{m-n-1} f(t)\, dt.$$

In view of the smoothness assumptions on f, we may integrate partially in this integral and find that

$$
\begin{aligned}
J_a^{m-n} f(x) &= \frac{1}{\Gamma(m-n+1)} \int_a^x (x-t)^{m-n} f'(t)\, dt \\
&\quad - \frac{1}{\Gamma(m-n+1)} (x-t)^{m-n} f(t) \Big|_{t=a}^{t=x} \\
&= \frac{1}{\Gamma(m-n+1)} (x-a)^{m-n} f(a) + J_a^{m-n+1} f'(x).
\end{aligned}
$$

The smoothness assumptions allow us to repeat this procedure for a total of m times; we find

$$J_a^{m-n} f(x) = \sum_{k=0}^{m-1} \frac{(x-a)^{k+m-n}}{\Gamma(k+m-n+1)} f^{(k)}(a) + J_a^{2m-n} f^{(m)}(x).$$

Thus,

$$D_a^n f(x) = D^m J_a^{m-n} f(x) = \sum_{k=0}^{m-1} \frac{(x-a)^{k-n}}{\Gamma(k-n+1)} f^{(k)}(a) + J_a^{m-n} f^{(m)}(x).$$

But according to Theorem D.14 the expression on the right-hand side of the equation is just $\int_a^x (x-t)^{-n-1} f(t)\, dt / \Gamma(-n)$. □

Remark 2.4. Another interesting feature of this representation appears if we formally assume n to be negative, $n = -n^*$ with some $n^* > 0$. Then the identity of Lemma 2.21 takes the shape of

$$D_a^{-n^*} f(x) = \frac{1}{\Gamma(n^*)} \int_a^x (x-t)^{n^*-1} f(t) \, dt,$$

and we find that the expression on the right-hand side is simply $J_a^{n^*} f(x)$.

2.3 Relations Between Riemann–Liouville Integrals and Derivatives

Having established a theory of Riemann–Liouville differential and integral operators separately, we now investigate how they interact. A very important first result in
this context has already been shown in Theorem 2.14 above: D_a^n is the left inverse of
J_a^n. Of course, we cannot claim that it is the right inverse. More precisely, we have
the following situation.

Theorem 2.22. *Let $n > 0$. If there exists some $\phi \in L_1[a,b]$ such that $f = J_a^n \phi$ then*

$$J_a^n D_a^n f = f$$

almost everywhere.

Proof. This is an immediate consequence of the previous result: We have, by definition of f and Theorem 2.14, that

$$J_a^n D_a^n f = J_a^n [D_a^n J_a^n \phi] = J_a^n \phi = f.$$ ⊓⊔

If f is not as required in the assumptions of Theorem 2.22, then we obtain a
different representation for $J_a^n D_a^n f$.

Theorem 2.23. *Let $n > 0$ and $m = \lfloor n \rfloor + 1$. Assume that f is such that $J_a^{m-n} f \in A^m[a,b]$. Then,*

$$J_a^n D_a^n f(x) = f(x) - \sum_{k=0}^{m-1} \frac{(x-a)^{n-k-1}}{\Gamma(n-k)} \lim_{z \to a+} D^{m-k-1} J_a^{m-n} f(z).$$

Specifically, for $0 < n < 1$ we have

$$J_a^n D_a^n f(x) = f(x) - \frac{(x-a)^{n-1}}{\Gamma(n)} \lim_{z \to a+} J_a^{1-n} f(z).$$

Proof. We first note that the limits on the right-hand side exist because of our assumption on f that implies the continuity of $D^{m-1} J_a^{m-n} f$. Moreover, because of this
assumption, there exists some $\phi \in L_1$ such that $D^{m-1} J_a^{m-n} f = D^{m-1} J_a^{m-n} f(a) + J_a^1 \phi$.
This is a classical differential equation of order $m-1$ for $J_a^{m-n} f$; its solution is easily
seen to be of the form

$$J_a^{m-n} f(x) = \sum_{k=0}^{m-1} \frac{(x-a)^k}{k!} \lim_{z \to a+} D^k J_a^{m-n} f(z) + J_a^m \phi(x). \tag{2.7}$$

Thus, by definition of D_a^n,

$$
\begin{aligned}
J_a^n D_a^n f(x) &= J_a^n D^m J_a^{m-n} f(x) \\
&= J_a^n D^m \left[\sum_{k=0}^{m-1} \frac{(\cdot - a)^k}{k!} \lim_{z \to a+} D^k J_a^{m-n} f(z) + J_a^m \phi \right](x) \\
&= J_a^n D^m J_a^m \phi(x) + \sum_{k=0}^{m-1} \frac{J_a^n D^m[(\cdot - a)^k](x)}{k!} \lim_{z \to a+} D^k J_a^{m-n} f(z) \\
&= J_a^n \phi(x) \tag{2.8}
\end{aligned}
$$

because of Theorem 2.14 (note that D^m annihilates every summand in the sum). Next we apply the operator D_a^{m-n} to both sides of (2.7) and find, in view of Theorem 2.14, that

$$
\begin{aligned}
f(x) &= \sum_{k=0}^{m-1} \frac{D_a^{m-n}[(\cdot - a)^k](x)}{k!} \lim_{z \to a+} D^k J_a^{m-n} f(z) + D_a^{m-n} J_a^m \phi(x) \\
&= \sum_{k=0}^{m-1} \frac{D_a^{m-n}[(\cdot - a)^k](x)}{k!} \lim_{z \to a+} D^k J_a^{m-n} f(z) + D_a^1 J_a^{1-m+n} J_a^m \phi(x).
\end{aligned}
$$

We now invoke Example 2.4 to evaluate the terms in the sum and the semigroup property of fractional integration and Theorem 2.14 to manipulate the remaining term. This yields

$$f(x) = \sum_{k=0}^{m-1} \frac{(x-a)^{k+n-m}}{\Gamma(k+n-m+1)} \lim_{z \to a+} D^k J_a^{m-n} f(z) + J_a^n \phi(x). \tag{2.9}$$

Finally we substitute k in the sum by $m - k - 1$, solve for $J_a^n \phi(x)$ and combine the result with (2.8) to obtain

$$J_a^n D_a^n f(x) = J_a^n \phi(x) = f(x) - \sum_{k=0}^{m-1} \frac{(x-a)^{n-k-1}}{\Gamma(n-k)} \lim_{z \to a+} D^{m-k-1} J_a^{m-n} f(z)$$

as desired.   □

Another important basic result in classical analysis is Taylor's theorem. We have already used its classical version in the proof of Theorem 2.20. It can be stated in the following way which is a bit more instructive than the standard formulation given in most textbooks in the sense that it gives some additional insight into the structure of the set A^m. The classical way to state this theorem can be obtained by considering only the implication (a) $\Rightarrow$ (b) and choosing $y = a$ there. For a proof of the general version presented here we refer to the monograph of Sard [168, §2] that also deals with related problems in a more general setting.

Theorem 2.C (Taylor expansion). *The following statements are equivalent:*

(a) $f \in A^m[a,b]$.
(b) *For every* $x, y \in [a,b]$,

$$f(x) = \sum_{k=0}^{m-1} \frac{(x-y)^k}{k!} D^k f(y) + J_y^m D^m f(x).$$

Based on the results derived so far we can find a fractional generalization of this statement.

Theorem 2.24 (Fractional Taylor expansion). *Under the assumptions of Theorem 2.23, we have*

$$f(x) = \frac{(x-a)^{n-m}}{\Gamma(n-m+1)} \lim_{z \to a+} J_a^{m-n} f(z)$$

$$+ \sum_{k=1}^{m-1} \frac{(x-a)^{k+n-m}}{\Gamma(k+n-m+1)} \lim_{z \to a+} D_a^{k+n-m} f(z) + J_a^n D_a^n f(x).$$

Note that in the case $n \in \mathbb{N}$ we have $m = n + 1$ and hence the limit outside the sum vanishes. We may thus retrieve the classical result (with m replaced by $m-1$).

Proof. From (2.8) and (2.9) we find

$$f(x) = \sum_{k=0}^{m-1} \frac{(x-a)^{k+n-m}}{\Gamma(k+n-m+1)} \lim_{z \to a+} D^k J_a^{m-n} f(z) + J_a^n D_a^n f(x).$$

We now move the summand for $k = 0$ out of the sum; for the remaining terms we apply Lemma 2.11. This gives the result.   □

2.4 Grünwald–Letnikov Operators

In the classical calculus it is well known that derivatives can be expressed as differential quotients, i.e. as limits of difference quotients. For example, we can use backward differences of order n with step size h, denoted and defined by

$$\Delta_h^n f(x) := \sum_{k=0}^{n} (-1)^k \binom{n}{k} f(x-kh), \tag{2.10}$$

to conclude the following classical result.

Theorem 2.D. *Let $n \in \mathbb{N}$, $f \in C^n[a,b]$ and $a < x \leq b$. Then*

$$D^n f(x) = \lim_{h \to 0} \frac{\Delta_h^n f(x)}{h^n}.$$

This result is actually not only useful for analytical investigations; by using a finite positive value for h instead of performing the limit operation $h \to 0$ it also gives us a straightforward numerical approximation for the derivative. In view of these advantages of this representation it is evidently desirable to have an analogue also for the fractional case. Such a construction is possible; it dates back to the work of Grünwald [85] and Letnikov [112]. Indeed all we have to do is give a meaning to the finite difference in (2.10) for $n \notin \mathbb{N}$. To this end we recall that the binomial coefficients with non-integer upper coefficient have been introduced in (2.5) above. We had already noted and used the property that $\binom{n}{k} = 0$ if $n \in \mathbb{N}$ and $n < k$. Thus, for $n \in \mathbb{N}$ (2.10) is equivalent to

$$\Delta_h^n f(x) := \sum_{k=0}^{\infty} (-1)^k \binom{n}{k} f(x - kh). \qquad (2.11)$$

Now recall that we want to have an expression for our function class, which is typically a subset of $C[a,b]$, i.e. a class of functions defined on the finite interval $[a,b]$. In this context we observe that the representation (2.11) introduces two problems in the case $n \notin \mathbb{N}$ where none of the binomial coefficients vanishes, so that this expression really represents an infinite series:

- In order to evaluate the expression in (2.11) for all $x \in (a,b]$, the function f needs to be defined on $(-\infty, b]$
- The function f must be such that the series converges

These two problems can be resolved simultaneously by a simple concept: Given a function $f : [a,b] \to \mathbb{R}$, define a new function

$$f^* : (-\infty, b] \to \mathbb{R}, \qquad x \mapsto \begin{cases} f(x) & \text{if } x \in [a,b], \\ 0 & \text{if } x \in (-\infty, a), \end{cases}$$

and use this function instead of the original f. In view of the fact that f and f^* coincide on the interval where both functions are defined, we interpret f^* as a continuation of f and, slightly abusing the notation, we will from now on write f instead of f^*.

This leads us to the required generalization of the concept of differential quotients. For the sake of simplicity, we impose a restriction in the way that $h \to 0$; specifically for the value of x under consideration we assume that h takes only the values $h_N = (x-a)/N$, $N = 1, 2, \ldots$. By a tedious analysis it is possible to show that this restriction is not necessary, but we will not go into details here.

Definition 2.3. Let $n > 0$, $f \in C^{\lceil n \rceil}[a,b]$ and $a < x \leq b$. Then

$$\widetilde{D}_a^n f(x) = \lim_{N \to \infty} \frac{\Delta_{h_N}^n f(x)}{h_N^n} = \lim_{N \to \infty} \frac{1}{h_N^n} \sum_{k=0}^{N} (-1)^k \binom{n}{k} f(x - kh_N)$$

with $h_N = (x - a)/N$ is called the *Grünwald–Letnikov fractional derivative* of order n of the function f.

The following result explains the relation between this new notion of a fractional derivative and the one that we already know.

Theorem 2.25. *Let $n > 0$, $m = \lceil n \rceil$ and $f \in C^m[a,b]$. Then, for $x \in (a,b]$,*

$$\widetilde{D}_a^n f(x) = D_a^n f(x).$$

Proof. If $n \in \mathbb{N}$ then this is evident because, as indicated above, the differential quotients reduce to the classical version which is covered by Theorem 2.D. Thus we now concentrate on the case $n \notin \mathbb{N}$.

We will follow Elliott [61]. First of all we note that it is no loss of generality to assume $a = 0$ and $x = 1$ because any other interval $[a,x]$ may be mapped to $[0,1]$ by an affine transformation, and the entire convergence analysis below will remain unchanged by this transformation. Thus we have to show that

$$\lim_{N \to \infty} N^n \sum_{k=0}^{N} (-1)^k \binom{n}{k} f\left(1 - \frac{k}{N}\right) = D_0^n f(1). \tag{2.12}$$

In this context let us write

$$D_0^n f(1) - N^n \sum_{k=0}^{N} (-1)^k \binom{n}{k} f\left(1 - \frac{k}{N}\right)$$

$$= \left[1 - \frac{N^n \Gamma(N+1-n)}{\Gamma(N+1)}\right] D_0^n f(1)$$

$$+ N^n \left(\frac{\Gamma(N+1-n)}{\Gamma(N+1)} D_0^n f(1) - \sum_{k=0}^{N} (-1)^k \binom{n}{k} f\left(1 - \frac{k}{N}\right)\right).$$

As $N \to \infty$, the expression in square brackets on the right-hand side of this equation converges to zero because, by Stirling's formula (Theorem D.5) and de l'Hospital's rule,

$$\frac{N^n \Gamma(N+1-n)}{\Gamma(N+1)} = N^n \left(\frac{N-n}{e}\right)^{N-n} \left(\frac{e}{N}\right)^N \frac{\sqrt{2\pi(N-n)}}{\sqrt{2\pi N}} (1 + o(1))$$

$$= e^n \left(1 - \frac{n}{N}\right)^{N-n} (1 + o(1)) = 1 + o(1).$$

Thus, in order to prove our desired result (2.12), it suffices to show that

$$N^n \left(\frac{\Gamma(N+1-n)}{\Gamma(N+1)} D_0^n f(1) - \sum_{k=0}^{N} (-1)^k \binom{n}{k} f\left(1 - \frac{k}{N}\right) \right) \xrightarrow{N \to \infty} 0. \tag{2.13}$$

To this end we introduce an auxiliary concept from approximation theory, the Nth *Bernstein polynomial* of the function f, denoted and defined by

$$B_N[f](t) := \sum_{k=0}^{N} \binom{N}{k} t^k (1-t)^{N-k} f\left(\frac{k}{N}\right)$$

(see Appendix D.5).

The connection to our approach is established as follows. For $N \in \mathbb{N}_0$ and $k \in \{0, 1, \ldots, N\}$ we define $b_{k,N}(t) := t^k (1-t)^{N-k}$. Note that these functions are related to the Bernstein polynomial via $B_N[f] = \sum_{k=0}^{N} \binom{N}{k} f(k/N) b_{k,N}$. An explicit calculation then yields, by Lemma 2.21,

$$D_0^n b_{k,N}(1) = \frac{1}{\Gamma(-n)} \int_0^1 (1-t)^{N-k-n-1} t^k \, dt = \frac{\Gamma(k+1)\Gamma(N-k-n)}{\Gamma(-n)\Gamma(N-n+1)}. \tag{2.14}$$

(This can be shown by using the Beta integral and its analytic continuation.) As a consequence of this relation we may express the sum on the left-hand side of (2.13) according to

$$\sum_{k=0}^{N} (-1)^k \binom{n}{k} f\left(1 - \frac{k}{N}\right) = \sum_{k=0}^{N} \frac{\Gamma(k-n)}{\Gamma(k+1)\Gamma(-n)} f\left(1 - \frac{k}{N}\right)$$

$$= \sum_{k=0}^{N} \frac{\Gamma(N-k-n)}{\Gamma(N-k+1)\Gamma(-n)} f\left(\frac{k}{N}\right)$$

$$= \frac{1}{\Gamma(N+1)} \sum_{k=0}^{N} \binom{N}{k} \frac{\Gamma(k+1)\Gamma(N-k-n)}{\Gamma(-n)} f\left(\frac{k}{N}\right)$$

$$= \frac{\Gamma(N+1-n)}{\Gamma(N+1)} \sum_{k=0}^{N} \binom{N}{k} D_0^n b_{k,N}(1) f\left(\frac{k}{N}\right)$$

$$= \frac{\Gamma(N+1-n)}{\Gamma(N+1)} D_0^n \left[\sum_{k=0}^{N} \binom{N}{k} f\left(\frac{k}{N}\right) b_{k,N} \right](1)$$

$$= \frac{\Gamma(N+1-n)}{\Gamma(N+1)} D_0^n B_N[f](1).$$

Hence our claim (2.13) reduces to

$$N^n \frac{\Gamma(N+1-n)}{\Gamma(N+1)} \left(D_0^n f(1) - D_0^n B_N[f](1) \right)$$

$$= N^n \frac{\Gamma(N+1-n)}{\Gamma(N+1)} D_0^n (f - B_N[f])(1) \to 0.$$

We had already seen above that

$$N^n \frac{\Gamma(N+1-n)}{\Gamma(N+1)} \to 1$$

as $N \to \infty$; thus it only remains to prove that $D_0^n(f - B_N[f])(1) \to 0$ as $N \to \infty$. For this last step we use the representation for D_0^n from Lemma 2.21 and the fundamental property of the finite-part integral described in Theorem D.14 which yields

$$D_0^n(f - B_N[f])(1)$$
$$= \frac{1}{\Gamma(-n)} \int_0^1 (1-t)^{-n-1}(f(t) - B_N[f](t))\,dt$$
$$= \sum_{k=0}^{\lceil n \rceil - 1} \frac{1}{\Gamma(k-n+1)} D^k[f - B_N[f]](0) + J_0^{\lceil n \rceil - n} D^{\lceil n \rceil}[f - B_N[f]](1).$$

We may now invoke Theorem D.16 which tells us that, under our assumptions, $D^k B_N[f]$ converges to $D^k f$ uniformly on $[0,1]$ for $k = 0, 1, \ldots, \lceil n \rceil$, and thus the entire expression on the right-hand side converges to zero as required. $\qquad \square$

Definition 2.3 immediately raises the question what happens if we replace n by $-n$. It turns out that this question has a simple answer.

Theorem 2.26. *Let $n > 0$, $f \in C[a,b]$ and $a \le x \le b$. Then, with $h_N = (x-a)/N$, we have*

$$J_a^n f(x) = \lim_{N \to \infty} h_N^n \sum_{k=0}^N (-1)^k \binom{-n}{k} f(x - kh_N).$$

Proof. For $x = a$, both sides of the equation vanish.

For $x > a$, we proceed formally as in the proof of Theorem 2.25, only substituting $-n$ for n throughout the entire argument and replacing (2.14) by

$$J_0^n b_{k,N}(1) = \frac{1}{\Gamma(n)} \int_0^1 (1-t)^{N-k+n-1} t^k\,dt = \frac{\Gamma(k+1)\Gamma(N-k+n)}{\Gamma(n)\Gamma(N+n+1)}.$$

which also can be deduced with the help of the Beta integral. We then arrive at the conclusion that our claim is equivalent to the statement

$$J_0^n(f - B_N[f])(1) \to 0 \qquad \text{as } N \to \infty$$

which follows from Theorem 2.7 since $B_N[f] \to f$ uniformly by Theorem D.16. $\qquad \square$

In view of this Theorem and the relation

$$(-1)^k \binom{-n}{k} = (-1)^k \frac{(-n)(-n-1)\cdots(-n-k+1)}{k!}$$
$$= \frac{n(n+1)\cdots(n+k-1)}{k!}$$

$$= \frac{(n+k-1)(n+k-2)\cdots n}{k!}$$

$$= \binom{n+k-1}{k} = \frac{\Gamma(n+k)}{\Gamma(n)\Gamma(k+1)},$$

the following formal definition is justified.

Definition 2.4. Let $n > 0$, $f \in C[a,b]$ and $a < x \leq b$. Then

$$\widetilde{J}_a^n f(x) := \frac{1}{\Gamma(n)} \lim_{N\to\infty} h_N^n \sum_{k=0}^{N} \frac{\Gamma(n+k)}{\Gamma(k+1)} f(x - kh_N)$$

with $h_N = (x-a)/N$ is called the *Grünwald–Letnikov fractional integral* of order n of the function f.

Indeed this is the historically correct definition introduced in the original papers of Grünwald [85] and Letnikov [112] in 1867 and 1868, respectively. Thus the Grünwald–Letnikov representation allows a formal unification of the concepts of fractional derivatives and fractional integrals by admitting both positive and negative values for n. Some authors have been motivated by this potential unification to introduce a unified notation concept for fractional integrals and derivatives, i.e. they would, e.g., write D_a^{-n} instead of J_a^n for $n > 0$. We have chosen to stick to separate symbols for differential and integral operators because it is then immediately clear from the notation whether the operator under consideration is of differential or integral type.

For a more detailed discussion of Grünwald–Letnikov differential and integral operators we refer to the monograph of Samko et al. [167, §20].

Exercises

Exercise 2.1. Compute the Riemann–Liouville integrals $J_0^n f(x)$ for $n > 0$ and the following functions f:

(a) $f(x) = \sin \omega x$, $\omega > 0$,
(b) $f(x) = \cos \omega x$, $\omega > 0$,
(c) $f(x) = (1+x)^{-1}$.

Hint: Proceed as in Example 2.2.

Exercise 2.2. Give a proof of relation (2.1).

Exercise 2.3. Prove the identity stated in Example 2.5.

Exercise 2.4. Prove the following identities that we required in the proof of Corollary 2.16:

(a)

$$(-1)^k \Gamma(k+1+n) = k!\Gamma(n)(n+k)\binom{-n}{k},$$

(b)

$$\sum_{k=0}^{\ell} \binom{n-\lceil n\rceil}{k}\binom{\lceil n\rceil}{\ell-k} = \binom{n}{\ell}.$$

Exercise 2.5. Compute the Riemann–Liouville derivatives $D_0^n f(x)$ for $n > 0$ and the following functions f:

(a) $f(x) = \sin \omega x,\ \omega > 0$,
(b) $f(x) = \cos \omega x,\ \omega > 0$,
(c) $f(x) = (1+x)^{-1}$.

Hint: Use the results obtained in Exercise 2.1.

Exercise 2.6. Determine the fractional Taylor expansions for the following cases:

(a) $f(x) = x^3$, $a = 0$, and $n = 2.8$,
(b) $f(x) = \sin x$, $a = 0$, and $n = 1.2$.

Chapter 3
Caputo's Approach

It turns out that the Riemann–Liouville derivatives have certain disadvantages when trying to model real-world phenomena with fractional differential equations. We shall therefore now discuss a modified concept of a fractional derivative. As we will see below when comparing the two ideas, this second one seems to be better suited to such tasks.

3.1 Definition and Basic Properties

We commence with a preliminary definition.

Definition 3.1. Let $n \geq 0$ and $m = \lceil n \rceil$. Then, we define the operator $\widehat{D}_a^n$ by

$$\widehat{D}_a^n f := J_a^{m-n} D^m f$$

whenever $D^m f \in L_1[a,b]$.

Let us start by looking at the case $n \in \mathbb{N}$. Here we have $m = n$ and hence our definition implies

$$\widehat{D}_a^n f = J_a^0 D^n f = D^n f,$$

i.e. we recover the standard definition in the classical case.

We begin the analysis of this operator in the strictly fractional case $n \notin \mathbb{N}$ with a simple example.

Example 3.1. Let $f(x) = (x-a)^\beta$ for some $\beta \geq 0$. Then,

$$\widehat{D}_a^n f(x) = \begin{cases} 0 & \text{if } \beta \in \{0,1,2,\ldots,m-1\}, \\ \dfrac{\Gamma(\beta+1)}{\Gamma(\beta+1-n)}(x-a)^{\beta-n} & \text{if } \beta \in \mathbb{N} \text{ and } \beta \geq m \\ & \text{or } \beta \notin \mathbb{N} \text{ and } \beta > m-1. \end{cases}$$

The reader is encouraged to compare this statement with the corresponding one for Riemann–Liouville operators (Example 2.4). Notice in particular that the two

K. Diethelm, *The Analysis of Fractional Differential Equations*, 49
Lecture Notes in Mathematics 2004, DOI 10.1007/978-3-642-14574-2_3,
© Springer-Verlag Berlin Heidelberg 2010

operators have different kernels, and that the domains of the two operators (exhibited here in terms of the allowed range of the parameter β) are also different.

A few additional examples of Caputo-type derivatives of certain important functions are collected for the reader's convenience in Appendix B.

Remark 3.1. We have required $m = \lceil n \rceil$ in Definition 3.1. The same condition has been imposed in the definition $D_a^n := D^m J_a^{m-n}$ of the Riemann–Liouville derivative (Definition 2.2). However, in the latter case we had seen in Lemma 2.11 that this restriction actually is not necessary; one may use any $m \in \mathbb{N}$ with $m \geq n$ in the Riemann–Liouville case. For the newly introduced operator $\widehat{D}_a^n := J_a^{m-n} D^m$ from Definition 3.1, the situation is different: Here we may not replace $m = \lceil n \rceil$ by some $m \in \mathbb{N}$ with $m > \lceil n \rceil$. This is evident by looking at the simple example $f(x) = (x - a)^{\lceil n \rceil}$. For such a function we have, according to Example 3.1,

$$\widehat{D}_a^n f(x) = \frac{\Gamma(\lceil n \rceil + 1)}{\Gamma(\lceil n \rceil + 1 - n)} (x - a)^{\lceil n \rceil - n}$$

but, for $m \in \mathbb{N}$ with $m > \lceil n \rceil$, we obtain $D^m f(x) = 0$ and hence $J_a^{m-n} D^m f(x) = 0$ too.

The key to the construction of the alternative differential operator that we are looking for is the following identity involving Riemann–Liouville derivatives on the one hand and the newly defined operator on the other hand.

Theorem 3.1. *Let $n \geq 0$ and $m = \lceil n \rceil$. Moreover assume that $f \in A^m[a,b]$. Then,*

$$\widehat{D}_a^n f = D_a^n[f - T_{m-1}[f;a]]$$

almost everywhere. Here, as in the proof of Theorem 2.20, $T_{m-1}[f;a]$ denotes the Taylor polynomial of degree $m - 1$ for the function f, centered at a; in the case $m = 0$ we define $T_{m-1}[f;a] := 0$.

Note that the expression on the right-hand side of the equation exists if $D_a^n f$ exists and f possesses $m - 1$ derivatives at a, the latter condition making sure that the Taylor polynomial exists. This condition is weaker than the previous condition that $f \in A^m$. (This follows since $f \in A^m$ implies (a) $f \in C^{m-1}$ and hence the existence of the required Taylor polynomial and its Riemann–Liouville derivative, and (b) the existence of $D_a^n f$ almost everywhere as can be seen by a repeated application of the ideas used in the proof of Lemma 2.12.) Therefore we will, from now on, use the latter expression. A formalization is given as follows.

Definition 3.2. Assume that $n \geq 0$ and that f is such that $D_a^n[f - T_{m-1}[f;a]]$ exists, where $m = \lceil n \rceil$. Then we define the function $D_{*a}^n f$ by

$$D_{*a}^n f := D_a^n[f - T_{m-1}[f;a]].$$

The operator D_{*a}^n is called the *Caputo differential operator of order n.*

Actually this concept has been introduced independently by many authors, including Caputo [23] and Rabotnov [157] who have based their developments on the approach given in Definition 3.1 and by Dzherbashyan and Nersesian [58] who have used Definition 3.2 as their starting point; other contributors who have dealt with such operators from various points of view include Gross [84] and Gerasimov [74], and it can even be found in a very old paper by Liouville [116, p. 10, formula (B)]. However it seems that Liouville did not see the difference between this operator and the Riemann–Liouville operator as he was mainly interested in those cases where the two operators coincide [Lützen, J., 2001, Private communication]. We follow the most common convention of naming it after Caputo only. The reader who is interested in a detailed historical account should consult the recent paper by Rossikhin [163] and the references cited therein. The notation that we have introduced here follows the generally accepted suggestion of Gorenflo and Mainardi [81].

Once again we note for $n \in \mathbb{N}$ that $m = n$ and hence

$$D_{*a}^n f = D_a^n [f - T_{n-1}[f;a]] = D^n f - D^n(T_{n-1}[f;a]) = D^n f$$

because $T_{n-1}[f;a]$ is a polynomial of degree $n - 1$ that is annihilated by the classical operator D^n. So in this case we recover the usual differential operator as well. In particular, D_{*a}^0 is once again the identity operator.

Various papers and books exist where some of the key properties of the Caputo operators have been described, see, e.g., [77, 81, 153]. Typically however they were, as stated explicitly in the abstract of [81], written "in a way accessible to applied scientists" and "avoiding unproductive generalities and excessive mathematical rigor". It seems that mathematically rigorous proofs of many important properties are not available in the literature. Therefore we try to give them here.

Proof (of Theorem 3.1). In the case $n \in \mathbb{N}$ the statement is trivial because, as we have seen above, both sides of the equation reduce to $D^n f$. We therefore only have to consider the case $n \notin \mathbb{N}$, which implies that $m > n$.

In this case we have

$$D_a^n[f - T_{m-1}[f;a]](x) = D^m J_a^{m-n}[f - T_{m-1}[f;a]](x)$$
$$= \frac{d^m}{dx^m} \int_a^x \frac{(x-t)^{m-n-1}}{\Gamma(m-n)} (f(t) - T_{m-1}[f;a](t)) \, dt. \quad (3.1)$$

A partial integration of the integral is permitted and yields

$$\int_a^x \frac{1}{\Gamma(m-n)} (f(t) - T_{m-1}[f;a](t))(x-t)^{m-n-1} \, dt$$
$$= -\frac{1}{\Gamma(m-n+1)} \left[(f(t) - T_{m-1}[f;a](t))(x-t)^{m-n} \right]_{t=a}^{t=x}$$
$$+ \frac{1}{\Gamma(m-n+1)} \int_a^x (Df(t) - DT_{m-1}[f;a](t))(x-t)^{m-n} \, dt.$$

The term outside the integral is zero (the first factor vanishes at the lower bound, the second vanishes at the upper bound). Thus,

$$J_a^{m-n}[f - T_{m-1}[f;a]] = J_a^{m-n+1}D[f - T_{m-1}[f;a]].$$

Under our assumptions, we may repeat this process a total number of m times, and this results in

$$J_a^{m-n}[f - T_{m-1}[f;a]] = J_a^{2m-n}D^m[f - T_{m-1}[f;a]] = J_a^m J_a^{m-n}D^m[f - T_{m-1}[f;a]].$$

We note that $D^m T_{m-1}[f;a] \equiv 0$ because $T_{m-1}[f;a]$ is a polynomial of degree $m-1$. Thus, the last identity can be simplified to

$$J_a^{m-n}[f - T_{m-1}[f;a]] = J_a^m J_a^{m-n}D^m f.$$

This may be combined with (3.1) to obtain

$$D_a^n[f - T_{m-1}[f;a]](x) = D^m J_a^m J_a^{m-n}D^m f = J_a^{m-n}D^m f = \widehat{D}_a^n f$$

in view of (1.1).   □

Taking into account the definition of the Caputo operator and Lemma 2.21, we obtain a direct consequence.

Lemma 3.2. *Under the assumptions of Lemma 2.21, we have*

$$D_{*a}^n f(x) = \frac{1}{\Gamma(-n)} \int_a^x (x-t)^{-n-1} (f(t) - T_{m-1}[f;a](t)) \, dt.$$

Remark 3.2. As in the case of the Riemann–Liouville operators, we see that the Caputo derivatives are not local either.

Yet another representation for the Caputo operator can be obtained by combining its definition with Theorem 2.25:

Lemma 3.3. *Let $n > 0$, $m = \lceil n \rceil$ and $f \in C^m[a,b]$. Then, for $x \in (a,b]$,*

$$D_{*a}^n f(x) = \lim_{N \to \infty} \frac{1}{h_N^n} \sum_{k=0}^N (-1)^k \binom{n}{k} [f(x - kh_N) - T_{m-1}[f;a](x - kh_N)]$$

with $h_N = (x-a)/N$.

The representations of these two Lemmas have proven to be useful for numerical work [34, 121]. Other representations are known as well; we shall present some of them in Sect. 3.2. However, we first continue the investigations of the analytical aspects of Caputo operators. In this context our next goal is to express the relation between the Riemann–Liouville operator and the Caputo operator in a different way.

Lemma 3.4. *Let $n \geq 0$ and $m = \lceil n \rceil$. Assume that f is such that both $D_{*a}^n f$ and $D_a^n f$ exist. Then,*

$$D_{*a}^n f(x) = D_a^n f(x) - \sum_{k=0}^{m-1} \frac{D^k f(a)}{\Gamma(k-n+1)} (x-a)^{k-n}.$$

Proof. In view of the definition of the Caputo derivative and Example 2.4,

$$D_{*a}^n f(x) = D_a^n f(x) - \sum_{k=0}^{m-1} \frac{D^k f(a)}{k!} D_a^n[(\cdot - a)^k](x)$$

$$= D_a^n f(x) - \sum_{k=0}^{m-1} \frac{D^k f(a)}{\Gamma(k-n+1)} (x-a)^{k-n}. \qquad \square$$

An immediate consequence of this Lemma is

Lemma 3.5. *Assume the hypotheses of Lemma 3.4. Then,*

$$D_a^n f = D_{*a}^n f$$

holds if and only if f has an m-fold zero at a, i.e. if and only if

$$D^k f(a) = 0 \quad \text{for} \quad k = 0, 1, \dots, m-1.$$

We may also combine Lemma 3.4 with Theorem 2.25 to deduce

Lemma 3.6. *Let $n > 0$, $m = \lceil n \rceil$ and $f \in C^m[a,b]$. Then, for $x \in (a,b)$,*

$$D_{*a}^n f(x) = \lim_{N \to \infty} \frac{1}{h_N^n} \sum_{k=0}^N (-1)^k \binom{n}{k} f(x - kh_N) - \sum_{k=0}^{m-1} \frac{D^k f(a)}{\Gamma(k-n+1)} (x-a)^{k-n}$$

with $h_N = (x-a)/N$.

When it comes to the composition of Riemann–Liouville integrals and Caputo differential operators, we find that the Caputo derivative is also a left inverse of the Riemann–Liouville integral:

Theorem 3.7. *If f is continuous and $n \geq 0$, then*

$$D_{*a}^n J_a^n f = f.$$

Proof. Let $\phi = J_a^n f$. By Theorem 2.5, we have $D^k \phi(a) = 0$ for $k = 0, 1, \dots, m-1$, and thus (in view of Lemma 3.5 and Theorem 2.14)

$$D_{*a}^n J_a^n f = D_{*a}^n \phi = D_a^n \phi = D_a^n J_a^n f = f. \qquad \square$$

Once again, we find that the Caputo derivative is not the right inverse of the Riemann–Liouville integral:

Theorem 3.8. *Assume that $n \geq 0$, $m = \lceil n \rceil$, and $f \in A^m[a,b]$. Then*

$$J_a^n D_{*a}^n f(x) = f(x) - \sum_{k=0}^{m-1} \frac{D^k f(a)}{k!} (x-a)^k.$$

Proof. By Theorem 3.1 and Definition 3.1, we have

$$D_{*a}^n f = \widehat{D}_a^n f = J_a^{m-n} D^m f.$$

Thus, applying the operator J_a^n to both sides of this equation and using the semigroup property of fractional integration, we obtain

$$J_a^n D_{*a}^n f = J_a^n J_a^{m-n} D^m f = J_a^m D^m f.$$

By the classical version of Taylor's theorem (cf. Theorem 2.C), we have that

$$f(x) = \sum_{k=0}^{m-1} \frac{D^k f(a)}{k!} (x-a)^k + J_a^m D^m f(x).$$

Combining these two equations we derive the claim. $\square$

A fractional analogue of Taylor's theorem follows immediately:

Corollary 3.9 (Taylor expansion for Caputo derivatives). *Under the assumptions of Theorem 3.8,*

$$f(x) = \sum_{k=0}^{m-1} \frac{D^k f(a)}{k!} (x-a)^k + J_a^n D_{*a}^n f(x).$$

The relations shown in Theorem 3.8 and Corollary 3.9 have major implications when it comes to the solution of differential equations involving the two types of differential operators. Specifically, assume that h is a given function with the property that there exists some function g such that $h = D_{*a}^n g$. Then, the solution of the Riemann–Liouville differential equation

$$D_a^n f = h$$

is given by

$$f(x) = g(x) + \sum_{j=1}^{\lceil n \rceil} c_j (x-a)^{n-j}$$

with arbitrary constants c_j. This follows by the same techniques that one would employ for differential equations of integer order because the equation is linear and inhomogeneous. By construction, g is a solution of the inhomogeneous equation, and by Example 2.4 each of the terms in the sum solves the corresponding homogeneous equation.

Similarly, if h_* is a given function with the property that $h_* = D^n_{*a} g_*$ and if we want to solve

$$D^n_{*a} f_* = h_*,$$

then we find

$$f_*(x) = g_*(x) + \sum_{j=1}^{\lceil n \rceil} c_j^*(x-a)^{\lceil n \rceil - j},$$

again with arbitrary constants c_j^*. Thus, in order to obtain a unique solution, it is most natural to prescribe the values $f_*(a), Df_*(a), \ldots, D^{\lceil n \rceil - 1} f_*(a)$ in the Caputo setting, whereas in the Riemann–Liouville case one would rather prescribe fractional derivatives of f at a. This will be explored in a more detailed fashion in the following chapters. For the moment we note that the Caputo version is usually preferred when physical models are described because the physical interpretation of the prescribed data is clear, and therefore it is in general possible to provide these data, e.g. by suitable measurements. This is not true for the fractional order initial conditions required for the Riemann–Liouville environment. For example, in applications like the modelling of viscoelastic materials in mechanics [184], $f(x)$ is typically a displacement at time x, and so $f'(x)$ and $f''(x)$ would be the corresponding velocity and acceleration, respectively – quantities that are well understood and easily measured. On the other hand, in spite of recently attempted explanations [154], a fractional derivative of a displacement remains an object whose physical nature is unclear, and so no measurement methods for such a quantity are readily available.

Apart from this reason (which is mainly motivated by arguments in connection with applications) there are actually other reasons coming from the "pure" side of mathematics for preferring the Caputo derivative over the Riemann–Liouville operator [97, 108], but we shall not dwell on this topic here. Rather, we shall continue by stating another representation for the Caputo derivative of a function under a quite natural assumption. To this end we require the following definition that is a special case of a concept established in [73] (see also [167, p. 426]).

Definition 3.3. Let $n > 0$ and let v be an entire function with the power series expansion $v(x) = \sum_{k=0}^{\infty} c_k x^k$. Then, the operator $\mathscr{D}^n$ that maps this function v to the function $\mathscr{D}^n v$ with

$$\mathscr{D}^n v(x) := \sum_{k=1}^{\infty} c_k \frac{\Gamma(kn+1)}{\Gamma(kn+1-n)} x^{k-1}$$

is called the *Gel'fond-Leont'ev operator* of order n.

The Caputo derivatives of certain functions can be expressed with the help of these operators in a very convenient way (see, e.g., [106, p. 139] or [167, p. 426]):

Theorem 3.10. *Let $n > 0$ and let v be an entire function with the power series expansion $v(x) = \sum_{k=0}^{\infty} c_k x^k$. Moreover let $f(x) := v(x^n)$ for $x \geq 0$. Then,*

$$D_{*0}^n f(x) = \mathscr{D}^n v(x^n).$$

Proof. This result follows using a straightforward computation using the defintions of the function f and the Caputo differential operator, the power series expansion of v and the fact that, because v is entire, we may interchange the infinite series operator and the Caputo differential operator (i.e., we may apply the Caputo operator to the series in a term-by-term manner). □

The following result establishes another significant difference between Riemann–Liouville and Caputo derivatives. A comparison with, e.g., Example 2.4 for $f(x) = 1$ and $n > 0$, $n \notin \mathbb{N}$, reveals that we are not allowed to replace D_{*a}^n by D_a^n here.

Lemma 3.11. *Let $n > 0$, $n \notin \mathbb{N}$ and $m = \lceil n \rceil$. Moreover assume that $f \in C^m[a,b]$. Then, $D_{*a}^n f \in C[a,b]$ and $D_{*a}^n f(a) = 0$.*

Proof. By definition and Theorem 3.1, $D_{*a}^n f = J_a^{m-n} D^m f$. The result follows from Theorem 2.5 because $D^m f$ is assumed to be continuous. □

We may relax the conditions on f slightly.

Lemma 3.12. *Let $n > 0$, $n \notin \mathbb{N}$ and $m = \lceil n \rceil$. Moreover let $f \in A^m[a,b]$ and assume that $D_{*a}^{\hat{n}} f \in C[a,b]$ for some $\hat{n} \in (n,m)$. Then, $D_{*a}^n f \in C[a,b]$ and $D_{*a}^n f(a) = 0$.*

Proof. By definition and Theorems 3.1 and 2.2,

$$D_{*a}^n f = J_a^{m-n} D^m f = J_a^{\hat{n}-n} J_a^{m-\hat{n}} D^m f = J_a^{\hat{n}-n} D_{*a}^{\hat{n}} f.$$

Thus the claim follows by virtue of Theorem 2.5. □

Remark 3.3. In order to assess the consequences of these two Lemmata, we point out the following fact. Assume, for example, the hypotheses of Lemma 3.11 and additionally that $n > 2$. Then the Lemma asserts that all derivatives $D_{*a}^n f$, $0 < n < 3$, are continuous, and thus, in particular, $D_{*a}^\ell f$ and $D_{*a}^{\ell+1} f$ are continuous whenever $0 < \ell < 2$. This does not mean, however, that $D_{*a}^\ell f \in C^1[a,b]$ for these ℓ. For a counterexample, we refer to Exercise 3.1. As a consequence of this observation, we obtain that we cannot deduce the identity $DD_{*a}^\ell f = D_{*a}^{\ell+1} f$ to be true under the assumptions of our Lemmata because the function on the right-hand side is continuous whereas the one on the left-hand side need not have this property. Hence we find that the Caputo differential operators do not form a semigroup in general.

In this context the following observation is important.

Lemma 3.13. *Let $f \in C^k[a,b]$ for some $a < b$ and some $k \in \mathbb{N}$. Moreover let $n, \varepsilon > 0$ be such that there exists some $\ell \in \mathbb{N}$ with $\ell \leq k$ and $n, n + \varepsilon \in [\ell - 1, \ell]$. Then,*

$$D_{*a}^\varepsilon D_{*a}^n f = D_{*a}^{n+\varepsilon} f.$$

Remark 3.4. Two comments concerning this Lemma need to be made:

(a) Such a result cannot be expected to hold in general if Riemann–Liouville derivatives were used instead of Caputo derivatives. As an example, consider the
function f with $f(x) = 1$ and let $a = 0$, $n = 1$ and $\varepsilon = 1/2$. If we were to
use Riemann–Liouville derivatives, the left-hand side would be $(D_0^{1/2} f')(x) =$
$D_0^{1/2} 0 = 0$, whereas the right-hand side is

$$D_0^{3/2} f(x) = D^2 J_0^{1/2} f(x) = \frac{1}{\Gamma(-1/2)} x^{-3/2}.$$

(b) The condition requiring the existence of the number ℓ with the properties mentioned in the Lemma is essential. To see what can happen without it, consider
the example $n = \varepsilon = 7/10$ (i.e. $7/10 = n < 1 < n + \varepsilon = 7/5$), $a = 0$ and $f(x) = x$.
Then, the right-hand side is $D_{*0}^{7/5} f(x) = (J_0^{3/5} f'')(x) = J_0^{3/5} 0 = 0$, but since

$$(D_{*0}^{7/10} f)(x) = \frac{1}{\Gamma(13/10)} x^{3/10},$$

the left-hand side takes the value

$$D_{*0}^{7/10} (D_{*0}^{7/10} f)(x) = \frac{1}{\Gamma(3/5)} x^{-2/5}.$$

Proof (of Lemma 3.13). The statement is trivial in the case $n = \ell - 1$ and $n + \varepsilon = \ell$,
so we only treat the other situations explicitly. Then we first observe that our assumptions imply $0 < \varepsilon < 1$. Thus, by Lemma 3.5 we find that

$$D_{*a}^{\varepsilon} z = D_a^{\varepsilon} z$$

whenever $z(a) = 0$. We consider three cases:

1. $n + \varepsilon \in \mathbb{N}$: In this case we have that $\lceil n \rceil = n + \varepsilon$ and hence $\lceil n \rceil - n = \varepsilon$. Moreover,
   by Lemma 3.11, $D_{*a}^n f(a) = 0$. Thus

$$D_{*a}^{\varepsilon} D_{*a}^n f = D_a^{\varepsilon} D_{*a}^n f = D_a^{\varepsilon} J_a^{\lceil n \rceil - n} D^{\lceil n \rceil} f$$
$$= D_a^{\varepsilon} J_a^{\varepsilon} D^{\lceil n \rceil} f = D^{\lceil n \rceil} f = D^{n+\varepsilon} f = D_{*a}^{n+\varepsilon} f.$$

2. $n \in \mathbb{N}$: Here we have, using Theorem 3.1,

$$D_{*a}^{\varepsilon} D_{*a}^n f = D_{*a}^{\varepsilon} D^n f = J_a^{1-\varepsilon} D^{n+1} f = D_{*a}^{n+\varepsilon} f.$$

3. Otherwise, we have $\lceil n \rceil = \lceil n + \varepsilon \rceil$, and thus we find by similar arguments that

$$D_{*a}^{\varepsilon} D_{*a}^n f = D_a^{\varepsilon} D_{*a}^n f = D_a^{\varepsilon} J_a^{\lceil n \rceil - n} D^{\lceil n \rceil} f$$
$$= D^1 J_a^{1-\varepsilon} J_a^{\lceil n \rceil - n} D^{\lceil n \rceil} f = D^1 J_a^1 J_a^{\lceil n + \varepsilon \rceil - (n+\varepsilon)} D^{\lceil n + \varepsilon \rceil} f$$
$$= D_{*a}^{n+\varepsilon} f. \qquad \square$$

The previous result has dealt with the concatenation of two Caputo differential operators. In some instances however it may also be useful to concatenate a Caputo operator with a differential operator of Riemann–Liouville type:

Theorem 3.14. *Let $f \in C^\mu[a,b]$ for some $\mu \in \mathbb{N}$. Moreover let $n \in [0,\mu]$. Then,*

$$D_a^{\mu-n} D_{*a}^n f = D^\mu f.$$

Notice that the operator D^μ appearing on the right-hand side of the claim is a classical (integer-order) differential operator.

Proof. If n is an integer then both differential operators on the left-hand side reduce to integer-order operators and hence we obtain the desired result by an application of the definition of the iterated operators, viz. Definition 1.1 (c).

If n is not an integer then we may invoke Theorem 3.1 to conclude that

$$D_{*a}^n f = \widehat{D}_a^n f = J_a^{\lceil n \rceil - n} D^{\lceil n \rceil} f.$$

Combining this with the definition of the Riemann–Liouville derivative and using the semigroup property of fractional integration and eq. (1.1) we find

$$D_a^{\mu-n} D_{*a}^n f = D^{\mu-\lceil n \rceil +1} J_a^{n+1-\lceil n \rceil} J_a^{\lceil n \rceil - n} D^{\lceil n \rceil} f = D^{\mu-\lceil n \rceil +1} J_a^1 D^{\lceil n \rceil} f$$
$$= D^{\mu-\lceil n \rceil} D^{\lceil n \rceil} f = D^\mu f. \qquad \square$$

It is actually possible to explore the smoothness properties of $D_{*0}^n f$ under smoothness assumptions on f in more detail than in Lemma 3.11:

Theorem 3.15. *If $f \in C^\mu[a,b]$ for some $\mu \in \mathbb{N}$ and $0 < n < \mu$ then*

$$D_{*a}^n f(x) = \sum_{\ell=0}^{\mu-\lceil n \rceil -1} \frac{f^{(\ell+\lceil n \rceil)}(a)}{\Gamma(\lceil n \rceil - n + \ell + 1)} (x-a)^{\lceil n \rceil - n + \ell} + g(x)$$

with some function $g \in C^{\mu-\lceil n \rceil}[a,b]$. Moreover, the $(\mu - \lceil n \rceil)$th derivative of g satisfies a Lipschitz condition of order $\lceil n \rceil - n$.

Proof. This is a direct consequence of the definition of the Caputo differential operator and Theorems 3.1 and 2.5. $\qquad \square$

The main computational rules for the Caputo derivative are similar, but not identical, to those for the Riemann–Liouville derivative.

Theorem 3.16. *Let $f_1, f_2 : [a,b] \to \mathbb{R}$ be such that $D_{*a}^n f_1$ and $D_{*a}^n f_2$ exist almost everywhere and let $c_1, c_2 \in \mathbb{R}$. Then, $D_{*a}^n(c_1 f_1 + c_2 f_2)$ exists almost everywhere, and*

$$D_{*a}^n(c_1 f_1 + c_2 f_2) = c_1 D_{*a}^n f_1 + c_2 D_{*a}^n f_2.$$

Proof. This linearity property of the fractional differential operator is an immediate consequence of the definition of D_{*a}^n. □

For the formula of Leibniz, we only state the case $0 < n < 1$ explicitly.

Theorem 3.17 (Leibniz' formula for Caputo operators). *Let $0 < n < 1$, and assume that f and g are analytic on $(a-h, a+h)$. Then,*

$$D_{*a}^n[fg](x) = \frac{(x-a)^{-n}}{\Gamma(1-n)}g(a)(f(x) - f(a)) + (D_{*a}^n g(x))\, f(x)$$

$$+ \sum_{k=1}^{\infty} \binom{n}{k} \left(J_a^{k-n}g(x) \right) D_{*a}^k f(x).$$

Proof. We apply the definition of the Caputo derivative and find

$$D_{*a}^n[fg] = D_a^n[fg - f(a)g(a)] = D_a^n[fg] - f(a)g(a)D_a^n[1].$$

Next we use Leibniz' formula for Riemann–Liouville derivatives and find

$$D_{*a}^n[fg] = f(D_a^n g) + \sum_{k=1}^{\infty} \binom{n}{k}(D_a^k f)(J_a^{k-n}g) - f(a)g(a)D_a^n[1].$$

Now we add and subtract $f \cdot g(a)(D_a^n[1])$ and rearrange to obtain

$$D_{*a}^n[fg] = f(D_a^n[g - g(a)]) + \sum_{k=1}^{\infty} \binom{n}{k}(D_a^k f)(J_a^{k-n}g)$$

$$+ g(a)(f - f(a))D_a^n[1]$$

$$= f \times (D_{*a}^n g) + \sum_{k=1}^{\infty} \binom{n}{k}(D_{*a}^k f)(J_a^{k-n}g) + g(a)(f - f(a)) \times D_a^n[1]$$

where we have used the fact that, for $k \in \mathbb{N}$, $D_a^k = D^k = D_{*a}^k$. To finally complete the proof it only remains to use the explicit expression for $D_a^n[1]$ from Example 2.4. □

Remark 3.5. For Faà di Bruno's formula (the chain rule) for Caputo operators we may combine eq. (2.6), i.e. the corresponding rule for Riemann–Liouville operators, and Lemma 3.4. This yields that, once again under suitable assumptions on the functions f and g that we shall not specify explicitly, it has the form

$$D_{*a}^n[f(g(\cdot))](x)$$

$$= \sum_{k=1}^{\infty} \binom{n}{k} \frac{k!(x-a)^{k-n}}{\Gamma(k-n+1)} \sum_{\ell=1}^{k}(D^\ell f)(g(x)) \sum_{(a_1,\ldots,a_k)\in A_{k,\ell}} \prod_{r=1}^{k} \frac{1}{a_r!} \left(\frac{D^r g(x)}{r!} \right)^{a_r}$$

$$+ \frac{(x-a)^{-n}}{\Gamma(1-n)}f(g(x)) - \sum_{k=0}^{\lceil n \rceil - 1} \frac{D^k[f(g(\cdot))](a)}{\Gamma(k-n+1)}(x-a)^{k-n}$$

where, as in the Riemann–Liouville case discussed in Theorem 2.19, $(a_1, \ldots, a_k) \in A_{k,\ell}$ means that

$$a_1, \ldots, a_k \in \mathbb{N}_0, \qquad \sum_{r=1}^{k} r a_r = k \quad \text{and} \quad \sum_{r=1}^{k} a_r = \ell.$$

3.2 Nonclassical Representations of Caputo Operators

In the previous section we have developed a number of different representations for Caputo differential operators under various assumptions on the function to be differentiated; see, e.g., Definition 3.2, Definition 3.1 in combination with Theorem 3.1, Lemma 3.2, Lemma 3.3, or Lemma 3.4. Essentially, all these representations consist of a combination of a convolution integral and some sort of a differential operator (or the limit of a discrete version of this). This approach immediately reveals the non-locality of the Caputo operator and provides a natural approach to handling this property. However, in some applications it has turned out that other ways to express the non-locality are more helpful. Two such alternative representations have recently been developed independently by Yuan and Agrawal [194] and Singh and Chatterjee [26, 177]. Our treatment of these two closely related methods is based on the generalizations provided and analyzed in [38].

We first recall the details of the method proposed by Yuan and Agrawal [194]. They have only discussed the case $0 < n < 1$. An extension to $1 < n < 2$ has been provided by Trinks and Ruge [185]. We will not impose any such restriction on the size of n. However, our techniques do require that $n \notin \mathbb{N}$ throughout this section. Since this only excludes cases that are not truly fractional anyway, this is not a substantial limitation. It can easily be seen that our approach reduces to the original scheme of Yuan and Agrawal if $0 < n < 1$.

The approach is tailored to functions $f \in C^{\lceil n \rceil}[a, b]$. In view of Theorem 3.1 this feature allows us to use the representation of Definition 3.1 for the Caputo derivative. Then we define an auxiliary bivariate function $\phi : (0, \infty) \times [a, b] \to \mathbb{R}$ by

$$\phi(w, x) := (-1)^{\lfloor n \rfloor} \frac{2 \sin \pi n}{\pi} w^{2n - 2\lceil n \rceil + 1} \int_a^x f^{(\lceil n \rceil)}(\tau) e^{-(x-\tau)w^2} \, d\tau. \qquad (3.2)$$

With this notation we obtain the following generalization of a result presented in [194, §2]:

Theorem 3.18. *Under the above assumptions,*

$$D_{*a}^n f(x) = \int_0^\infty \phi(w, x) \, dw. \qquad (3.3)$$

In addition, for fixed $w > 0$ the function $\phi(w, \cdot)$ satisfies the differential equation

$$\frac{\partial}{\partial x}\phi(w,x) = -w^2\phi(w,x) + (-1)^{\lfloor n \rfloor}\frac{2\sin \pi n}{\pi}w^{2n-2\lceil n \rceil + 1}f^{(\lceil n \rceil)}(x) \quad (3.4)$$

subject to the initial condition $\phi(w, a) = 0$.

Notice that the differential equation (3.4) is effectively an ordinary differential equation since we assume w to be a fixed parameter. Moreover it is a differential equation of order 1 and hence of classical (not fractional in the strict sense) type. In addition we note that the differential equation is linear and inhomogeneous and that it has constant coefficients. Therefore it is a simple matter to compute the solution of the initial value problem explicitly, and of course this computation reproduces the representation (3.2).

Proof. Bearing in mind the definition of the Gamma function (Definition 1.2), Theorem D.3 and the obvious identity

$$\sin \pi(n - \lceil n \rceil + 1) = (-1)^{\lfloor n \rfloor}\sin \pi n \quad (n \notin \mathbb{N})$$

we obtain that

$$\begin{aligned}
D_{*a}^n f(x) &= \frac{1}{\Gamma(\lceil n \rceil - n)}\int_a^x (x-\tau)^{\lceil n \rceil - n - 1}f^{(\lceil n \rceil)}(\tau)\,d\tau \\
&= \frac{1}{\Gamma(n - \lceil n \rceil + 1)\Gamma(\lceil n \rceil - n)} \\
&\quad \times \int_a^x \int_0^\infty e^{-z}z^{n - \lceil n \rceil}\,dz\,(x-\tau)^{\lceil n \rceil - n - 1}f^{(\lceil n \rceil)}(\tau)\,d\tau \\
&= \frac{\sin \pi(n - \lceil n \rceil + 1)}{\pi}\int_a^x \int_0^\infty e^{-z}\left(\frac{z}{x-\tau}\right)^{n - \lceil n \rceil + 1}\frac{1}{z}f^{(\lceil n \rceil)}(\tau)\,dz\,d\tau \\
&= (-1)^{\lfloor n \rfloor}\frac{\sin \pi n}{\pi}\int_a^x \int_0^\infty e^{-z}\left(\frac{z}{x-\tau}\right)^{n - \lceil n \rceil + 1}\frac{1}{z}f^{(\lceil n \rceil)}(\tau)\,dz\,d\tau.
\end{aligned}$$

We may now apply the substitution $z = (x - \tau)w^2$ in the inner integral and note that Fubini's Theorem allows us to interchange the order of the integrations since $f^{(\lceil n \rceil)}$ is assumed to be continuous. This yields

$$\begin{aligned}
D_{*a}^n f(x) &= (-1)^{\lfloor n \rfloor}\frac{2\sin \pi n}{\pi}\int_a^x \int_0^\infty e^{-(x-\tau)w^2}w^{2n-2\lceil n \rceil + 1}f^{(\lceil n \rceil)}(\tau)\,dw\,d\tau \\
&= \int_0^\infty (-1)^{\lfloor n \rfloor}\frac{2\sin \pi n}{\pi}w^{2n-2\lceil n \rceil + 1}\int_a^x e^{-(x-\tau)w^2}f^{(\lceil n \rceil)}(\tau)\,d\tau\,dw
\end{aligned}$$

and, recalling the definition (3.2) of ϕ, we deduce (3.3).

Next we differentiate the definition (3.2) of ϕ with respect to x. The classical rules for the differentiation of parameter integrals with respect to the parameter immediately give (3.4). Finally the fact that $\phi(w,a) = 0$ for $w > 0$ is also a direct consequence of (3.2) because the integrand of the integral on the right-hand side of (3.2) is continuous.   □

The approach proposed by Chatterjee [26] and investigated further in [177] is also based on expressing the fractional derivative of the given function f in the form of an integral over $(0,\infty)$ whose integrand can be computed as the solution of a first-order initial value problem. Specifically, it is based on the following analogue of Theorem 3.18. The result essentially states that we may replace the integrand ϕ by a function ϕ^* which can be characterized as the solution of a different first-order initial value problem.

Theorem 3.19. *Let* $f \in C^{\lceil n \rceil}[a,b]$. *Moreover, for fixed* $w > 0$, *let* $\phi^*(w,\cdot)$ *be the solution of the differential equation*

$$\frac{\partial}{\partial x}\phi^*(w,x) = -w^{1/(n-\lceil n \rceil-1)}\phi^*(w,x) + \frac{(-1)^{\lfloor n \rfloor}\sin\pi n}{\pi(n-\lceil n \rceil+1)}f^{(\lceil n \rceil)}(x) \qquad (3.5)$$

subject to the initial condition $\phi^*(w,a) = 0$. *Then, we have*

$$\phi^*(w,x) = \frac{(-1)^{\lfloor n \rfloor}\sin\pi n}{\pi(n-\lceil n \rceil+1)}\int_0^x f^{(\lceil n \rceil)}(\tau)\exp\left(-(x-\tau)w^{1/(n-\lceil n \rceil+1)}\right)d\tau \qquad (3.6)$$

and

$$D_{*a}^n f(x) = \int_0^\infty \phi^*(w,x)\,dw. \qquad (3.7)$$

Proof. The proof of this result is almost identical to the proof of Theorem 3.18; one only needs to replace the substitution $z = (x-\tau)w^2$ by $z = (x-\tau)w^{1/(n-\lceil n \rceil+1)}$ and use the functional equation of the Gamma function, $u\Gamma(u) = \Gamma(u+1)$. We leave the details to the reader.   □

In many applications of these representations [41, 118, 171, 177, 185] it is important to have some additional knowledge about the behaviour of the function ϕ in eq. (3.2) or the function ϕ^* in eq. (3.6), respectively. The most important of these properties are summarized in the following theorems. Here, the symbol $\alpha(v) \sim \beta(v)$ means that there exist two strictly positive constants A and B such that $|\alpha(v)/\beta(v)| \in [A,B]$ as v tends to the indicated limit. We begin with the function ϕ arising in the original Yuan-Agrawal representation that we had given in our Theorem 3.18.

Theorem 3.20. *Let* $x \in (a,b)$ *be fixed and* $0 < n \notin \mathbb{N}$, *and assume that there exists some* $C > 0$ *such that* $|f^{(\lceil n \rceil)}(\tilde{x})| > C$ *for all* $\tilde{x} \in [a,b]$.

(a) The function $\phi(\cdot,x)$ *defined in (3.2) behaves as*

$$\phi(w,x) \sim w^{2n-2\lceil n \rceil+1} \qquad as\ w \to 0. \qquad (3.8)$$

(b) Moreover,

$$\phi(w,x) \sim w^{2n-2\lceil n \rceil - 1} \qquad as \ w \to \infty. \tag{3.9}$$

(c) We have $\phi(\cdot,x) \in C^{\infty}(0,\infty)$.

Remark 3.6. The condition that $f^{(\lceil n \rceil)}$ be bounded away from zero is a technical condition required in order to keep the proof simple and to keep the result valid for *all* $x \in (a,b)$. Using more complicated techniques, one could show that the same asymptotic behaviour is present for *almost all* $x \in (a,b)$ under substantially weaker conditions. Thus it is justified to say that the asymptotic behaviour described in Theorem 3.20 is the behaviour that one may reasonably expect for the function ϕ unless the given function f is of a highly exceptional nature.

Proof. For part (a), a partial integration gives

$$\int_a^x f^{(\lceil n \rceil)}(\tau) e^{-(x-\tau)w^2} \, d\tau$$

$$= f^{(\lceil n \rceil - 1)}(\tau) e^{-(x-\tau)w^2} \Big|_{\tau=a}^{\tau=x} - w^2 \int_a^x f^{(\lceil n \rceil - 1)}(\tau) e^{-(x-\tau)w^2} \, d\tau$$

$$= f^{(\lceil n \rceil - 1)}(x) - f^{(\lceil n \rceil - 1)}(a) e^{-xw^2} - w^2 \int_0^x f^{(\lceil n \rceil - 1)}(\tau) e^{-(x-\tau)w^2} \, d\tau.$$

Since x is fixed, the rightmost integral obviously remains bounded as $w \to 0$, and hence we conclude

$$\lim_{w \to 0} \int_a^x f^{(\lceil n \rceil)}(\tau) e^{-(x-\tau)w^2} \, d\tau = f^{(\lceil n \rceil - 1)}(x) - f^{(\lceil n \rceil - 1)}(0). \tag{3.10}$$

Inserting this relation into the definition (3.2) of ϕ we obtain the first claim.

For the proof of (b), we write

$$w^2 \int_a^x f^{(\lceil n \rceil)}(\tau) e^{-(x-\tau)w^2} \, d\tau$$

$$= w^2 \int_a^{x-w^{-1}} f^{(\lceil n \rceil)}(\tau) e^{-(x-\tau)w^2} \, d\tau + w^2 \int_{x-w^{-1}}^x f^{(\lceil n \rceil)}(\tau) e^{-(x-\tau)w^2} \, d\tau$$

$$= f^{(\lceil n \rceil)}(\xi_1) w^2 \int_a^{x-w^{-1}} e^{-(x-\tau)w^2} \, d\tau + f^{(\lceil n \rceil)}(\xi_2) w^2 \int_{x-w^{-1}}^x e^{-(x-\tau)w^2} \, d\tau$$

$$= f^{(\lceil n \rceil)}(\xi_1)(e^{-w} - e^{-w^2}) + f^{(\lceil n \rceil)}(\xi_2)(1 - e^{-w})$$

with some $\xi_1 \in [a, x - w^{-1}]$ and $\xi_2 \in [x - w^{-1}, x]$ because of the Mean Value Theorem. Now, as $w \to \infty$, $f^{(\lceil n \rceil)}(\xi_1)$ remains bounded whereas $e^{-w} - e^{-w^2} \to 0$. Thus the first summand on the right-hand side vanishes. For the second summand we have $1 - e^{-w} \to 1$ and $f^{(\lceil n \rceil)}(\xi_2) \to f^{(\lceil n \rceil)}(x)$ because $\xi_2 \in [x - w^{-1}, x]$. Thus, we conclude

$$\lim_{w \to \infty} w^2 \int_0^x f^{(\lceil n \rceil)}(\tau) e^{-(x-\tau)w^2} \, d\tau = f^{(\lceil n \rceil)}(x).$$

Inserting this relation into the definition of ϕ, we arrive at

$$\phi(w,x) = (-1)^{\lfloor n \rfloor} \frac{2\sin \pi n}{\pi} w^{2n-2\lceil n \rceil -1}[f^{(\lceil n \rceil)}(x) + o(1)] \qquad (3.11)$$

which completes the proof of (b).

Finally, part (c) follows directly from the definition of ϕ that we had given in eq. (3.2). $\qquad\qquad\qquad\qquad\qquad\qquad\qquad\qquad\qquad\qquad\qquad\qquad\qquad\qquad\qquad$ □

We can also provide a corresponding result for the function ϕ^* used in Chatterjee's representation (Theorem 3.19). The behaviour of this function ϕ^*, which is defined in eq. (3.6), can be described as follows.

Theorem 3.21. *Let $x \in (a,b)$ be fixed and $0 < n \notin \mathbb{N}$, and assume that there exists some $C > 0$ such that $|f^{(\lceil n \rceil)}(\tilde{x})| > C$ for all $\tilde{x} \in [0,X]$.*

(a) The function $\phi^*(\cdot,x)$ described in eq. (3.6) behaves as

$$\phi^*(w,x) \sim 1 \quad as\ w \to 0. \qquad (3.12)$$

(b) Moreover,

$$\phi^*(w,x) \sim w^{-1/(n-\lceil n \rceil +1)} \quad as\ w \to \infty. \qquad (3.13)$$

(c) We have $\phi^*(\cdot,x) \in C^\infty(0,\infty)$.

The proof proceeds along the same lines as the proof of Theorem 3.20.

Theorems 3.20 and 3.21 allow us to compare the analytical properties of the function ϕ used by Yuan and Agrawal and the function ϕ^* proposed by Chatterjee. First of all we note that both functions possess infinitely many derivatives (in the classical sense) with respect to the first variable. However, there are significant differences in the asymptotic behaviour of the functions as the first variable tends to either end of the interval $(0,\infty)$ over which the functions need to be integrated in order to compute the Caputo derivative $D_{*a}^n f$.

To be precise, for $w \to 0$ the function $\phi(w,x)$ exhibits an asymptotic behaviour of the form $w^{2n-2\lceil n \rceil -1}$ according to Theorem 3.20 (a). The exponent of w here is always strictly between -1 and $+1$. This asserts the integrability of $\phi(w,x)$ with respect to w near $w = 0$ at least in the improper sense. However, we can expect a smooth behaviour near this end point of the integration interval only if the exponent is an integer, and this is the case if and only if $n = k + 1/2$ with some $k \in \mathbb{N}_0$. For all other values of n the behaviour is less regular. This irregularity needs to be taken into account carefully when one tries to use this approach in a numerical algorithm [38, 118]. Theorem 3.21 (a) demonstrates that the function ϕ^* is easier to handle in this respect since here we always have that $\phi^*(w,x)$ remains bounded by nonzero constants from above and below.

The behaviour of the integrands $\phi(w,x)$ and $\phi^*(w,x)$ for $w \to \infty$ also exhibits substantial differences. As shown in Theorem 3.20 (b), the Yuan-Agrawal integrand $\phi(w,x)$ behaves as $w^{2n-2\lceil n \rceil -1}$. The exponent of w here is always contained in the interval $(-3,-1)$. This is just about fast enough to make sure that the improper integral exists. On the other hand, according to Theorem 3.21 (b), the Chatterjee integrand $\phi^*(w,x)$ behaves in a way that depends on n in a somewhat more complicated fashion: If $n = k + \varepsilon$ with some $k \in \mathbb{N}_0$ and $0 < \varepsilon < 1$ then the exponent in question is $-1/\varepsilon$. This is always less than -1, and hence the improper integral converges. In this respect we have no difference to the Yuan-Agrawal method. However, if ε is close to 0 then the exponent of course remains negative but it may be arbitrarily large in modulus, leading to a much faster (but still algebraic) decay of the integrand. In particular, in contrast to the Yuan-Agrawal method there is no lower bound on the exponent as n runs through all the admissible numbers. For numerical work, a rapidly decaying integrand is preferable, so at least for $n = k + \varepsilon$ with $k \in \mathbb{N}$ and ε close to 0 the approach via Theorem 3.19 has some advantages over the path via Theorem 3.18. It should be pointed out though that, from the point of view of approximation theory (see, e.g., the survey article [122]), an ideal integrand (i.e. an integrand that can be handled very nicely by a numerical algorithm) would decay exponentially as $w \to \infty$, i.e. much faster than we can ever hope even for ϕ^*.

Exercises

Exercise 3.1. Let $f(x) = \cos \lambda x$ for some $\lambda > 0$.

(a) Determine the functions $D_0^n f(x)$ and $D_{*0}^n f(x)$ for arbitrary $n > 0$.
(b) For which values of x are these derivatives defined?
(c) Investigate these derivatives with respect to continuity and differentiability.
(d) Draw a sketch of the derivatives for some values of n.

Exercise 3.2. Prove the identities stated in Appendix B.

Exercise 3.3. Work out the details of the proofs of Theorems 3.19 and 3.21.

Chapter 4
Mittag-Leffler Functions

Before we can come to the core of this text, i.e. to the discussion of fractional differential equations, we need to introduce two classes of functions (one of which may be considered to be a special case of the other) and investigate their basic properties. These functions will turn out to be of fundamental importance in our context, and they will be used in many places throughout the second part of this book. We begin with the more restrictive of the two concepts.

Definition 4.1. Let $n > 0$. The function E_n defined by

$$E_n(z) := \sum_{j=0}^{\infty} \frac{z^j}{\Gamma(jn+1)}$$

whenever the series converges is called the *Mittag-Leffler function* of order n.

This function has been introduced by Mittag-Leffler [136, 137]. We immediately notice that

$$E_1(z) = \sum_{j=0}^{\infty} \frac{z^j}{\Gamma(j+1)} = \sum_{j=0}^{\infty} \frac{z^j}{j!} = \exp(z) \tag{4.1}$$

is just the well known exponential function.

The more general class of functions is defined as follows.

Definition 4.2. Let $n_1, n_2 > 0$. The function E_{n_1,n_2} defined by

$$E_{n_1,n_2}(z) := \sum_{j=0}^{\infty} \frac{z^j}{\Gamma(jn_1+n_2)}$$

whenever the series converges is called the *two-parameter Mittag-Leffler function* with parameters n_1 and n_2.

Remark 4.1. It is evident that the one-parameter Mittag-Leffler functions may be defined in terms of their two-parameter counterparts via the relation $E_n(z) = E_{n,1}(z)$.

The naming of the latter functions after Mittag-Leffler is due to the fact that they are a very simple and obvious generalization of the functions originally introduced

K. Diethelm, *The Analysis of Fractional Differential Equations*,
Lecture Notes in Mathematics 2004, DOI 10.1007/978-3-642-14574-2_4,
© Springer-Verlag Berlin Heidelberg 2010

by him and recalled here in Definition 4.1. For the sake of historical correctness one should however mention that the two-parameter functions were actually first discussed by Wiman [191] shortly after the publication of Mittag-Leffler's original work.

Remark 4.2. Most of the results for the one-parameter Mittag-Leffler function given below will remain valid if the restriction $n > 0$ is replaced by $n \in \mathbb{C}$ with $\operatorname{Re} n > 0$. Similarly, the conditions $n_1, n_2 > 0$ for the two-parameter Mittag-Leffler function may be relaxed to $n_1, n_2 \in \mathbb{C}$ with $\operatorname{Re} n_1 > 0$ and $\operatorname{Re} n_2 > 0$. For our purposes however it will be sufficient to work with real parameters.

First of all, we need to discuss the radius of convergence of the power series given above, i.e. the domains of definition of the Mittag-Leffler functions. It turns out that we can give a full result for the two-parameter version. In view of Remark 4.1 this then of course holds a forteriori for the one-parameter Mittag-Leffler functions too.

Theorem 4.1. *Consider the two-parameter Mittag-Leffler function E_{n_1,n_2} for some $n_1, n_2 > 0$. The power series defining $E_{n_1,n_2}(z)$ is convergent for all $z \in \mathbb{C}$. In other words, E_{n_1,n_2} is an entire function.*

Proof. By definition, the two-parameter Mittag-Leffler function has the power series representation

$$\sum_{j=0}^{\infty} a_j z^j \quad \text{with} \quad a_j = \frac{1}{\Gamma(jn_1 + n_2)}.$$

In view of Stirling's formula (see Theorem D.5) we find that

$$a_j^{1/j} = \left(\frac{e}{jn_1 + n_2} \right)^{n_1 + n_2/j} (2\pi(jn_1 + n_2))^{-1/(2j)} (1 + o(1)) \to 0$$

as $j \to \infty$ since $n_1 > 0$. Thus, by the root criterion, the radius of convergence of the power series is infinite. □

Remark 4.3. It is possible to investigate the Mittag-Leffler function E_{n_1,n_2} also for $n_1 = 0$. In this case, the power series has a finite convergence radius. If, for example, $n_2 = 1$ then the convergence radius is 1. Indeed a close inspection of the power series representation yields

$$E_{0,1}(z) = \sum_{j=0}^{\infty} \frac{1}{\Gamma(1)} z^j = \sum_{j=0}^{\infty} z^j = \frac{1}{1-z}.$$

However, in the context of fractional differential equations, the Mittag-Leffler functions E_{0,n_2} are not very important, and so we shall not discuss them any further.

Remark 4.4. The so-called *multi-index Mittag-Leffler functions*

$$E_{(n_{11},n_{12},\ldots,n_{1k}),(n_{21},n_{22},\ldots,n_{2k})}(z) := \sum_{j=0}^{\infty} z^j \prod_{\mu=1}^{k} \frac{1}{\Gamma(jn_{1\mu}+n_{2\mu})}$$

form an even more general class of functions. It turns out that methods using concepts from fractional calculus can be conveniently used to analyze these functions. However, since these functions do not play a significant role in the context of this book, we shall not go into detail with respect to these functions here and only refer to the survey article [103] and the references cited therein.

Example 4.1. For some special choices of the parameters n_1 and n_2, we can recover certain well known functions:

(a) For $x \in \mathbb{C}$, $E_2(-x^2) = E_{2,1}(-x^2) = \cos x$.
(b) For $x \in \mathbb{C}$, $E_2(x^2) = E_{2,1}(x^2) = \cosh x$.
(c) For $x > 0$, $E_{1/2}(x^{1/2}) = E_{1/2,1}(x^{1/2}) = (1 + \mathrm{erf}(x)) \exp(x^2)$. (This relation can be extended to $x \in \mathbb{C}$ if $x^{1/2}$ is interpreted as the principal branch of the complex square root function.)
(d) For $x \in \mathbb{C}$ and $r \in \mathbb{N}$,

$$E_{1,r}(x) = \frac{1}{x^{r-1}} \left(\exp(x) - \sum_{k=0}^{r-2} \frac{x^k}{k!} \right).$$

(In the case $x = 0$, appropriate limits need to be taken on the right-hand side.)

In (c), erf denotes the error function defined by

$$\mathrm{erf}(x) = \frac{2}{\sqrt{\pi}} \int_0^x \exp(-t^2)\, dt.$$

We leave the proof of these identities as an exercise for the reader.

One last property of Mittag-Leffler functions that we mention before building the bridge to fractional calculus is a relation between two Mittag-Leffler functions with different parameters.

Theorem 4.2. *Let $n_1, n_2 > 0$ and $x \in \mathbb{C}$. Then,*

$$E_{n_1,n_2}(x) = xE_{n_1,n_1+n_2}(x) + \frac{1}{\Gamma(n_2)}.$$

Proof. This can be shown by explicitly writing down the power series on either side of the claimed identity and by comparing the coefficients. We omit the details. □

The key result that indicates why Mittag-Leffler functions (in particular those with one parameter) are so important in fractional calculus is the following theorem.

It essentially states that the eigenfunctions of Caputo differential operators may be expressed in terms of Mittag-Leffler functions.

Theorem 4.3. *Let $n > 0$ and $\lambda \in \mathbb{R}$. Moreover define*

$$y(x) := E_n(\lambda x^n), \qquad x \geq 0.$$

Then,

$$D_{*0}^n y(x) = \lambda y(x).$$

Proof. We first look at the case $\lambda = 0$ and note that in this case $y(x) = E_n(0) = 1$. Hence, $D_{*0}^n y(x) = 0 = \lambda y(x)$ as required. If, on the other hand, $\lambda \neq 0$, then (using the notation $p_k(x) := x^k$)

$$D_{*0}^n y(x) = D_{*0}^n \left[\sum_{j=0}^{\infty} \frac{(\lambda p_n)^j}{\Gamma(1+jn)} \right](x) = J_0^{m-n} D^m \left[\sum_{j=0}^{\infty} \frac{\lambda^j p_{nj}}{\Gamma(1+jn)} \right](x)$$

$$= J_0^{m-n} \left[\sum_{j=0}^{\infty} \frac{\lambda^j D^m p_{nj}}{\Gamma(1+jn)} \right](x) = J_0^{m-n} \left[\sum_{j=1}^{\infty} \frac{\lambda^j D^m p_{nj}}{\Gamma(1+jn)} \right](x)$$

$$= J_0^{m-n} \left[\sum_{j=1}^{\infty} \frac{\lambda^j p_{nj-m}}{\Gamma(1+jn-m)} \right](x) = \sum_{j=1}^{\infty} \frac{\lambda^j J_0^{m-n} p_{nj-m}(x)}{\Gamma(1+jn-m)}$$

$$= \sum_{j=1}^{\infty} \frac{\lambda^j p_{nj-n}(x)}{\Gamma(1+jn-n)} = \sum_{j=1}^{\infty} \frac{\lambda^j x^{nj-n}}{\Gamma(1+jn-n)}$$

$$= \sum_{j=0}^{\infty} \frac{\lambda^{j+1} x^{nj}}{\Gamma(1+jn)} = \lambda \sum_{j=0}^{\infty} \frac{(\lambda x^n)^j}{\Gamma(1+jn)} = \lambda y(x).$$

Here we have used the fact that, in view of the convergence properties of the series defining the Mittag-Leffler function, we may interchange first summation and differentiation and later summation and integration.   □

Remark 4.5. It is evident from (4.1) that the Mittag-Leffler function E_1 satisfies the functional equation

$$E_1(x-y) = \frac{E_1(x)}{E_1(y)}. \tag{4.2}$$

A generalization of this result to Mittag-Leffler functions E_n with $n \notin \mathbb{N}$ is not known and probably such a relation does not exist. The functional equation (4.2) plays a very important role in the analysis of first-order differential equations (in particular in the theory of linear equations) because it allows to write a convolution kernel that arises, e.g., in the variation-of-constants approach, in the form of a product of a fundamental solution at one point and the inverse of the fundamental solution at some other point. The fact that a fractional generalization of this feature is not available is a major obstacle in the development of a comprehensive theory for linear fractional differential equations.

It is frequently of interest to have some knowledge about the asymptotic behaviour of these functions. In this context we recall the following result on the one-parameter Mittag-Leffler functions that will be seen to be important later on in Chap. 6.

Theorem 4.4. *Let $n > 0$. The Mittag-Leffler function E_n behaves as follows:*

(a) $E_n(re^{i\phi}) \to 0$ *for* $r \to \infty$ *if* $|\phi| > n\pi/2$,
(b) $E_n(re^{i\phi})$ *remains bounded for* $r \to \infty$ *if* $|\phi| = n\pi/2$,
(c) $|E_n(re^{i\phi})| \to \infty$ *for* $r \to \infty$ *if* $|\phi| < n\pi/2$.

Obviously, in the classical case $n = 1$ this reduces to the well known fact that, as $|z| \to \infty$, $\exp(z)$ (a) goes to zero if $\arg z > \pi/2$, (b) remains bounded if $\arg z = \pi/2$ and (c) grows without bound if $\arg z < \pi/2$.

Proof. We shall outline the proof given by Wiman [191].

In the first step let us consider the case that $n = 1/k$ with some $k \in \mathbb{N}$. In this case we can see that $E_n = E_{1/k}$ satisfies the first-order linear differential equation

$$E'_{1/k}(x) = kx^{k-1}E_{1/k}(x) + k\sum_{\mu=1}^{k-1}\frac{x^{\mu-1}}{\Gamma(\mu/k)}.$$

This can be shown easily using the power series representation of $E_{1/k}$. Since we also know that $E_{1/k}$ satisfies the initial condition $E_{1/k}(0) = 1$ that corresponds to this differential equation, we can use the standard methods for the solution of such differential equations and find the representation

$$E_{1/k}(x) = \exp(x^k) + \exp(x^k)\int_0^x k\exp(-z^k)\sum_{\mu=1}^{k-1}\frac{z^{\mu-1}}{\Gamma(\mu/k)}\,dz.$$

For each of the k summands on the right-hand side of this equation one can then set up asymptotic expressions that allow to conclude the desired result.

If now n is a positive rational number, say $n = \ell/k$ with relatively prime ℓ and k, then we can invoke the identity

$$E_{\ell/k}(x) = \frac{1}{\ell}\sum_{\mu=0}^{\ell-1}E_{1/k}\left(x^{1/\ell}\exp\frac{2\mu\pi i}{\ell}\right) \tag{4.3}$$

which, using the result for $E_{1/k}$ that we have already shown, implies the claim also for $n = \ell/k$.

Finally, for irrational values of n we may choose a sequence $(n_j)_{j=0}^{\infty}$ of rational numbers that converges to n. For each of the E_{n_j} the result is already in place, and by taking appropriate limits we then derive the result also for E_n. $\square$

In the final results for this short chapter, we shall describe the interconnection between a one-parameter Mittag-Leffler function and the Laplace transform operation and an important consequence of this theorem. Further information on the Laplace transform is given in Appendix D.3; at this point we only note that it is a very useful tool for the solution of certain classes of fractional differential equations. We shall deal with such an approach in Sect. 7.1.

Theorem 4.5. *Let $n > 0$ and $\lambda \in \mathbb{C}$ and define $y(x) := E_n(-\lambda x^n)$. Then, the Laplace transform of y is given by*

$$\mathscr{L}y(s) = \frac{s^{n-1}}{s^n + \lambda}. \tag{4.4}$$

Proof. This can be shown by explicitly writing down the series expansion of $y(x)$ in powers of x^n and applying the Laplace transform in a termwise manner. We leave the details to the reader.   □

In conjunction with the Final Value Theorem for the Laplace transform (Theorem D.13) we obtain a statement on the asymptotic behaviour of the function y mentioned in Theorem 4.5 as its argument tends to infinity:

Theorem 4.6. *Let $n > 0$, $r > 0$, $\varphi \in [-\pi, \pi]$ and $\lambda = r\exp(i\varphi)$. Denote $y(x) := E_n(-\lambda x^n)$. Then,*

(a) $\lim_{x \to \infty} y(x) = 0$ if $|\varphi| < n\pi/2$,
(b) $y(x)$ is unbounded as $x \to \infty$ if $|\varphi| > n\pi/2$.

Proof. This is an immediate consequence of Theorem 4.5, Theorem D.13 and Remark D.1. Alternatively, we may also deduce this result from Theorem 4.4.   □

It is worth pointing out that the numerical evaluation of the Mittag-Leffler function $E_{n_1, n_2}(x)$ may, depending on the precise values of the parameters n_1 and n_2 and the argument x, be an extremely difficult task. A useful algorithm for the solution of this problem has been provided by Gorenflo et al. [79]. More recently, an alternative numerical method has been developed by Seybold and Hilfer [175].

We will frequently use Mittag-Leffler functions in the following chapters. For additional general results on Mittag-Leffler functions we refer to [63, Chap. 18], [81, Appendix A], [131, §2] and the detailed survey paper [90]; further information describing their connection to fractional calculus is given in [80, 127]. Some more results about the long-term behaviour of certain special Mittag-Leffler functions and about the number of their zeros will also be given later in this book; see Theorems 7.3–7.8.

Exercises

Exercise 4.1. Give an explicit proof of the identities mentioned in Example 4.1.

Exercise 4.2. Give an explicit proof of Theorem 4.2.

Exercise 4.3. Fill in the details of the proof of Theorem 4.4. In particular, prove the identity (4.3).

Exercise 4.4. Give an explicit proof of Theorem 4.5.

Part II
Theory of Fractional
Differential Equations

Chapter 5
Existence and Uniqueness Results for Riemann–Liouville Fractional Differential Equations

In this part of the text we now discuss the classical questions concerning ordinary differential equations involving fractional derivatives, i.e. the questions of existence and uniqueness of solutions. We shall mainly be interested in initial value problems (Cauchy problems), and in particular in global results. For a discussion of other types of conditions we refer to Samko, Kilbas and Marichev [167, §42.3], Kilbas and Trujillo [102] or Agarwal, Benchohra and Hamani [4]. Additional aspects of the problems to be treated here may also be found in [101] and [167, §42.4]. Some local results are derived in [87] and [138]. Other useful references are [135] and [153]. The present chapter will be focused on equations with Riemann–Liouville differential operators; Caputo derivatives are the topic of the following chapters.

The fundamental result is an existence and uniqueness theorem. Without loss of generality, we assume in this result and in the ensuing developments that the fractional derivatives are developed at the point 0.

Theorem 5.1. *Let $n > 0$, $n \notin \mathbb{N}$ and $m = \lceil n \rceil$. Moreover let $K > 0$, $h^* > 0$, and $b_1, \ldots, b_m \in \mathbb{R}$. Define*

$$G := \{(x,y) \in \mathbb{R}^2 : 0 \le x \le h^* \, , \, y \in \mathbb{R} \text{ for } x = 0 \text{ and}$$

$$\left| x^{m-n}y - \sum_{k=1}^{m} b_k x^{m-k}/\Gamma(n-k+1) \right| < K \text{ else} \},$$

and assume that the function $f : G \to \mathbb{R}$ is continuous and bounded in G and that it fulfils a Lipschitz condition with respect to the second variable, i.e. there exists a constant $L > 0$ such that, for all (x, y_1) and $(x, y_2) \in G$, we have

$$|f(x, y_1) - f(x, y_2)| < L|y_1 - y_2|.$$

Then the differential equation

$$D_0^n y(x) = f(x, y(x))$$

equipped with the initial conditions

$$D_0^{n-k} y(0) = b_k \quad (k = 1, 2, \ldots, m-1), \quad \lim_{z \to 0+} J_0^{m-n} y(z) = b_m$$

K. Diethelm, *The Analysis of Fractional Differential Equations*,
Lecture Notes in Mathematics 2004, DOI 10.1007/978-3-642-14574-2_5,
© Springer-Verlag Berlin Heidelberg 2010

has a uniquely defined continuous solution $y \in C(0,h]$ where

$$h := \min\left\{ h^*, \tilde{h}, \left(\frac{\Gamma(n+1)K}{M} \right)^{1/m} \right\}$$

with $M := \sup_{(x,z) \in G} |f(x,z)|$ and $\tilde{h}$ being an arbitrary positive number satisfying the constraint

$$\tilde{h} < \left(\frac{\Gamma(2n-m+1)}{\Gamma(n-m+1)L} \right)^{1/n}.$$

The result is very similar to the known classical results for first-order equations. Therefore it is probably not surprising to find that the proof is analogous as well. Specifically we shall first transform the initial value problem into an equivalent Volterra integral equation (Lemma 5.2), and then we are going to prove the existence and uniqueness of the solution of this integral equation by a Picard-type iteration process (i.e. by using a variant of Banach's fixed point theorem in a suitably chosen complete metric space), cf. Lemma 5.3. Theorem 5.1 is thus an immediate consequence of these two lemmas.

Lemma 5.2. *Assume the hypotheses of Theorem 5.1 and let $h > 0$. The function $y \in C(0,h]$ is a solution of the differential equation*

$$D_0^n y(x) = f(x, y(x)),$$

equipped with the initial conditions

$$D_0^{n-k} y(0) = b_k \qquad (k = 1, 2, \ldots, m-1), \qquad \lim_{z \to 0+} J_0^{m-n} y(z) = b_m,$$

if and only if it is a solution of the Volterra integral equation

$$y(x) = \sum_{k=1}^{m} \frac{b_k x^{n-k}}{\Gamma(n-k+1)} + \frac{1}{\Gamma(n)} \int_0^x (x-t)^{n-1} f(t, y(t)) \, dt.$$

Remark 5.1. A look at the integral equation reveals why we have only assumed y to be continuous on the half-open interval $(0,h]$ and not on the closed interval $[0,h]$ as we could have done for equations of integer order: If y were continuous throughout $[0,h]$ then the left-hand side of the integral equation would be continuous in this interval, and so would be the integral on the right-hand side (because of the continuity of f). Therefore the sum must be continuous on $[0,h]$ too. In view of the definition of m, we easily see that the summands are indeed continuous on $[0,h]$ for $k = 1, 2, \ldots, m-1$, but the remaining one ($k = m$) is unbounded as $x \to 0$ because $m > n$ if $n \notin \mathbb{N}$, unless $b_m = 0$.

Proof. Assume first that y is a solution of the integral equation. We can rewrite this equation in the shorter form

$$y(x) = \sum_{k=1}^{m} \frac{b_k x^{n-k}}{\Gamma(n-k+1)} + J_0^n f(\cdot, y(\cdot))(x).$$

Now we apply the differential operator D_0^n to both sides of this relation and immediately obtain, in view of Example 2.4 and Theorem 2.14, that y also solves the differential equation. With respect to the initial conditions, we look at the case $1 \le k \le m-1$ first and find, by an application of D_0^{n-k} to the Volterra equation, that

$$D_0^{n-k} y(x) = \sum_{j=1}^{m} \frac{b_j D_0^{n-k}(\cdot)^{n-j}(x)}{\Gamma(n-j+1)} + D_0^{n-k} J_0^{n-k} J_0^k f(\cdot, y(\cdot))(x)$$

in view of the semigroup property of fractional integration. By Example 2.4 we find that the summands vanish identically for $j > k$. Moreover, by the same example, the summands for $j < k$ vanish if $x = 0$. Thus, according to Theorem 2.14,

$$D_0^{n-k} y(0) = \frac{b_k D_0^{n-k}(\cdot)^{n-k}(0)}{\Gamma(n-k+1)} + J_0^k f(\cdot, y(\cdot))(0).$$

Since $k \ge 1$, the integral vanishes, and once again applying Example 2.4 we find that $D_0^{n-k}(\cdot)^{n-k}(x) = \Gamma(n-k+1)$. Thus $D_0^{n-k} y(0) = b_k$ as required by the initial condition. Finally for $k = m$ we apply the operator J_0^{m-n} to both sides of the integral equation and find that, in the limit $z \to 0$, all the summands of the sum vanish except for the mth. The integral $J_0^{m-n} J_0^n f(\cdot, y(\cdot))(z) = J_0^m f(\cdot, y(\cdot))(z)$ also vanishes as $z \to 0$. Thus we find

$$\lim_{z \to 0+} J_0^{m-n} y(z) = \lim_{z \to 0+} J_0^{m-n} \frac{b_m J_0^{m-n}(\cdot)^{n-m}(z)}{\Gamma(n-m+1)} = b_m$$

because of Example 2.1. Hence y solves the given initial value problem.

If y is a continuous solution of the initial value problem then we define $z(x) := f(x, y(x))$. By assumption, z is a continuous function and $z(x) = f(x, y(x)) = D_0^n y(x) = D^m J_0^{m-n} y(x)$. Thus, $D^m J_0^{m-n} y$ is continuous too, i.e. $J_0^{m-n} y \in C^m(0, h]$. We may therefore apply Theorem 2.23 to derive

$$y(x) = J_0^n D_0^n y(x) + \sum_{k=1}^{m} c_k x^{n-k} = J_0^n f(\cdot, y(\cdot))(x) + \sum_{k=1}^{m} c_k x^{n-k}$$

with certain constants $c_1, \ldots, c_m$. Introducing the initial conditions as indicated above, we can determine these constants c_k as $c_k = b_k / \Gamma(n-k+1)$. $\qquad \square$

Lemma 5.3. *Under the assumptions of Theorem 5.1, the Volterra equation*

$$y(x) = \sum_{k=1}^{m} \frac{b_k x^{n-k}}{\Gamma(n-k+1)} + \frac{1}{\Gamma(n)} \int_0^x (x-t)^{n-1} f(t,y(t)) \, dt$$

possesses a uniquely determined solution $y \in C(0,h]$.

Proof. We define the set

$$B := \left\{ y \in C(0,h] : \sup_{0<x\leq h} \left| x^{m-n} y(x) - \sum_{k=1}^{m} \frac{b_k x^{m-k}}{\Gamma(n-k+1)} \right| \leq K \right\}$$

and on this set we define the operator A by

$$Ay(x) := \sum_{k=1}^{m} \frac{b_k x^{n-k}}{\Gamma(n-k+1)} + \frac{1}{\Gamma(n)} \int_0^x (x-t)^{n-1} f(t,y(t)) \, dt.$$

Then we note that, for $y \in B$, Ay is also a continuous function on $(0,h]$. Moreover,

$$\left| x^{m-n} Ay(x) - \sum_{k=1}^{m} \frac{b_k x^{m-k}}{\Gamma(n-k+1)} \right| = \left| \frac{x^{m-n}}{\Gamma(n)} \int_0^x (x-t)^{n-1} f(t,y(t)) \, dt \right|$$

$$\leq \frac{x^{m-n}}{\Gamma(n)} M \int_0^x (x-t)^{n-1} \, dt$$

$$\leq \frac{x^{m-n}}{\Gamma(n)} M \frac{x^n}{n} = \frac{x^m M}{\Gamma(n+1)} \leq K$$

for $x \in (0,h]$, where the last inequality follows from the definition of h. This shows that $Ay \in B$ if $y \in B$, i.e. the operator A maps the set B into itself.

Next we introduce a new set

$$\widehat{B} := \left\{ y \in C(0,h] : \sup_{0<x\leq h} |x^{m-n} y(x)| < \infty \right\},$$

and on this set we define a norm $\|\cdot\|_{\widehat{B}}$ by

$$\|y\|_{\widehat{B}} := \sup_{0<x\leq h} |x^{m-n} y(x)|.$$

It is easily seen that $\widehat{B}$, equipped with this norm, is a normed linear space, and that B is a complete subset of this space.

We use the definition of A to rewrite the Volterra equation more compactly as

$$y = Ay.$$

Hence, in order to prove the desired result, it is sufficient to show that the operator A has a unique fixed point. For this purpose, we shall employ Weissinger's fixed point theorem (Theorem D.7). In this context we prove, for $y, \tilde{y} \in B$,

$$\left\| A^j y - A^j \tilde{y} \right\|_{\hat{B}} \le \left(\frac{Lh^n \Gamma(n-m+1)}{\Gamma(2n-m+1)} \right)^j \|y - \tilde{y}\|_{\hat{B}}. \tag{5.1}$$

This can be shown by induction: In the case $j = 0$, the statement is trivially true. For the induction step $j - 1 \mapsto j$, we proceed as follows. We write

$$
\begin{aligned}
\left\| A^j y - A^j \tilde{y} \right\|_{\hat{B}} &= \sup_{0<x\le h} \left| x^{m-n} (A^j y(x) - A^j \tilde{y}(x)) \right| \\
&= \sup_{0<x\le h} \left| x^{m-n} (AA^{j-1} y(x) - AA^{j-1} \tilde{y}(x)) \right| \\
&= \sup_{0<x\le h} \frac{x^{m-n}}{\Gamma(n)} \left| \int_0^x (x-t)^{n-1} \left[f(t, A^{j-1} y(t)) - f(t, A^{j-1} \tilde{y}(t)) \right] dt \right| \\
&\le \sup_{0<x\le h} \frac{x^{m-n}}{\Gamma(n)} \int_0^x (x-t)^{n-1} \left| f(t, A^{j-1} y(t)) - f(t, A^{j-1} \tilde{y}(t)) \right| dt \\
&\le \frac{L}{\Gamma(n)} \sup_{0<x\le h} x^{m-n} \int_0^x (x-t)^{n-1} \left| A^{j-1} y(t) - A^{j-1} \tilde{y}(t) \right| dt
\end{aligned}
$$

by definition of the operator A and the Lipschitz condition on f. In the next step we estimate further to find

$$
\begin{aligned}
\left\| A^j y - A^j \tilde{y} \right\|_{\hat{B}} &< \frac{L}{\Gamma(n)} \sup_{0<x\le h} x^{m-n} \int_0^x (x-t)^{n-1} \left| A^{j-1} y(t) - A^{j-1} \tilde{y}(t) \right| dt \\
&\le \frac{L}{\Gamma(n)} \sup_{0<x\le h} x^{m-n} \int_0^x (x-t)^{n-1} t^{n-m} t^{m-n} \left| A^{j-1} y(t) - A^{j-1} \tilde{y}(t) \right| dt \\
&\le \frac{L}{\Gamma(n)} \left\| A^{j-1} y - A^{j-1} \tilde{y} \right\|_{\hat{B}} \sup_{0<x\le h} x^{m-n} \int_0^x (x-t)^{n-1} t^{n-m} dt \\
&= \frac{L}{\Gamma(n)} \left\| A^{j-1} y - A^{j-1} \tilde{y} \right\|_{\hat{B}} \sup_{0<x\le h} \frac{\Gamma(n)\Gamma(n-m+1)}{\Gamma(2n-m+1)} x^n \\
&= \frac{Lh^n \Gamma(n-m+1)}{\Gamma(2n-m+1)} \left\| A^{j-1} y - A^{j-1} \tilde{y} \right\|_{\hat{B}}.
\end{aligned}
$$

Now we use the induction hypothesis, proving (5.1). Therefore we may apply Theorem D.7 with $\alpha_j = \gamma^j$ where $\gamma = (Lh^n \Gamma(n-m+1)/\Gamma(2n-m+1))$. It remains to prove that the series $\sum_{j=0}^{\infty} \alpha_j$ is convergent. This, however, is trivial in view of the fact that $h \le \tilde{h}$ and the definition of $\tilde{h}$ that implies $\gamma < 1$. Thus an application of the fixed point theorem yields the existence and the uniqueness of the solution of our integral equation. $\square$

Remark 5.2. The proof of Lemma 5.3 also gives us, at least in theory, a constructive method to find the solution of the initial value problem by means of the calculation of the sequence $(A^j y_0)_{j=0}^\infty$, where y_0 in an arbitrary element of B. The limit of this sequence is the desired solution. Typically one chooses

$$y_0(x) = \sum_{k=1}^m \frac{b_k x^{n-k}}{\Gamma(n-k+1)}.$$

In this case we call the sequence $(A^j y_0)_{j=1}^\infty$ the *Picard iteration sequence* corresponding to the given initial value problem.

Remark 5.3. Theorem 5.1 can be interpreted as an analogue of the Picard–Lindelöf theorem for first-order differential equations. We may ask ourselves whether the conditions are too sharp in the fractional setting. It turns out that it is possible to prove a weaker result under weaker assumptions: If we drop the Lipschitz condition on f, the existence of the solution can still be shown. This corresponds to Peano's existence theorem in the classical theory. The proof is essentially similar, just replacing Weissinger's theorem by Schauder's fixed point theorem. For the corresponding problem involving Caputo operators instead of Riemann–Liouville derivatives, we shall investigate this explicitly in Sect. 6.1.

For a further illustration of this remark, we discuss a very simple example of a fractional differential equation with a right-hand side that does not fulfil a Lipschitz condition.

Example 5.1. Consider the differential equation

$$D_0^n y(x) = [y(x)]^\mu$$

where $0 < \mu < 1$. In this case the right-hand side of the equation in continuous but the Lipschitz condition is violated. If we select the initial condition corresponding to this differential equation as

$$\lim_{z \to 0+} J_0^{1-n} y(z) = 0 \text{ and } D_0^{n-k} y(0) = 0 \qquad (k = 1, 2, \ldots, \lceil n \rceil - 1),$$

we easily see that one solution is $y \equiv 0$. However, an explicit calculation reveals that the function y given by

$$y(x) = {}^{\mu-1}\sqrt{\frac{\Gamma(j+1)}{\Gamma(j+1-n)}} x^j$$

with $j = n/(1-\mu)$ also solves the initial value problem. Thus, we indeed see that, in general, the uniqueness of the solution cannot be expected without the Lipschitz condition.

Exercises

Exercise 5.1. Show that the set $\widehat{B}$ given in the proof of Lemma 5.3, equipped with the norm indicated there, is a normed linear space, and that B is a complete subset of $\widehat{B}$.

Exercise 5.2. Complete the details of the proof of the existence theorem mentioned in Remark 5.3.
Hint: Follow the structure of the proof of Theorem 6.1.

Exercise 5.3. Consider the fractional differential equation

$$D_0^{1/3} y(x) = x^2 + xy(x)$$

with initial condition $\lim_{z \to 0+} J_0^{2/3} y(z) = -1$. For this initial value problem, construct the operator A from the proof of Lemma 5.3 and determine the first five elements of the corresponding Picard iteration sequence.

Exercise 5.4. Consider the fractional differential equation

$$D_0^{3/2} y(x) = x + (xy(x))^2$$

with initial conditions $\lim_{z \to 0+} J_0^{1/2} y(z) = 3$ and $D_0^{1/2} y(0) = 1$. For this initial value problem, construct the operator A from the proof of Lemma 5.3 and determine the first three elements of the corresponding Picard iteration sequence.

Chapter 6
Single-Term Caputo Fractional Differential Equations: Basic Theory and Fundamental Results

Having established the fundamentals of a theory for fractional differential equations with Riemann–Liouville derivatives, we now come to the corresponding problem for Caputo operators. In view of the fact that the latter seem to be much more important than the former as far as applications outside of mathematics are concerned, we shall discuss this problem in a more detailed fashion. The main emphasis will be on initial value problems. In particular, the first two sections of this chapter will be devoted to existence and uniqueness questions, respectively, for a most general class of equations whereas in the third section we shall deal with structural stability of the solutions: How do they depend on the given data? The smoothness properties of the solutions will then be discussed in Sect. 6.4. Finally, in Sect. 6.5, we will leave the area of initial value problems and provide some fundamental results about boundary value problems.

The results of this chapter will be used in Chap. 7 to thoroughly establish more specific theorems describing the properties of certain practically very important special cases. Later, in Chap. 8, the considerations will be extended to a more general class of equations, namely equations containing more than one differential operator. Corresponding results for Riemann–Liouville equations are available, e.g., in the publications mentioned at the beginning of Chap. 5.

6.1 Existence of Solutions

We begin once again with equations of the form

$$D_{*0}^n y(x) = f(x, y(x)), \tag{6.1a}$$

combined with appropriate initial conditions. As indicated in Chap. 3 these conditions have the form

$$D^k y(0) = y_0^{(k)}, \qquad k = 0, 1, \ldots, m-1, \tag{6.1b}$$

K. Diethelm, *The Analysis of Fractional Differential Equations*,
Lecture Notes in Mathematics 2004, DOI 10.1007/978-3-642-14574-2_6,
© Springer-Verlag Berlin Heidelberg 2010

where as usual we have set $m = \lceil n \rceil$. In Chap. 8, a more general problem will be considered. The results presented in this section and in the first part of Sect. 6.2 are mainly slightly modified versions of the findings of Diethelm and Ford [43].

The first result is an existence result that corresponds to the classical Peano existence theorem for first order equations.

Theorem 6.1. *Let $0 < n$ and $m = \lceil n \rceil$. Moreover let $y_0^{(0)}, \ldots, y_0^{(m-1)} \in \mathbb{R}$, $K > 0$ and $h^* > 0$. Define $G := \{(x,y) : x \in [0,h^*], |y - \sum_{k=0}^{m-1} x^k y_0^{(k)}/k!| \leq K\}$, and let the function $f : G \to \mathbb{R}$ be continuous. Furthermore, define $M := \sup_{(x,z) \in G} |f(x,z)|$ and*

$$h := \begin{cases} h^* & \text{if } M = 0, \\ \min\{h^*, (K\Gamma(n+1)/M)^{1/n}\} & \text{else.} \end{cases}$$

Then, there exists a function $y \in C[0,h]$ solving the initial value problem (6.1).

Remark 6.1. For the sake of simplicity of the presentation we only treat the scalar case explicitly here. However, all the results in this and the following chapter can be extended to vector-valued functions y (i.e. systems of equations) without any problems.

Remark 6.2. In many applications in science and engineering, we have $0 < n \leq 1$. In this case, the set G defined in Theorem 6.1 is just the simple rectangle $G = [0,h^*] \times [y_0^{(0)} - K, y_0^{(0)} + K]$.

For the proofs of most of the theorems in this and the next section, we will use the following lemma that adapts the statement of Lemma 5.2 to the present situation.

Lemma 6.2. *Assume the hypotheses of Theorem 6.1. The function $y \in C[0,h]$ is a solution of the initial value problem (6.1) if and only if it is a solution of the nonlinear Volterra integral equation of the second kind*

$$y(x) = \sum_{k=0}^{m-1} \frac{x^k}{k!} y_0^{(k)} + \frac{1}{\Gamma(n)} \int_0^x (x-t)^{n-1} f(t, y(t)) \, dt \tag{6.2}$$

with $m = \lceil n \rceil$.

Before we come to the proof of this lemma, we would like to make some comments concerning its relation to the corresponding results for equations with Riemann–Liouville operators and equations of integer order, respectively.

Remark 6.3. Recalling Remark 5.1 and using Lemmas 5.2 and 6.2, we can compare the behaviour of the solutions of Caputo-type fractional differential equations with those of Riemann–Liouville type. Doing so we find that the continuity problem at the origin that we found for the Riemann–Liouville setting does not arise in the Caputo environment. Rather, it turns out that continuity of the function f implies continuity of the solution y throughout the *closed* interval $[0,h]$.

Remark 6.4. Let us look at (6.2) for some $n \in (0, 1]$ and for two different values of x, say x_1 and x_2 with $x_1 < x_2$, and subtract the second of these equations from the first. This yields

$$y(x_2) - y(x_1) = \frac{1}{\Gamma(n)} \int_0^{x_2} (x_2 - t)^{n-1} f(t, y(t)) \, dt$$

$$- \frac{1}{\Gamma(n)} \int_0^{x_1} (x_1 - t)^{n-1} f(t, y(t)) \, dt$$

$$= \frac{1}{\Gamma(n)} \int_0^{x_1} \left[(x_2 - t)^{n-1} - (x_1 - t)^{n-1} \right] f(t, y(t)) \, dt$$

$$+ \frac{1}{\Gamma(n)} \int_{x_1}^{x_2} (x_2 - t)^{n-1} f(t, y(t)) \, dt. \tag{6.3}$$

Now let us first consider the classical (non-fractional) case $n - 1$. Here the term in brackets on the right-hand side of (6.3) is zero, and hence the entire first integral vanishes. This equation then implies the well known fact that, if we already know the solution $y(x_1)$ of our given initial value problem (6.1) at the point $x_1 > 0$, then we may compute the solution at the point $x_2 > x_1$ exclusively on the basis of $y(x_1)$ and the function f. We do not need to use any information on $y(x)$ for $x \in [0, x_1)$. This observation, which is just another way of expressing the locality of the integer-order differential operator, is the basis of almost all classical methods for the numerical solution of first-order differential equations, and it is also of fundamental significance in the mathematical modelling of many systems in physics, engineering, and other sciences because it states that it is sufficient to observe the state of a first-order system at an arbitrary point in time to compute its behaviour in the future.

When we look at the fractional case $0 < n < 1$ however, the situation is fundamentally different. Here the first integral on the right-hand side of (6.3) does not vanish in general. Hence, whenever we want to compute the solution $y(x_2)$ at some point x_2 it is necessary to take into account the entire history of y from the starting point 0 up to the point of interest x_2. This reflects the non-locality of the Caputo fractional differential operators that we had already observed in Remark 3.2. Obviously, this observation has a substantial influence on the construction of numerical methods for such equations, Moreover, for a fractional-order system modelling some real-world phenomena one may be drawn into the conclusion that here one would be forced to measure the state of the system at the initial point and would not be allowed to measure at an arbitrary point. This latter conclusion, however, is not correct as we shall see in Theorem 6.17.

It thus follows that integer-order equations are appropriate tools for the modelling of systems without memory whereas fractional-order equations are the method of choice for the description of systems with memory.

Of course, since we had noted that Riemann–Liouville derivatives are not local either, this remark applies to Riemann–Liouville fractional differential equations as well.

Remark 6.5. As stated in the previous remark, (6.3) has a major impact on the theoretical basis for numerical methods for fractional differential equations. The standard methods usually take this into account properly, but they often do not make use of the full power of this equation. Specifically, as noted by Deng [33], even though the term in brackets on the right-hand side of (6.3) is not zero, it will be quite small in magnitude in certain situations (e.g. if x_1 and x_2 are rather large compared to $x_2 - x_1$, a situation that is quite common for numerical methods where $x_2 - x_1$ may be a small step size). Thus one may be able to approximate the first integral on the right-hand side of (6.3) by a less accurate but cheap method, thus reducing the total complexity without significantly losing accuracy.

Proof (of Lemma 6.2). The proof that every continuous solution of the Volterra equation also solves the initial value problem is very close to the proof of the corresponding part of Lemma 5.2; we therefore leave the details to the reader.

   For the other direction, we define $z(x) := f(x, y(x))$ and once again note that $z \in C[0, h]$ by our assumptions on y and f. Then, using the definition of the Caputo differential operator, the differential equation can be rewritten as

$$z(x) = f(x, y(x)) = D_{*0}^n y(x) = D_0^n (y - T_{m-1}[y; 0])(x)$$
$$= D^m J_0^{m-n} (y - T_{m-1}[y; 0])(x).$$

Since we are dealing with continuous functions, we may apply the operator J_0^m to both sides of the equation and find

$$J_0^m z(x) = J_0^{m-n} (y - T_{m-1}[y; 0])(x) + q(x)$$

with some polynomial q of degree not exceeding $m - 1$. Since z is continuous, the function $J_0^m z$ on the left-hand side of this equation has a zero of order (at least) m at the origin. Moreover, the difference $y - T_{m-1}[y; 0]$ has the same property by construction, and therefore the function $J_0^{m-n}(y - T_{m-1}[y; 0])$ on the right-hand side of our equation must have such an mth order zero too. Thus, the polynomial q has the same property, and we immediately deduce (since its degree is not more than $m - 1$) that $q = 0$. Consequently,

$$J_0^m z(x) = J_0^{m-n} (y - T_{m-1}[y; 0])(x),$$

and by applying the Riemann–Liouville differential operator D_0^{m-n} to this equation we find

$$y(x) - T_{m-1}[y; 0](x) = D_0^{m-n} J_0^m z(x) = D^1 J_0^{1+n-m} J_0^m z(x) = D J_0^{1+n} z(x)$$
$$= J_0^n z(x).$$

Recalling the definitions of z and the Taylor polynomial $T_{m-1}[y; 0]$, this is just the required Volterra equation.   □

Proof (of Theorem 6.1). If $M = 0$ then $f(x,y) = 0$ for all $(x,y) \in G$. In this case it is evident that the function $y : [0,h] \to \mathbb{R}$ with $y(x) = \sum_{k=0}^{m-1} y_0^{(k)} x^k / k!$ is a solution of the initial value problem (6.1). Hence we conclude, as required, that a solution exists in this case.

Otherwise, we apply Lemma 6.2 and see that our initial value problem (6.1) is equivalent to the Volterra equation (6.2). We thus introduce the polynomial T that satisfies the initial conditions, viz.

$$T(x) := \sum_{k=0}^{m-1} \frac{x^k}{k!} y_0^{(k)}, \tag{6.4}$$

and the set $U := \{y \in C[0,h] : \|y - T\|_\infty \leq K\}$. It is evident that U is a closed and convex subset of the Banach space of all continuous functions on $[0,h]$, equipped with the Chebyshev norm. Hence, U is a Banach space too. Since the polynomial T is an element of U, we also see that U is not empty. On this set U we define the operator A by

$$(Ay)(x) := T(x) + \frac{1}{\Gamma(n)} \int_0^x (x-t)^{n-1} f(t, y(t)) \, dt. \tag{6.5}$$

Using this operator, the equation whose solvability we need to prove, viz. the Volterra equation (6.2), can be rewritten as

$$y = Ay,$$

and thus, in order to prove our desired existence result, we have to show that A has a fixed point. We therefore proceed by investigating the properties of the operator A more closely.

Our first goal in this context is to show that $Ay \in U$ for $y \in U$. To this end we begin by noting that, for $0 \leq x_1 \leq x_2 \leq h$,

$$|(Ay)(x_1) - (Ay)(x_2)|$$

$$= \frac{1}{\Gamma(n)} \left| \int_0^{x_1} (x_1 - t)^{n-1} f(t, y(t)) \, dt - \int_0^{x_2} (x_2 - t)^{n-1} f(t, y(t)) \, dt \right|$$

$$= \frac{1}{\Gamma(n)} \left| \int_0^{x_1} \left((x_1 - t)^{n-1} - (x_2 - t)^{n-1} \right) f(t, y(t)) \, dt + \int_{x_1}^{x_2} (x_2 - t)^{n-1} f(t, y(t)) \, dt \right|$$

$$\leq \frac{M}{\Gamma(n)} \left(\int_0^{x_1} \left| (x_1 - t)^{n-1} - (x_2 - t)^{n-1} \right| \, dt + \int_{x_1}^{x_2} (x_2 - t)^{n-1} \, dt \right). \tag{6.6}$$

The second integral in the right-hand side of (6.6) has the value $(x_2 - x_1)^n / n$. For the first integral, we look at the three cases $n = 1$, $n < 1$ and $n > 1$ separately. In the first case $n = 1$, the integrand vanishes identically, and hence the integral has the value zero. Secondly, for $n < 1$, we have $n - 1 < 0$, and hence $(x_1 - t)^{n-1} \geq (x_2 - t)^{n-1}$. Thus,

$$\int_0^{x_1} \left| (x_1 - t)^{n-1} - (x_2 - t)^{n-1} \right| dt = \int_0^{x_1} \left((x_1 - t)^{n-1} - (x_2 - t)^{n-1} \right) dt$$

$$= \frac{1}{n} \left(x_1^n - x_2^n + (x_2 - x_1)^n \right) \leq \frac{1}{n} (x_2 - x_1)^n.$$

Finally, if $n > 1$ then $(x_1 - t)^{n-1} \leq (x_2 - t)^{n-1}$, and hence

$$\int_0^{x_1} \left| (x_1 - t)^{n-1} - (x_2 - t)^{n-1} \right| dt = \int_0^{x_1} \left((x_2 - t)^{n-1} - (x_1 - t)^{n-1} \right) dt$$

$$= \frac{1}{n} \left(-x_1^n + x_2^n - (x_2 - x_1)^n \right) \leq \frac{1}{n} (x_2^n - x_1^n).$$

A combination of these results yields

$$|(Ay)(x_1) - (Ay)(x_2)| \leq \begin{cases} \dfrac{2M}{\Gamma(n+1)} (x_2 - x_1)^n & \text{if } n \leq 1, \\[2ex] \dfrac{M}{\Gamma(n+1)} \left((x_2 - x_1)^n + x_2^n - x_1^n \right) & \text{if } n > 1. \end{cases} \tag{6.7}$$

In either case, the expression on the right-hand side of (6.7) converges to 0 as $x_2 \to x_1$ which proves that Ay is a continuous function. Moreover, for $y \in U$ and $x \in [0, h]$, we find

$$|(Ay)(x) - T(x)| = \frac{1}{\Gamma(n)} \left| \int_0^x (x - t)^{n-1} f(t, y(t)) dt \right| \leq \frac{1}{\Gamma(n+1)} M x^n$$

$$\leq \frac{1}{\Gamma(n+1)} M h^n \leq \frac{1}{\Gamma(n+1)} M \frac{K \Gamma(n+1)}{M} = K.$$

Thus, we have shown that $Ay \in U$ if $y \in U$, i.e. A maps the set U to itself.

Since we want to apply Schauder's Fixed Point Theorem (Theorem D.9), all that remains now is to show that $A(U) := \{Au : u \in U\}$ is a relatively compact set. This can be done by means of the Arzelà–Ascoli Theorem (Theorem D.10). For $z \in A(U)$ we find that, for all $x \in [0, h]$,

$$|z(x)| = |(Ay)(x)| \leq \|T\|_\infty + \frac{1}{\Gamma(n)} \int_0^x (x - t)^{n-1} |f(t, y(t))| dt$$

$$\leq \|T\|_\infty + \frac{1}{\Gamma(n+1)} M h^n \leq \|T\|_\infty + K,$$

which is the required boundedness property. Moreover, the equicontinuity property can be derived from (6.7) above. Specifically, for $0 \leq x_1 \leq x_2 \leq h$, we have found in the case $n \leq 1$ that

$$|(Ay)(x_1) - (Ay)(x_2)| \leq \frac{2M}{\Gamma(n+1)} (x_2 - x_1)^n.$$

Thus, if $|x_2 - x_1| < \delta$, then

$$|(Ay)(x_1) - (Ay)(x_2)| \leq 2 \frac{M}{\Gamma(n+1)} \delta^n.$$

Noting that the expression on the right-hand side is independent of y, x_1 and x_2, we see that the set $A(U)$ is equicontinuous. Similarly, in the case $n > 1$ we may use the Mean Value Theorem to conclude that

$$\begin{aligned}
|(Ay)(x_1) - (Ay)(x_2)| &\leq \frac{M}{\Gamma(n+1)} \left((x_2 - x_1)^n + x_2^n - x_1^n \right) \\
&= \frac{M}{\Gamma(n+1)} \left((x_2 - x_1)^n + n(x_2 - x_1) \xi^{n-1} \right) \\
&\leq \frac{M}{\Gamma(n+1)} \left((x_2 - x_1)^n + n(x_2 - x_1) h^{n-1} \right)
\end{aligned}$$

with some $\xi \in [x_1, x_2] \subseteq [0, h]$. Hence, if once again $|x_2 - x_1| < \delta$, then

$$|(Ay)(x_1) - (Ay)(x_2)| \leq \frac{M}{\Gamma(n+1)} \left(\delta^n + n\delta h^{n-1} \right)$$

and the right-hand side is once more independent of y, x_1 and x_2, proving the equicontinuity. In either case the Arzelà–Ascoli Theorem yields that $A(U)$ is relatively compact, and hence Schauder's Fixed Point Theorem asserts that A has a fixed point. By construction, a fixed point of A is a solution of our initial value problem. $\qquad\square$

We note two important special cases of Theorem 6.1. The first of these states that, under certain assumptions, the solution exists on the entire interval $[0, h^*]$ (and so for all x for which $f(x, y)$ is defined) and not only for a subinterval $[0, h]$ with some $h \leq h^*$.

Corollary 6.3. *Assume the hypotheses of Theorem 6.1, except that the set G, i.e. the domain of definition of the function f on the right-hand side of the differential equation (6.1a), is now taken to be $G := [0, h^*] \times \mathbb{R}$. Moreover we assume that f is continuous and that there exist constants $c_1 \geq 0$, $c_2 \geq 0$ and $0 \leq \mu < 1$ such that*

$$|f(x, y)| \leq c_1 + c_2 |y|^\mu \quad \textit{for all } (x, y) \in G. \tag{6.8}$$

Then, there exists a function $y \in C[0, h^*]$ solving the initial value problem (6.1).

Proof. We use the polynomial T defined in (6.4) in the previous proof. Since $\mu < 1$ we may find some $K > 0$ such that

$$c_1 + c_2 (K + \max_{x \in [0,h^*]} |T(x)|)^\mu \leq \frac{K \Gamma(n+1)}{h^{*n}}.$$

Using this value of K, we then restrict our function f to the set $G_K := \{(x,y) : x \in [0,h^*], |y - T(x)| \leq K\}$ (this is the set that was denoted by G in Theorem 6.1). Then we see that

$$M := \sup_{(x,y) \in G_K} |f(x,y)| \leq c_1 + c_2 \sup_{(x,y) \in G_K} |y|^\mu$$

$$\leq c_1 + c_2 (K + \max_{x \in [0,h^*]} |T(x)|)^\mu \leq \frac{K \Gamma(n+1)}{h^{*n}}.$$

Thus we may apply Theorem 6.1 with this value of K and the given h^* and see that $(K\Gamma(n+1)/M)^{1/n} \geq h^*$ which implies that

$$h^* = \min \left\{ h^*, \left(\frac{K\Gamma(n+1)}{M} \right)^{1/n} \right\} = h. \qquad \square$$

Remark 6.6. Three comments with respect to condition (6.8) are in order:

(a) A sufficient condition on f for (6.8) to be satisfied is that f is continuous and bounded on G.
(b) In Theorem 6.1, f was required to be a continuous function on a compact set, and hence it was automatically bounded. In this corollary, f is still continuous but now it is defined on a non-compact set, and hence we have to demand a suitable bound explicitly.
(c) Condition (6.8) can be considered to be a quite severe restriction since it is violated even by some very elementary and practically important types of equations like, e.g., linear equations. Therefore, additional investigations are necessary.

The second corollary to Theorem 6.1 asserts the existence of a solution on the entire half-axis $[0, \infty)$ under appropriate conditions.

Corollary 6.4. *Assume the hypotheses of Corollary 6.3, except that the set G, i.e. the domain of definition of the function f on the right-hand side of the differential equation (6.1a), is now taken to be $G := \mathbb{R}^2$. Then, there exists a function $y \in C[0, \infty)$ solving the initial value problem (6.1).*

Of course, Remark 6.6 applies to this corollary too.

Proof. Let $h^* > 0$. Under our assumptions, we may apply Corollary 6.3 for this h^* and conclude that a continuous solution exists on $[0, h^*]$. Since h^* can be chosen arbitrarily large, we find that a continuous solution exists on $[0, \infty)$. $\qquad \square$

6.2   Uniqueness of Solutions

Next we come to a uniqueness theorem that corresponds to the well-known Picard–Lindelöf result. It can be seen as an analogue to the statement shown for Riemann–Liouville operators in the previous chapter (Theorem 5.1).

Theorem 6.5. *Let $0 < n$ and $m = \lceil n \rceil$. Moreover let $y_0^{(0)}, \ldots, y_0^{(m-1)} \in \mathbb{R}$, $K > 0$ and $h^* > 0$. Define the set G as in Theorem 6.1 and let the function $f : G \to \mathbb{R}$ be continuous and fulfil a Lipschitz condition with respect to the second variable, i.e.*

$$|f(x, y_1) - f(x, y_2)| \le L|y_1 - y_2|$$

with some constant $L > 0$ independent of x, y_1, and y_2. Then, denoting h as in Theorem 6.1, there exists a uniquely defined function $y \in C[0, h]$ solving the initial value problem (6.1).

Remark 6.7. Remark 6.3 that we had stated above in connection with the existence of solutions applies here in the discussion of uniqueness questions too.

Proof (of Theorem 6.5). We first note that Theorem 6.1 asserts that the initial value problem has a solution. In order to prove the uniqueness of this solution, we start with arguments similar to those of the proof of Theorem 6.1. In particular, we use the same polynomial T (defined in (6.4)) and the same operator A (defined in (6.5)) and recall that it maps the nonempty, convex and closed set $U = \{y \in C[0, h] : \|y - T\|_\infty \le K\}$ to itself. We now have to prove that A has a unique fixed point. In order to do this, we shall first prove that, for every $j \in \mathbb{N}_0$, every $x \in [0, h]$ and all $y, \tilde{y} \in U$, we have

$$\left\| A^j y - A^j \tilde{y} \right\|_{L_\infty[0, x]} \le \frac{(Lx^n)^j}{\Gamma(1 + nj)} \|y - \tilde{y}\|_{L_\infty[0, x]}. \tag{6.9}$$

This can be seen by induction. In the case $j = 0$, the statement is trivially true. For the induction step $j - 1 \mapsto j$, we write

$$\left\| A^j y - A^j \tilde{y} \right\|_{L_\infty[0, x]}$$
$$= \left\| A(A^{j-1} y) - A(A^{j-1} \tilde{y}) \right\|_{L_\infty[0, x]}$$
$$= \frac{1}{\Gamma(n)} \sup_{0 \le w \le x} \left| \int_0^w (w - t)^{n-1} \left[f(t, A^{j-1} y(t)) - f(t, A^{j-1} \tilde{y}(t)) \right] dt \right|.$$

We proceed in the induction step by using the Lipschitz assumption on f and the induction hypothesis. This allows us to estimate the quantities under consideration in the following way:

$$\left\| A^j y - A^j \tilde{y} \right\|_{L_\infty[0, x]}$$
$$\le \frac{L}{\Gamma(n)} \sup_{0 \le w \le x} \int_0^w (w - t)^{n-1} \left| A^{j-1} y(t) - A^{j-1} \tilde{y}(t) \right| dt$$

$$\leq \frac{L}{\Gamma(n)} \int_0^x (x-t)^{n-1} \sup_{0 \leq w \leq t} \left| A^{j-1} y(w) - A^{j-1} \tilde{y}(w) \right| dt$$

$$\leq \frac{L^j}{\Gamma(n) \Gamma(1 + n(j-1))} \int_0^x (x-t)^{n-1} t^{n(j-1)} \sup_{0 \leq w \leq t} |y(w) - \tilde{y}(w)| \, dt$$

$$\leq \frac{L^j}{\Gamma(n) \Gamma(1 + n(j-1))} \sup_{0 \leq w \leq x} |y(w) - \tilde{y}(w)| \int_0^x (x-t)^{n-1} t^{n(j-1)} \, dt$$

$$= \frac{L^j}{\Gamma(n) \Gamma(1 + n(j-1))} \|y - \tilde{y}\|_{L_\infty[0,x]} \frac{\Gamma(n) \Gamma(1 + n(j-1))}{\Gamma(1 + nj)} x^{nj}.$$

This is our desired result (6.9). As a consequence, we find, taking Chebyshev norms on our fundamental interval $[0, h]$,

$$\left\| A^j y - A^j \tilde{y} \right\|_\infty \leq \frac{(Lh^n)^j}{\Gamma(1 + nj)} \|y - \tilde{y}\|_\infty.$$

We have now shown that the operator A fulfils the assumptions of Theorem D.7 with $\alpha_j = (Lh^n)^j / \Gamma(1 + nj)$. In order to apply that theorem, we only need to verify that the series $\sum_{j=0}^\infty \alpha_j$ converges. To this end we notice that $\sum_{j=0}^\infty \alpha_j$ with α_j as above is simply the power series representation of the Mittag-Leffler function $E_n(Lh^n)$, and hence the required convergence of the series follows immediately from Theorem 4.1. Therefore, we may apply Weissinger's Fixed Point Theorem and deduce the uniqueness of the solution of our differential equation.   □

Remark 6.8. An observation similar to the one made for the Riemann–Liouville case in Remark 5.2 holds here too: The proofs of the uniqueness theorem 6.5 once again give a constructive method to find a sequence of Picard iterations that converges against the exact solution of the initial value problem. In particular, the individual elements of this sequence can be considered as approximations for the exact solution, at least in theory. Since it is not necessarily possible to evaluate the required integrals in closed form, these approximations are unlikely to be useful numerically.

Remark 6.9. Concerning the connection between the uniqueness statement and the Lipschitz condition we look at the differential equation

$$D_{*0}^n y = y^\mu$$

with $0 < \mu < 1$. We had discussed the Riemann–Liouville analogue of this equation in Example 5.1. In the Caputo version we again use homogeneous initial conditions

$$y(0) = 0 \quad \text{and} \quad D_0^{n-k} y(0) = 0 \qquad (k = 1, 2, \ldots, \lceil n \rceil - 1)$$

and find the same two solutions as in the Riemann–Liouville case, viz.

$$y(x) = 0 \quad \text{and} \quad y(x) = {}^{\mu-1}\sqrt{\frac{\Gamma(j+1)}{\Gamma(j+1-n)}} x^j$$

with $j = n/(1-\mu)$.

An apparent weakness in the statement of Theorem 6.5 is that it yields the existence and uniqueness of the solution of the initial value problem not on the interval $[0, h^*]$ where the first argument of the given function f was allowed to come from, but only on the possibly smaller interval $[0, h]$ with $h = \min\{h^*, (K\Gamma(n+1)/M)^{1/n}\}$. In this respect, Theorem 6.5 is completely analogous to the Peano-type existence Theorem 6.1. For the latter, we had derived Corollaries 6.3 and 6.4 that gave sufficient conditions for the solution to exist on the complete interval $[0, h^*]$ or even on $[0, \infty)$. Corresponding results can be shown for the uniqueness question too:

Corollary 6.6. *Assume the hypotheses of Theorem 6.5, except that the set G, i.e. the domain of definition of the function f on the right-hand side of the differential equation (6.1a), is now taken to be $G := [0, h^*] \times \mathbb{R}$. Moreover we assume that f is continuous and that there exist constants $c_1 \geq 0$, $c_2 \geq 0$ and $0 \leq \mu < 1$ such that*

$$|f(x,y)| \leq c_1 + c_2|y|^{\mu} \quad \text{for all } (x,y) \in G. \tag{6.10}$$

Then, there exists a function $y \in C[0, h^*]$ solving the initial value problem (6.1).

Corollary 6.7. *Assume the hypotheses of Corollary 6.6, except that the set G, i.e. the domain of definition of the function f on the right-hand side of the differential equation (6.1a), is now taken to be $G := \mathbb{R}^2$. Then, there exists a function $y \in C[0, \infty)$ solving the initial value problem (6.1).*

The proofs of these two corollaries exactly follow the lines of the proofs of Corollaries 6.3 and 6.4, respectively. We omit the details.

Remark 6.10. As in Remark 6.6 we want to comment on condition (6.10):

(a) Since we have to assume the continuity of f anyway, a sufficient condition on f for (6.10) to be satisfied is that f is bounded on G. As G is unbounded now, the boundedness of f is not an automatic consequence of its continuity.

(b) Condition (6.10) can be considered to be a quite severe restriction since it is violated even by some very elementary and practically important types of equations like, e.g., linear equations.

The last remark here motivates us to include another existence and uniqueness theorem. It uses a slightly different set of assumptions that actually allow us to derive a global existence and uniqueness result for a large class of equations that now includes the linear equations.

Theorem 6.8. *Let $0 < n$ and $m = \lceil n \rceil$. Moreover let $y_0^{(0)}, \ldots, y_0^{(m-1)} \in \mathbb{R}$ and $h^* > 0$. Define the set $G := [0, h^*] \times \mathbb{R}$ and let the function $f : G \to \mathbb{R}$ be continuous and fulfil a Lipschitz condition with respect to the second variable with a Lipschitz constant $L > 0$ that is independent of x, y_1, and y_2. Then there exists a uniquely defined function $y \in C[0, h^*]$ solving the initial value problem (6.1).*

In particular, this theorem is applicable to linear equations, i.e. equations of the form

$$D_{*0}^n y(x) = f(x) y(x) + g(x)$$

with certain functions $f, g \in C[0, h^*]$, because here we may choose $L = \|f\|_\infty < \infty$.

We obtain an immediate consequence:

Corollary 6.9. *Let $0 < n$ and $m = \lceil n \rceil$. Moreover let $y_0^{(0)}, \ldots, y_0^{(m-1)} \in \mathbb{R}$ and $h^* > 0$. Define the set $G := [0, \infty) \times \mathbb{R}$ and let the function $f : G \to \mathbb{R}$ be continuous and fulfil a Lipschitz condition with respect to the second variable with a Lipschitz constant $L > 0$ that is independent of x, y_1, and y_2. Then there exists a uniquely defined function $y \in C[0, \infty)$ solving the initial value problem (6.1).*

Proof. Let $h^* > 0$. Under the assumptions of the corollary, we may apply Theorem 6.8 on the interval $[0, h^*]$ and conclude the existence of a unique continuous solution on this interval. Since h^* may be chosen arbitrarily large, we derive the existence and uniqueness on the entire half-axis $[0, \infty)$. $\qquad \square$

For the proof of Theorem 6.8 we collect some auxiliary results.

Lemma 6.10. *Assume the hypotheses of Theorem 6.8. Moreover denote $p(x, t) := (x - t)^{n-1} / \Gamma(n)$. This function p has the following properties:*

(a) For every $x \geq 0$, $p(x, \cdot)$ is absolutely integrable on $[0, x]$.

(b) For every function $k \in C[0, h^*]$ and all $\xi_1, \xi_2 \in [0, h^*]$, the expressions

$$\int_{\xi_1}^{\xi_2} p(x, t) f(t, k(t)) \, dt \quad and \quad \int_0^x p(x, t) f(t, k(t)) \, dt$$

are continuous with respect to x.

(c) There exist numbers $0 = h_0 < h_1 < h_2 < \ldots < h_N = h^*$ such that for all $i \in \{0, 1, \ldots, N-1\}$ and all $x \in [h_i, h^*]$ we have

$$L \int_{h_i}^{\min\{x, h_{i+1}\}} |p(x, t)| \, dt \leq \frac{1}{2}.$$

(d) For every $x \geq 0$,

$$\lim_{\delta \to 0+} \int_x^{x+\delta} |p(x+\delta, t)| \, dt = 0.$$

Proof. Part (a) is an immediate consequence of the definition of p and the fact that $n > 0$.

For (b) we will consider the two cases $n \geq 1$ and $n < 1$. The former is very simple since in this case all the functions under the integral operation are continuous in their respective domains of definition, and hence the continuity of the integrals is trivial. In the other case we may proceed as in the final part of the proof of Theorem 2.2 and use the facts that p is integrable and $f(\cdot, k(\cdot))$ is continuous to conclude the required result.

For part (c) we distinguish the same two cases. In the case $n \geq 1$, p is continuous and hence bounded on the compact set $\{(x,t) : 0 \leq t \leq x \leq h^*\}$. Thus, choosing $N := \lceil 2h^* L \|p\|_\infty \rceil$, where the Chebyshev norm of p is taken over the above mentioned set, we may define $h_i := ih^*/N$. This implies $0 = h_0 < h_1 < \ldots < h_N = h^*$ and

$$
L \int_{h_i}^{\min\{x, h_{i+1}\}} |p(x,t)| \, dt \leq L \int_{h_i}^{h_{i+1}} |p(x,t)| \, dt \leq L \|p\|_\infty (h_{i+1} - h_i)
$$
$$
= \frac{L\|p\|_\infty h^*}{N} \leq \frac{L\|p\|_\infty h^*}{2L\|p\|_\infty h^*} = \frac{1}{2}
$$

as required. In the case $n < 1$ we proceed in a slightly different manner. Here we use $N := \lceil h^* (2L/\Gamma(n+1))^{1/n} \rceil$ and $h_i := ih^*/N$. This once again implies $0 = h_0 < h_1 < \ldots < h_N = h^*$. Moreover we have, for $x \leq h_{i+1}$,

$$
L \int_{h_i}^{\min\{x, h_{i+1}\}} |p(x,t)| \, dt = \frac{L}{\Gamma(n)} \int_{h_i}^{x} (x-t)^{n-1} \, dt = \frac{L}{\Gamma(n+1)} (x - h_i)^n
$$
$$
\leq \frac{L}{\Gamma(n+1)} (h_{i+1} - h_i)^n = \frac{L}{\Gamma(n+1)} (h^*/N)^n
$$
$$
\leq \frac{L}{\Gamma(n+1)} \frac{\Gamma(n+1)}{2L} = \frac{1}{2}.
$$

In addition, for $x \geq h_{i+1}$, we find

$$
\Psi(x) := L \int_{h_i}^{\min\{x, h_{i+1}\}} |p(x,t)| \, dt = \frac{L}{\Gamma(n)} \int_{h_i}^{h_{i+1}} (x-t)^{n-1} \, dt
$$
$$
= \frac{L}{\Gamma(n+1)} [(x - h_i)^n - (x - h_{i+1})^n].
$$

Thus,

$$
\Psi'(x) = \frac{L}{\Gamma(n)} [(x - h_i)^{n-1} - (x - h_{i+1})^{n-1}] < 0
$$

since $n < 0$. It follows that, for $x \geq h_{i+1}$, $\Psi(x) \leq \Psi(h_{i+1}) \leq 1/2$ in view of our result above. This completes the proof of (c) for $n < 1$ too.

Finally, for (d) we see that

$$\int_x^{x+\delta} |p(x+\delta,t)|\,dt = \frac{1}{\Gamma(n+1)}\delta^n$$

which, since $n > 0$, implies the desired result. $\square$

Proof (of Theorem 6.8). Our approach is inspired by the sketched proof of [115, Theorem 4.8]. Let us recall the points h_i of Lemma 6.10 (c). We first concentrate on the interval $[h_0,h_1]$. We begin by defining the functions

$$y_0(x) := T(x) := \sum_{k=0}^{m-1} y_0^{(k)}\frac{x^k}{k!}$$

and

$$y_j(x) := T(x) + \frac{1}{\Gamma(n)}\int_0^x (x-t)^{n-1} f(t,y_{j-1}(t))\,dt, \quad j=1,2,\ldots,$$

and we note that Lemma 6.10 (b) implies the continuity of these functions on $[h_0,h_1]$. Moreover we define

$$\phi_j(x) := y_j(x) - y_{j-1}(x), \quad j=1,2,\ldots,$$

and

$$\phi_0(x) := T(x) = y_0(x).$$

Evidently these functions are continuous on $[h_0,h_1]$ too, and for $j=0,1,\ldots$ it is obvious that

$$y_j(x) = \sum_{\mu=0}^{j} \phi_\mu(x).$$

Moreover, for $j=2,3,\ldots$ we have

$$\phi_j(x) = \frac{1}{\Gamma(n)}\int_0^x (x-t)^{n-1}(f(t,y_{j-1}(t)) - f(t,y_{j-2}(t)))\,dt. \tag{6.11}$$

For $x \in [h_0,h_1]$ we then find, using (6.11), the Lipschitz condition on f and the statement of Lemma 6.10 (c),

$$\begin{aligned}
|\phi_j(x)| &\le \frac{1}{\Gamma(n)}\int_0^x (x-t)^{n-1}|f(t,y_{j-1}(t)) - f(t,y_{j-2}(t))|\,dt \\
&\le \frac{L}{\Gamma(n)}\int_0^x (x-t)^{n-1}|y_{j-1}(t) - y_{j-2}(t)|\,dt \\
&\le \frac{L}{\Gamma(n)}\max_{t\in[h_0,h_1]}|y_{j-1}(t) - y_{j-2}(t)|\int_0^x (x-t)^{n-1}\,dt \\
&\le \frac{1}{2}\max_{t\in[h_0,h_1]}|y_{j-1}(t) - y_{j-2}(t)| = \frac{1}{2}\max_{t\in[h_0,h_1]}|\phi_{j-1}(t)|
\end{aligned}$$

for $j = 2, 3, \ldots$, which implies that

$$\max_{x \in [h_0, h_1]} |\phi_j(x)| \leq \frac{1}{2^{j-1}} \max_{t \in [h_0, h_1]} |\phi_1(t)|.$$

Thus, we have found a convergent majorant for the series $\sum_{\mu=0}^{\infty} \phi_\mu$ on the interval $[h_0, h_1]$, and hence the series is uniformly convergent there. Since y_j is the jth partial sum of this series, it follows that the sequence $(y_j)_{j=1}^{\infty}$ is also uniformly convergent on $[h_0, h_1]$, and the limits coincide. Let us denote the limit of this sequence by y. We had seen above that y_j is continuous on $[0, h^*]$, and hence on $[h_0, h_1]$, for all j. Hence, in view of the uniform convergence, $y \in C[h_0, h_1]$.

Our next goal is to show that this function y solves the Volterra equation (6.2), and thus the initial value problem (6.1). This will then, in particular, imply the existence of a continuous solution. To this end we note that the uniform convergence of the sequence $(y_j)_{j=1}^{\infty}$ and the Lipschitz property of f imply $|f(x, y(x)) - f(x, y_j(x))| \leq L|y(x) - y_j(x)| \to 0$ uniformly for $x \in [h_0, h_1]$. In other words, the sequence $(f(\cdot, y_j(\cdot)))_{j=0}^{\infty}$ converges uniformly against $f(\cdot, y(\cdot))$. Thus we may interchange the limit operation and the fractional integration. This yields

$$y(x) = \lim_{j \to \infty} y_j(x) = \lim_{j \to \infty} \left(T(x) + J_0^n f(\cdot, y_{j-1}(\cdot))(x) \right)$$

$$= T(x) + J_0^n \left(\lim_{j \to \infty} f(\cdot, y_{j-1}(\cdot)) \right)(x)$$

$$= \sum_{k=0}^{m-1} y_0^{(k)} \frac{x^k}{k!} + J_0^n f(\cdot, y(\cdot))(x),$$

which is the required relation (6.2).

For the interval $[h_0, h_1]$ it remains to prove that this solution is unique. We therefore assume that $\tilde{y}$ is a solution of (6.2) as well. It then follows that

$$|y(x) - \tilde{y}(x)| \leq \frac{1}{\Gamma(n)} \int_0^x (x - t)^{n-1} |f(t, y(t)) - f(t, \tilde{y}(t))| \, dt$$

$$\leq \frac{L}{\Gamma(n)} \int_0^x (x - t)^{n-1} |y(t) - \tilde{y}(t)| \, dt$$

$$\leq \frac{1}{2} \max_{t \in [h_0, h_1]} |y(t) - \tilde{y}(t)|$$

by Lemma 6.10 (c). Since this inequality holds uniformly for all $x \in [h_0, h_1]$, we deduce

$$\max_{x \in [h_0, h_1]} |y(x) - \tilde{y}(x)| \leq \frac{1}{2} \max_{t \in [h_0, h_1]} |y(t) - \tilde{y}(t)|$$

which implies the required uniqueness statement $y \equiv \tilde{y}$.

Now we need to extend this existence and uniqueness result from the initial subinterval $[h_0, h_1]$ to the remaining subintervals $[h_{i-1}, h_i]$ $(i = 2, 3, \ldots, N)$. We will do this in an inductive manner, using our result for the interval $[h_0, h_1]$ as the basis. Thus we assume that the claim holds on $[h_{i-1}, h_i]$ for some i and we shall show that, if $i < N$, it also holds on $[h_i, h_{i+1}]$. We need to prove that, for x in this interval, (6.2) has a unique continuous solution. In this context it is convenient to rewrite the Volterra equation in question in the form

$$y(x) = T_i(x) + \frac{1}{\Gamma(n)} \int_{h_i}^{x} (x-t)^{n-1} f(t, y(t)) \, dt \qquad (6.12a)$$

with

$$T_i(x) = \sum_{k=0}^{m-1} y_0^{(k)} \frac{x^k}{k!} + \frac{1}{\Gamma(n)} \int_0^{h_i} (x-t)^{n-1} f(t, y(t)) \, dt. \qquad (6.12b)$$

The essential observation here is that the function T_i is a known function because its representation contains only given data and the values of the solution y on the interval $[0, h_i]$ that has already been computed. Moreover, Lemma 6.10 (b) implies that T_i is continuous on $[h_i, h_{i+1}]$. Recalling the definitions of the functions y_j and ϕ_j, we can then proceed in a way that is very similar to our approach above. Specifically we define

$$y_0^{(i)}(x) := T_i(x)$$

and

$$y_j^{(i)}(x) := T_i(x) + \frac{1}{\Gamma(n)} \int_{h_i}^{x} (x-t)^{n-1} f(t, y_{j-1}^{(i)}(t)) \, dt, \quad j = 1, 2, \ldots,$$

$$\phi_j^{(i)}(x) := y_j^{(i)}(x) - y_{j-1}^{(i)}(x), \quad j = 1, 2, \ldots,$$

and

$$\phi_0^{(i)}(x) := T_i(x) = y_0^{(i)}(x).$$

Evidently, by Lemma 6.10 (b), all these functions are continuous on $[h_i, h_{i+1}]$ too, and for $j = 0, 1, \ldots$ it is obvious that

$$y_j^{(i)}(x) = \sum_{\mu=0}^{j} \phi_\mu^{(i)}(x).$$

A convergent majorant for $\sum_{\mu=0}^{\infty} \phi_\mu^{(i)}$ can then be found as above which provides the existence of the uniform limit $y := \lim_{j \to \infty} y_j^{(i)}$ on the interval $[h_i, h_{i+1}]$. Lemma 6.10 (d) implies that the function y, which so far has been defined in a piecewise manner on the intervals $[0, h_i]$ and $[h_i, h_{i+1}]$, respectively, is actually continuous at the point h_i and hence throughout the entire interval $[0, h_{i+1}]$.

The proof that this function solves the Volterra equation and that any other continuous solution of the Volterra equation must be identical to y then follows the same
lines as above.   □

We may actually find the solution of a simple class of Caputo-type fractional
differential equations explicitly, namely for the class of homogeneous linear equations with constant coefficients where at most the lowest-order initial condition is
inhomogeneous.

Theorem 6.11. *Let $n > 0$, $m = \lceil n \rceil$ and $\lambda \in \mathbb{R}$. The solution of the initial value
problem*

$$D_{*0}^n y(x) = \lambda y(x), \qquad y(0) = y_0, \quad y^{(k)}(0) = 0 \quad (k = 1, 2, \ldots, m-1)$$

is given by

$$y(x) = y_0 E_n(\lambda x^n), \qquad x \geq 0.$$

The differential equation under consideration in Theorem 6.11 is a very simple
example of a linear fractional differential equation. We shall provide a more detailed
investigation of more general linear equations in Sect. 7.1.

Proof. It is evident from our existence and uniqueness result in Corollary 6.9 above
that the initial value problem has a unique solution. Therefore we only have to verify
that the function y stated above is a solution. For the initial condition, we indeed see
that $y(0) = y_0 E_n(0) = y_0$ since

$$E_n(z) = 1 + \frac{z}{\Gamma(1+n)} + \frac{z^2}{\Gamma(1+2n)} + \cdots;$$

moreover in the case $m \geq 2$ (i.e. $n > 1$) we have $y^{(k)}(0) = 0$ for $k = 1, 2, \ldots, m-1$
since

$$y(x) = 1 + \frac{\lambda x^n}{\Gamma(1+n)} + \frac{\lambda^2 x^{2n}}{\Gamma(1+2n)} + \cdots$$

which implies that

$$y^{(k)}(x) = \frac{\lambda x^{n-k}}{\Gamma(1+n-k)} + \frac{\lambda^2 x^{2n-k}}{\Gamma(1+2n-k)} + \cdots$$

for $k = 1, 2, \ldots, m-1 < n$.

Finally, the fact that our function y really solves the differential equation is an
immediate consequence of Theorem 4.3.   □

To conclude this section, we recall one more classical result from the theory of
first order differential equations and look at possible generalizations of this result
to the fractional setting. It is a well known theorem closely related to uniqueness
questions.

Theorem 6.A. *Let* $f : [a,b] \times [c,d] \to \mathbb{R}$ *be continuous and satisfy a Lipschitz condition with respect to the second variable. Consider two solutions* y_1 *and* y_2 *of the differential equation*

$$D^1 y_j(x) = f(x, y_j(x)) \cdot \quad (j = 1, 2) \tag{6.13}$$

subject to the initial conditions $y_j(x_j) = y_{j0}$, *respectively. Then the functions* y_1 *and* y_2 *coincide either everywhere or nowhere.*

Proof. The proof is very simple: Assume that the functions y_1 and y_2 coincide at some point x^*, i.e. $y_1(x^*) = y_2(x^*) =: y^*$, say. Then, both functions solve the initial value problem $D^1 y_j(x) = f(x, y_j(x))$, $y_j(x^*) = y^*$. Since the assumptions assert that this problem has a unique solution, y_1 and y_2 must be identical, i.e. they coincide everywhere.   □

Let us interpret the statement of Theorem 6.A. To this end we start with one solution y_1 of the differential equation (6.13) with some initial condition $y_1(x_1) = y_{10}$. Then we may choose an arbitrary abscissa x_2 and prescribe an initial condition $y_2(x_2) = y_{20}$ at this abscissa. It may happen that the point (x_2, y_{20}) is located on the graph of y_1. In this case, the functions y_1 and y_2 are identical. Otherwise, the graphs of y_1 and y_2 will never meet or even cross each other.

We specifically draw the reader's attention to the fact that in Theorem 6.A it does not matter whether the values x_1 and x_2, i.e. the abscissas where the initial conditions are specified, coincide with each other or not.

The question for a fractional generalization of this theorem has been raised and partially answered in [40]. An attempt to give a full answer has been made in [100, Theorem 3.27]. However, a close inspection of the theory developed there reveals a gap in the proof. Therefore, an alternative approach has been developed in a slightly more general setting in [47]. We shall now describe the specialization of the results of that paper to our class of problems. As it turns out, this allows us to provide a quite satisfactory answer for our question.

As a first step towards the indicated answer to our question we notice that the proof of Theorem 6.A strongly depends on the locality of the differential operator D^1. We had already noted in Remark 6.4 that such a property is not available in the fractional case, and hence the proof cannot be carried over directly. Moreover, as a consequence of this non-locality, we find that now it does matter whether the abscissas of the initial conditions coincide or not. Indeed the following example, taken from [40], shows that we cannot expect a result comparable to Theorem 6.A to hold if $x_1 \neq x_2$:

Example 6.1. Let $0 < n < 1$ and consider the fractional differential equations

$$D^n_{*0} y_1(x) = \Gamma(n+1), \qquad y_1(0) = 0, \tag{6.14}$$

and

$$D^n_{*1} y_2(x) = \Gamma(n+1), \qquad y_2(1) = 1. \tag{6.15}$$

We have here two differential equations with identical right-hand sides but initial conditions specified at different points. The solutions are easily seen to be $y_1(x) = x^n$ and $y_2(x) = 1 + (x-1)^n$. It is obvious that these two functions coincide at $x = 1$ but nowhere else.

Theorem 6.A deals with differential equations of order 1, i.e. with equations subject to exactly one initial condition. It is well known that a similar statement does not hold for equations of higher order (for example, the equation $D^2 y(x) = -y(x)$ has solutions $y(x) = \cos x$, $y(x) = \sin x$ and $y(x) = 0$ the first two of which oscillate and the graphs of which cross each other). Similar effects arise in the context of fractional differential equations with more than one initial condition (i.e. equations of order $n > 1$); see Sect. 7.1 below. We can therefore not expect more than the following result to hold:

Theorem 6.12. *Let $0 < n < 1$ and assume $f : [0,b] \times [c,d] \to \mathbb{R}$ to be continuous and satisfy a Lipschitz condition with respect to the second variable. Consider two solutions y_1 and y_2 of the differential equation*

$$D_{*0}^n y_j(x) = f(x, y_j(x)) \qquad (j = 1, 2) \tag{6.16}$$

subject to the initial conditions $y_j(0) = y_{j0}$, respectively, where $y_{10} \neq y_{20}$. Then, for all x where both $y_1(x)$ and $y_2(x)$ exist, we have $y_1(x) \neq y_2(x)$.

Proof. Let us assume that there exists some x^* such that $y_1(x^*) = y_2(x^*)$. We then have to show that $y_{10} = y_{20}$. To this end, let L denote the Lipschitz constant of f with respect to the second variable. We may then choose a number $\tau > 0$ that satisfies the conditions

$$\gamma := \frac{2L\tau^n}{\Gamma(n+1)} < 1 \tag{6.17}$$

and $N := x^*/\tau \in \mathbb{N}$ and split the fundamental interval $[0, x^*] = [0, N\tau]$ into the subintervals $[(j-1)\tau, j\tau]$, $j = 1, 2, \ldots, N$.

We begin by looking at the first of these subintervals, i.e. on the interval $[0, \tau]$. Our goal here is to show that the problem consisting of the differential equation (6.16) and the additional condition $y(\tau) = y^*$ cannot have more than one solution. To prove this, we recall that any solution of the differential equation must satisfy the Volterra equation

$$y(x) = Y_0 + \frac{1}{\Gamma(n)} \int_0^x (x-t)^{n-1} f(t, y(t)) \, dt \tag{6.18}$$

with a suitable number Y_0. The additional condition at the point τ then implies that

$$y^* = y(\tau) = Y_0 + \frac{1}{\Gamma(n)} \int_0^\tau (\tau - t)^{n-1} f(t, y(t)) \, dt,$$

i.e.

$$Y_0 = y^* - \frac{1}{\Gamma(n)} \int_0^\tau (\tau - t)^{n-1} f(t, y(t)) \, dt.$$

Inserting this into (6.18) we obtain

$$y(x) = Ay(x) := y^* + \frac{1}{\Gamma(n)} \left(\int_0^x (x - t)^{n-1} f(t, y(t)) \, dt - \int_0^\tau (\tau - t)^{n-1} f(t, y(t)) \, dt \right).$$

Thus we must now demonstrate that the operator A defined here does not have more than one fixed point. It is clear that A maps $C[0, \tau]$ to itself, and moreover we find that

$$|A\tilde{y}(x) - A\bar{y}(x)| \leq \frac{1}{\Gamma(n)} \left(\int_0^x (x - t)^{n-1} |f(t, \tilde{y}(t)) - f(t, \bar{y}(t))| \, dt \right.$$

$$\left. + \int_0^\tau (\tau - t)^{n-1} |f(t, \tilde{y}(t)) - f(t, \bar{y}(t))| \, dt \right)$$

$$\leq \frac{L}{\Gamma(n)} \|\tilde{y} - \bar{y}\|_{L_\infty[0,\tau]} \left(\int_0^x (x - t)^{n-1} \, dt + \int_0^\tau (\tau - t)^{n-1} \, dt \right)$$

$$\leq \frac{2L\tau^n}{\Gamma(n+1)} \|\tilde{y} - \bar{y}\|_{L_\infty[0,\tau]}$$

$$= \gamma \|\tilde{y} - \bar{y}\|_{L_\infty[0,\tau]},$$

and from (6.17) we see that A is a contraction. Thus, by Banach's fixed point theorem, we obtain the desired uniqueness of the solution of the differential equation (6.16) subject to a condition of the form $y(\tau) = y^*$. As a consequence of this observation, we conclude that two solutions of (6.16) that coincide at the point τ must also coincide on the entire interval $[0, \tau]$ and hence, in particular, at the point 0.

Thus, if $N = 1$ then the proof is complete. In the case $N > 1$ it remains to show for $j = 2, 3, \ldots, N$ that two solutions of (6.16) that coincide at $j\tau$ must actually be identical on the preceding interval $[(j-1)\tau, j\tau]$ and hence at the previous grid point $(j-1)\tau$. Once we have shown this to be true, we can say that our assumption $y_1(N\tau) = y_1(x^*) = y_2(x^*) = y_2(N\tau)$ implies that the solutions y_1 and y_2 coincide on $[(N-1)\tau, N\tau], [(N-2)\tau, (N-1)\tau], \ldots, [0, \tau]$ which is our desired result.

For this remaining part of the proof, we look at the given differential equation (6.16) on the interval $[0, j\tau]$, subject to the condition $y(j\tau) = y^*$. As above, we find that this is equivalent to writing

$$y(x) = y^* + \frac{1}{\Gamma(n)} \left(\int_0^x (x - t)^{n-1} f(t, y(t)) \, dt - \int_0^{j\tau} (j\tau - t)^{n-1} f(t, y(t)) \, dt \right).$$

As indicated above, we are interested in this problem on the interval $[(j-1)\tau, j\tau]$. Thus, for x in this interval, we rewrite the identity as

$$y(x) = B_j(x) := g_j(x) + \frac{1}{\Gamma(n)} \left(\int_{(j-1)\tau}^{x} (x-t)^{n-1} f(t,y(t))\, dt \right.$$
$$\left. - \int_{(j-1)\tau}^{j\tau} (j\tau - t)^{n-1} f(t,y(t))\, dt \right)$$

where

$$g_j(x) := y^* + \frac{1}{\Gamma(n)} \int_0^{(j-1)\tau} \left((x-t)^{n-1} - (j\tau - t)^{n-1} \right) f(t,y(t))\, dt$$

(note that this quantity is well defined because it only uses values of y on the interval $[0,(j-1)\tau]$ and we already know that y is uniquely determined there). We can then again proceed as in the first step of our argumentation and conclude that B_j is a contraction which leads to the required uniqueness property.   □

Example 6.2. We verify the statement of Theorem 6.12 by looking at the differential equation

$$D_{*0}^{0.28} y_j(x) = (0.5 - x)\sin y_j(x) + 0.8 x^3$$

with initial conditions

$$y_1(0) = 1.6, \quad y_2(0) = 1.5, \quad y_3(0) = 1.4, \quad y_4(0) = 1.3, \quad y_5(0) = 1.2.$$

Since a generally applicable method to determine the analytical solutions of our initial value problems is not readily available, we have to revert to some numerically computed approximate solutions. To this end, we can use the Adams–Bashforth Moulton method described in Appendix C.1. The resulting solutions for the five initial value problems, obtained using a step size of $1/200$, are shown in Fig. 6.1. The property predicted by Theorem 6.12 is evident; in view of the convergence

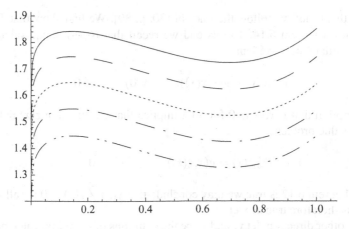

Fig. 6.1 Plots of the solutions of the five initial value problems from Example 6.2

analysis for the numerical method provided in Appendix C.1 we may be confident that this shows a qualitatively correct picture of the exact solutions too.

An obvious alternative formulation of Theorem 6.12 reads as follows.

Theorem 6.13. *Let $0 < n < 1$ and assume $f : [0,b] \times [c,d] \to \mathbb{R}$ to be continuous and satisfy a Lipschitz condition with respect to the second variable. Consider two solutions y_1 and y_2 of the differential equation*

$$D_{*0}^n y_j(x) = f(x, y_j(x)) \qquad (j = 1,2)$$

subject to the initial conditions $y_j(0) = y_{j0}$, respectively. Let $b^* \in (0,b]$ be such that both solutions y_1 and y_2 exist on $[0,b^*]$. Then, either $y_1(x) \neq y_2(x)$ for all $x \in [0,b^*]$ or $y_1(x) = y_2(x)$ for all $x \in [0,b^*]$.

As it turns out, we may also reformulate this result in a slightly different way. We present this reformulation here because it is also of interest in a different context later on.

Theorem 6.14. *Let $0 < n < 1$ and assume $g : [0,b] \times [y^* - K, y^* + K] \to \mathbb{R}$ to be continuous and satisfy a Lipschitz condition with respect to the second variable. Moreover, let $g(x,0) = 0$ for all $x \in [0,b]$, and let $y^* \neq 0$. Then, the solution y of the Volterra equation*

$$y(x) = y^* + \frac{1}{\Gamma(n)} \int_0^x (x-t)^{n-1} g(t, y(t)) \, dt \tag{6.19}$$

satisfies $y(x) \neq 0$ for all x.

The relation between Theorems 6.12 and 6.14 is easily described:

Theorem 6.15. *Theorem 6.12 holds if and only if Theorem 6.14 holds.*

Proof. In this proof we follow the lines of [30, p. 89]. We first show that Theorem 6.12 implies Theorem 6.14. To this end we recall that (6.19) is equivalent to the fractional initial value problem

$$D_{*0}^n y(x) = g(x, y(x)), \qquad y(0) = y^* \neq 0.$$

Our assumption that $g(x,0) = 0$ for all x implies that the function $z(x) = 0$ solves the initial value problem

$$D_{*0}^n z(x) = g(x, z(x)), \qquad z(0) = 0.$$

Thus, if Theorem 6.12 is true we may conclude that $y(x) \neq z(x) = 0$ for all x. Theorem 6.14 is therefore true as well.

For the other direction, let y_1 and y_2 be the solutions of the initial value problems mentioned in Theorem 6.12, and define

$$g(x,z) := f(x,z+y_1(x)) - f(x,y_1(x)).$$

Moreover we set $y(x) := y_2(x) - y_1(x)$ and $y^* := y_{20} - y_{10} \neq 0$. Now we rewrite the two initial value problems from Theorem 6.12 in their corresponding Volterra forms, viz.

$$y_j(x) = y_{j0} + \frac{1}{\Gamma(n)} \int_0^x (x-t)^{n-1} f(t,y_j(t)) \, dt \qquad (j = 1,2).$$

Subtracting these two equations, we find

$$\begin{aligned}
y(x) &= y_2(x) - y_1(x) \\
&= y_{20} - y_{10} + \frac{1}{\Gamma(n)} \int_0^x (x-t)^{n-1} [f(t,y_2(t)) - f(t,y_1(t))] \, dt \\
&= y^* + \frac{1}{\Gamma(n)} \int_0^x (x-t)^{n-1} g(t,y_2(t) - y_1(t)) \, dt \\
&= y^* + \frac{1}{\Gamma(n)} \int_0^x (x-t)^{n-1} g(t,y(t)) \, dt,
\end{aligned}$$

i.e. y solves the Volterra equation (6.19). Since, for all x, $g(x,0) = 0$, we may apply Theorem 6.14 to conclude that $0 \neq y(x) = y_2(x) - y_1(x)$ for all x. □

A trivial consequence of our results above is

Corollary 6.16. *Assume the hypotheses of Theorem 6.12. Moreover, let $y_{10} < y_{20}$. Then, $y_1(x) < y_2(x)$ for all x for which both solutions exist.*

We end this section by taking a look at this problem from a different point of view. Specifically, from Theorem 6.12 we may deduce the following result.

Theorem 6.17. *Let $0 < n < 1$ and assume $f : [0,b] \times [c,d] \to \mathbb{R}$ to be continuous and satisfy a Lipschitz condition with respect to the second variable. Then, for each $x^* \in [0,b]$ and each $y^* \in [c,d]$, the differential equation*

$$D_{*0}^n y(x) = f(x,y(x)) \tag{6.20}$$

subject to the condition

$$y(x^*) = y^* \tag{6.21}$$

has at most one solution.

Thus we have a uniqueness theorem for the solutions of a fractional differential equation of the usual form combined with a prescribed value of the unknown solution at a point that may differ form the starting point of the fractional differential operator. In the case $x^* = 0$ this is just the standard initial condition that we had discussed thoroughly at the beginning of this section, but if $x^* > 0$ then we have a significantly different problem that is sometimes called a *terminal value problem* because one usually is interested in the solution on the interval $[0, x^*]$, i.e. one provides a condition on the unknown solution at the terminal point of the interval of interest. The following simple equivalence theorem elucidates the difference in character between initial and terminal value problems.

Theorem 6.18. *Assume the hypotheses of Theorem 6.17, and let $0 < x^* \leq b$. Then, the fractional differential equation (6.20) combined with condition (6.21) is equivalent to the weakly singular integral equation*

$$y(x) = y^* + \frac{1}{\Gamma(n)} \int_0^{x^*} G(x,t) f(t, y(t)) \, dt \qquad (6.22)$$

where

$$G(x,t) = \begin{cases} -(x^* - t)^{n-1} & \text{for } t > x, \\ (x-t)^{n-1} - (x^* - t)^{n-1} & \text{for } t \leq x. \end{cases} \qquad (6.23)$$

The substantial difference between the integral equation (6.22) and the integral equation (6.2) derived in Lemma 6.2 in the case $x^* = 0$ is that we now have a *Fredholm* integral equation of Hammerstein type whereas (6.2) was a *Volterra* equation. Thus, in analogy with the corresponding results for integer-order equations, it would be natural to interpret condition (6.21) as a boundary condition and not an initial condition. Hence, in contrast to the situation observed for first-order differential equations, the terminal value problem consisting of eqs. (6.20) and (6.21) is much more closely related to a boundary value problem than it is to an initial value problem. This is also quite natural since we need to find a solution on the interval $[0, x^*]$ whose end points both play a major role in the definition of the problem – the left end point is used to define the differential operator and the right end point provides the additional condition that asserts the uniqueness of the solution.

We shall provide a few additional results about boundary value problems for Caputo-type fractional differential equations in Sect. 6.5.

Proof. Let us first assume that the fractional differential equation (6.20) combined with condition (6.21) has a solution $y \in C[0, x^*]$. Then, we may apply the integral operator J_0^n to both sides of (6.20), which yields

$$y(x) = y(0) + \frac{1}{\Gamma(n)} \int_0^x (x-t)^{n-1} f(t, y(t)) \, dt \qquad (6.24)$$

for all $x \in [0, x^*]$. (Notice that the quantity $y(0)$ is unknown.) Specifically setting $x = x^*$ and using the condition (6.21), we obtain

$$y^* = y(x^*) = y(0) + \frac{1}{\Gamma(n)} \int_0^{x^*} (x^* - t)^{n-1} f(t, y(t)) \, dt$$

Table 6.1 Results of bisection search for Example 6.3

y_0	1	2	1.5	1.75	1.625	1.6875	1.71875
$y(1)$	2.0556	2.63485	2.37728	2.51106	2.44567	2.47871	2.49496

which implies

$$y(0) = y^* - \frac{1}{\Gamma(n)} \int_0^{x^*} (x^* - t)^{n-1} f(t, y(t)) \, dt.$$

Inserting this relation into (6.24), we obtain (6.22) and (6.23).

On the other hand, if (6.22) with the kernel G being defined as in (6.23) has a continuous solution then we may apply the Caputo operator D_{*0}^n to both sides of (6.22) which yields the fractional differential equation (6.20). Finally, we immediately obtain (6.21) by setting $x = x^*$ in (6.22). □

To conclude this section, let us look at a simple example of a terminal value problem.

Example 6.3. Find a solution of the terminal value problem

$$D_{*0}^{1/2} y(x) = \sin y(x), \qquad y(1) = 2.5.$$

Once again we cannot offer an analytic method to find the exact solution; therefore we shall revert to an approximation technique. We start by choosing two arbitrary initial values, say $y_0 = 1$ and $y_0 = 2$, for the given differential equation and compute the corresponding solutions numerically on the interval $[0, 1]$ by means of the Adams method of Appendix C.1 with a step size of $1/200$. It turns out (see Table 6.1) that one of these choices for the initial values leads to a value of $y(1)$ that is smaller than desired whereas the other one produces a too large value. We thus employ a simple bisection technique to find a new initial value, $y_0 = 1.5$, and compute $y(1)$ again. Proceeding in an iterative manner as indicated in Table 6.1 we find that we can get as close to the desired exact solution as we wish.

Figure 6.2 shows a visualization of this procedure. We have displayed five of the neighbouring solutions of our exact solution, namely those obtained for $y_0 = 2$ (dashed line; top), $y_0 = 1.75$ (dash-dotted line; second from top), $y_0 = 1.5$ (dotted line, bottom), $y_0 = 1.625$ (dashed and double dotted line; second from bottom) and $y_0 = 1.71875$ (continuous line; centre).

6.3 Influence of Perturbed Data

Having established criteria for the existence and uniqueness of solutions to initial value problems for Caputo-type fractional differential equations, in this and the following sections we come to the classical questions concerning the most important

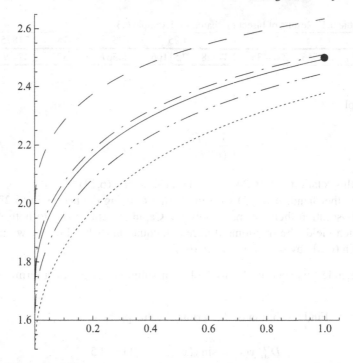

Fig. 6.2 Graphs of neighbouring approximate solutions for the terminal value problem from Example 6.3

properties of the solutions. These include, in particular, questions for smoothness (continuity and differentiability) of the solutions under various assumptions on the given data as well as investigations concerning the well-posedness of the initial value problems. We begin with the latter and recall the key results in this direction from [43].

Traditionally, a problem is called *well-posed* if it has the following three properties:

- A solution exists
- the solution is unique
- the solution depends on the given data in a continuous way

The first two aspects have already been discussed in the previous sections; the third one requires further attention. In particular we note one important difference between the fractional and the classical setting, and that is the precise meaning of the expression "the given data": In the classical theory of differential equations (of integer order), one usually assumes the initial values and the function f on the right-hand side of the differential equation

$$D^k y(x) = f(x, y(x), Dy(x), \ldots, D^{k-1} y(x))$$

to be given, and then the behaviour of the solution under perturbations of these expressions is discussed. Of course, we have the same given data in the fractional setting, but here one additional problem needs to be taken into account: In a typical application (see, for example, [69]), a differential equation of the form

$$D_{*0}^n y(x) = f(x, y(x))$$

arises. Here the main parameters of the equation, namely the initial value(s), possibly also the right-hand side f, *and the order of the differential operator*, depend on material constants that are only known up to a certain, usually moderate, accuracy. For example, in the problems arising in viscoelasticity considered briefly in [50, 51] and in a more detailed fashion in [69], the knowledge of the values of n is only imprecise and typically restricted to about two decimal digits. Therefore, it is important to investigate how the solution depends on this parameter too.

Throughout this section, we assume y to be the exact solution of the initial value problem

$$D_{*0}^n y(x) = f(x, y(x)), \tag{6.25a}$$

$$D^k y(0) = y_0^{(k)}, \qquad k = 0, 1, \ldots, m-1, \tag{6.25b}$$

where as usual we have set $m = \lceil n \rceil$. Moreover we assume f to be such that the hypotheses of Theorem 6.5 are satisfied, so that we can be sure that a continuous solution exists on some interval $[0, h]$, and that this solution is unique. We then have to compare this solution y with the solution of another initial value problem involving perturbed given data. It turns out that we will indeed be able to prove a positive result in all cases: Small perturbations in any of the given data yield small perturbations in the solution.

As a prerequisite that we shall find useful in most of the proofs below, we present a Gronwall-type inequality. For the case $0 < n < 1$ this result is contained in [55, Theorem 3.1], but we shall look at the fully general case here.

Lemma 6.19. *Let $n, T, \varepsilon_1, \varepsilon_2 \in \mathbb{R}_+$. Moreover assume that $\delta : [0, T] \to \mathbb{R}$ is a continuous function satisfying the inequality*

$$|\delta(x)| \le \varepsilon_1 + \frac{\varepsilon_2}{\Gamma(n)} \int_0^x (x-t)^{n-1} |\delta(t)| \, dt$$

for all $x \in [0, T]$. Then

$$|\delta(x)| \le \varepsilon_1 E_n(\varepsilon_2 x^n)$$

for $x \in [0, T]$.

Here, E_n once again denotes the Mittag-Leffler function of order n.

Proof. Let $\varepsilon_3 > 0$ and introduce the function Φ with $\Phi(x) := (\varepsilon_1 + \varepsilon_3) E_n(\varepsilon_2 x^n)$. Applying Theorem 6.11 and using the linearity of the initial value problem considered there, this function Φ is seen to be the solution of the initial value problem $D_{*0}^n \Phi(x) = \varepsilon_2 \Phi(x)$ with $\Phi(0) = \varepsilon_1 + \varepsilon_3$ and $D^k \Phi(0) = 0$ for $k = 1, 2, \ldots, \lceil n \rceil - 1$. In view of Lemma 6.2 we immediately deduce that Φ satisfies the integral equation

$$\Phi(x) = \varepsilon_1 + \varepsilon_3 + \frac{\varepsilon_2}{\Gamma(n)} \int_0^x (x-t)^{n-1} \Phi(t)\,dt.$$

By our assumption on δ, we find that

$$|\delta(0)| \leq \varepsilon_1 < \varepsilon_1 + \varepsilon_3 = \Phi(0).$$

Standard continuity arguments thus yield that $|\delta(x)| < \Phi(x)$ for all $x \in [0, \eta]$ with some $\eta > 0$. To prove that this inequality holds throughout $[0, T]$, we first assume the contrary and denote by x_0 the smallest positive number with the property that $|\delta(x_0)| = \Phi(x_0)$. Then, for $0 \leq x \leq x_0$ we have $|\delta(x)| \leq \Phi(x)$ and thus

$$|\delta(x_0)| \leq \varepsilon_1 + \frac{\varepsilon_2}{\Gamma(n)} \int_0^{x_0} (x_0 - t)^{n-1} |\delta(t)|\,dt$$

$$\leq \varepsilon_1 + \frac{\varepsilon_2}{\Gamma(n)} \int_0^{x_0} (x_0 - t)^{n-1} \Phi(t)\,dt$$

$$< \varepsilon_1 + \varepsilon_3 + \frac{\varepsilon_2}{\Gamma(n)} \int_0^{x_0} (x_0 - t)^{n-1} \Phi(t)\,dt = \Phi(x_0)$$

which cannot be true in view of our choice of x_0. Thus the assumption must be false, and we find that indeed

$$|\delta(x)| < \Phi(x) = (\varepsilon_1 + \varepsilon_3) E_n(\varepsilon_2 x^n)$$

for all $x \in [0, T]$. Since this holds for every $\varepsilon_3 > 0$, we derive the desired result. □

In our first main result, we investigate the dependence of the solution of a fractional differential equation on the initial values.

Theorem 6.20. *Let y be the solution of the initial value problem (6.25), and let z be the solution of the initial value problem*

$$D_{*0}^n z(x) = f(x, z(x)), \tag{6.26a}$$

$$D^k z(0) = z_0^{(k)}, \qquad k = 0, 1, \ldots, m-1. \tag{6.26b}$$

Moreover let $\varepsilon := \max_{k=0,1,\ldots,m-1} |y_0^{(k)} - z_0^{(k)}|$. Then, if ε is sufficiently small, there exists some $h > 0$ such that both the functions y and z are defined on $[0, h]$, and

$$\sup_{0 \leq x \leq h} |y(x) - z(x)| = O\left(\max_{k=0,1,\ldots,m-1} \left| y_0^{(k)} - z_0^{(k)} \right| \right).$$

Proof. By construction, a unique solution of (6.25) exists over some nonempty interval. It is clear from Theorem 6.5 that the set G corresponding to the problem (6.26) is not empty and thus this problem also has a uniquely determined solution on some interval. We may now choose $[0, h]$ to be the smaller of these two intervals. Defining $\delta(x) := y(x) - z(x)$, we find that δ is a solution of the initial value problem

$$D_{*0}^n \delta(x) = f(x, y(x)) - f(x, z(x)), \quad D^k \delta(0) = y_0^{(k)} - z_0^{(k)}, \quad k = 0, \ldots, m-1.$$

In view of Lemma 6.2, this initial value problem is equivalent to the integral equation

$$\delta(x) = \sum_{k=0}^{m-1} \frac{x^k}{k!} \left(y_0^{(k)} - z_0^{(k)} \right) + \frac{1}{\Gamma(n)} \int_0^x (x-t)^{n-1} (f(t, y(t)) - f(t, z(t))) \, dt.$$

Taking absolute values and using Hölder's inequality for the sum and the Lipschitz condition on f for the integral, we find

$$|\delta(x)| \leq \varepsilon \sum_{k=0}^{m-1} \frac{h^k}{k!} + \frac{L}{\Gamma(n)} \int_0^x (x-t)^{n-1} |\delta(t)| \, dt.$$

where L is the Lipschitz constant of f. Thus, by Lemma 6.19,

$$|\delta(x)| \leq O(\varepsilon) E_n(Lh^n) = O(\varepsilon)$$

as desired.   □

Next we look at the influence of changes in the given function on the right-hand side of the differential equation.

Theorem 6.21. *Let y be the solution of the initial value problem (6.25), and let z be the solution of the initial value problem*

$$D_{*0}^n z(x) = \tilde{f}(x, z(x)), \tag{6.27a}$$

$$D^k z(0) = y_0^{(k)}, \qquad k = 0, 1, \ldots, m-1, \tag{6.27b}$$

where $\tilde{f}$ is supposed to satisfy the same hypotheses as f. Moreover let $\varepsilon := \max_{(x_1, x_2) \in G} |f(x_1, x_2) - \tilde{f}(x_1, x_2)|$. Then, if ε is sufficiently small, there exists some $h > 0$ such that both the functions y and z are defined on $[0, h]$, and we have that

$$\sup_{0 \leq x \leq h} |y(x) - z(x)| = O\left(\max_{(x_1, x_2) \in G} |f(x_1, x_2) - \tilde{f}(x_1, x_2)| \right).$$

Proof. Existence and uniqueness of the solutions of both initial value problems follow as in the previous theorem. To prove the required inequality, we also proceed in a similar way. Once again we define $\delta(x) := y(x) - z(x)$ and find that δ solves the initial value problem

$$D_{*0}^n \delta(x) = f(x, y(x)) - \tilde{f}(x, z(x)), \qquad D^k \delta(0) = 0, \quad k = 0, 1, \ldots, m - 1$$

that is equivalent to the integral equation

$$\delta(x) = \frac{1}{\Gamma(n)} \int_0^x (x - t)^{n-1} (f(t, y(t)) - \tilde{f}(t, z(t))) \, dt.$$

Taking absolute values and using the Lipschitz assumptions on f and $\tilde{f}$, we deduce

$$\begin{aligned}
|\delta(x)| &\leq \int_0^x \frac{(x - t)^{n-1}}{\Gamma(n)} \left(|f(t, y(t)) - f(t, z(t))| + |f(t, z(t)) - \tilde{f}(t, z(t))| \right) dt \\
&\leq \frac{L}{\Gamma(n)} \int_0^x (x - t)^{n-1} |\delta(t)| \, dt + \varepsilon \frac{1}{\Gamma(n)} \int_0^x (x - t)^{n-1} \, dt \\
&\leq \frac{L}{\Gamma(n)} \int_0^x (x - t)^{n-1} |\delta(t)| \, dt + \varepsilon \frac{h^n}{\Gamma(n+1)}.
\end{aligned}$$

We may now apply Lemma 6.19 once again and find

$$|\delta(x)| \leq O(\varepsilon) E_n(Lh^n) = O(\varepsilon)$$

as required.   □

Finally we discuss the consequences of a modification of the order of the differential equation. Here we need to be particularly careful because a change in the order of the differential equation may lead to a change in the number of prescribed initial values.

Theorem 6.22. *Let y be the solution of the initial value problem (6.25), and let z be the solution of the initial value problem*

$$D_{*0}^{\tilde{n}} z(x) = f(x, z(x)), \tag{6.28a}$$

$$D^k z(0) = y_0^{(k)}, \qquad k = 0, 1, \ldots, \tilde{m} - 1, \tag{6.28b}$$

where $\tilde{n} > n$ and $\tilde{m} := \lceil \tilde{n} \rceil$. Moreover let $\varepsilon := \tilde{n} - n$ and

$$\varepsilon^* := \begin{cases} 0 & \text{if } m = \tilde{m}, \\ \max\left\{ \left| y_0^{(k)} \right| : m \leq k \leq \tilde{m} - 1 \right\} & \text{else.} \end{cases}$$

Then, if ε and ε^* are sufficiently small, there exists some $h > 0$ such that both the functions y and z are defined on $[0, h]$, and we have that

$$\sup_{0 \leq x \leq h} |y(x) - z(x)| = O(\tilde{n} - n) + O\left(\max\left\{ 0, \max\left\{ \left| y_0^{(k)} \right| : m \leq k \leq \tilde{m} - 1 \right\} \right\} \right).$$

Proof. Existence and uniqueness of the solutions can be deduced as above. For the bound on the difference, we rewrite the initial value problems (6.25) and (6.28) as equivalent integral equations according to Lemma 6.2 and subtract the two resulting equations from each other. Thus,

$$
\begin{aligned}
\delta(x) &:= y(x) - z(x) \\
&= -\sum_{k=m}^{\tilde{m}-1} \frac{x^k}{k!} y_0^{(k)} + \frac{1}{\Gamma(n)} \int_0^x (x-t)^{n-1} f(t, y(t)) \, dt \\
&\quad - \frac{1}{\Gamma(\tilde{n})} \int_0^x (x-t)^{\tilde{n}-1} f(t, z(t)) \, dt \\
&= -\sum_{k=m}^{\tilde{m}-1} \frac{x^k}{k!} y_0^{(k)} + \frac{1}{\Gamma(n)} \int_0^x (x-t)^{n-1} (f(t, y(t)) - f(t, z(t))) \, dt \\
&\quad + \int_0^x \left(\frac{(x-t)^{n-1}}{\Gamma(n)} - \frac{(x-t)^{\tilde{n}-1}}{\Gamma(\tilde{n})} \right) f(t, z(t)) \, dt.
\end{aligned}
$$

In view of the Lipschitz property of f, this implies

$$
\begin{aligned}
|\delta(x)| \leq {}& \sum_{k=m}^{\tilde{m}-1} \frac{h^k}{k!} \left| y_0^{(k)} \right| + \frac{L}{\Gamma(n)} \int_0^x (x-t)^{n-1} |\delta(t)| \, dt \\
&+ \max_{(x_1, x_2) \in G} |f(x_1, x_2)| \int_0^x \left| \frac{(x-t)^{n-1}}{\Gamma(n)} - \frac{(x-t)^{\tilde{n}-1}}{\Gamma(\tilde{n})} \right| dt.
\end{aligned}
$$

Obviously the sum is $O(\varepsilon^*)$. Moreover, we can bound the second integral according to

$$
\begin{aligned}
\int_0^x \left| \frac{(x-t)^{n-1}}{\Gamma(n)} - \frac{(x-t)^{\tilde{n}-1}}{\Gamma(\tilde{n})} \right| dt &= \int_0^x \left| \frac{u^{n-1}}{\Gamma(n)} - \frac{u^{\tilde{n}-1}}{\Gamma(\tilde{n})} \right| dt \\
&\leq \int_0^h \left| \frac{u^{n-1}}{\Gamma(n)} - \frac{u^{\tilde{n}-1}}{\Gamma(\tilde{n})} \right| dt = O(\varepsilon).
\end{aligned}
$$

To see this, one can compute the integral explicitly: One first has to find the zero of the integrand, which is located at $(\Gamma(\tilde{n})/\Gamma(n))^{1/(\tilde{n}-n)}$ (a quantity that converges to $\exp(\psi(n))$ as $\tilde{n} \to n$, where $\psi = \Gamma'/\Gamma$ denotes the Digamma function). If h is smaller than this value, then the integrand has no change of sign, and we can move the absolute value operation outside the integral. Otherwise we must split the interval of integration at this point, and each integral can be handled in this way. In either case, we may use the Mean Value Theorem of Differential Calculus to see that the resulting expression is bounded by $O(\tilde{n} - n) = O(\varepsilon)$.

Summing up, we have found that

$$
|\delta(x)| \leq O(\varepsilon) + O(\varepsilon^*) + \frac{L}{\Gamma(n)} \int_0^x (x-t)^{n-1} |\delta(t)| \, dt
$$

and so Lemma 6.19 yields the desired result.   □

There are two important special cases of Theorem 6.22:

Corollary 6.23. *Assume the hypotheses of Theorem 6.22. Moreover, let $\tilde{m} = m$. Then,*

$$\sup_{0 \le x \le h} |y(x) - z(x)| = O(\tilde{n} - n).$$

Corollary 6.24. *Assume the hypotheses of Theorem 6.22. Moreover, let $\tilde{m} > m$ and $y_0^{(k)} = 0$ for $k = m, m+1, \ldots, \tilde{m} - 1$. Then,*

$$\sup_{0 \le x \le h} |y(x) - z(x)| = O(\tilde{n} - n).$$

Proof (of Corollaries 6.23 and 6.24). Under the assumptions of either of the corollaries, the quantity ε^* in Theorem 6.22 vanishes and thus we immediately obtain the simpler bound.   □

Remark 6.11. Let z be a given continuous function. Since the function $y := J_0^n z$ is the unique solution of the differential equation $D_{*0}^n y = z$ with homogeneous initial conditions, Corollary 6.24 gives an alternative proof for the second statement of Theorem 2.10.

6.4 Smoothness of the Solutions

An interesting question that frequently arises is the question for the smoothness of the solution of a differential equation under certain assumptions on the given data (mainly on the given function f on the right-hand side of the equation). A typical result in the classical setting is the following well known theorem.

Theorem 6.B. *Let $k \in \mathbb{N}$ and $f \in C^{k-1}(G)$, where $G = [y_0 - K, y_0 + K] \times \mathbb{R}$. Then, the solution y of the initial value problem*

$$Dy(x) = f(x, y(x)), \qquad y(0) = y_0,$$

is k times continuously differentiable.

A simple example shows that we cannot expect this result to be true in general for fractional-order equations: Even for $f \in C^\infty$ it may happen that $y \notin C^1$.

Example 6.4. From Example 3.1 we know that $D_{*0}^{1/2}(\cdot)^{1/2} = \Gamma(3/2)$. Therefore the non-differentiable function y given by $y(x) = x^{1/2}$ is the unique solution of the initial value problem $D_{*0}^{1/2} y(x) = \Gamma(3/2)$, $y(0) = 0$, whose given function f (the right-hand side of the differential equation) is analytic.

We now try to find out some positive statements that can be said about equations of fractional order. As in the previous section, we assume in the present section that y is the exact solution of the initial value problem

$$D_{*0}^n y(x) = f(x, y(x)), \tag{6.29a}$$

$$D^k y(0) = y_0^{(k)}, \qquad k = 0, 1, \ldots, m-1, \tag{6.29b}$$

where once more we have set $m = \lceil n \rceil$. Additionally we assume f to be such that the hypotheses of Theorem 6.5 are satisfied, so that we can be sure that a continuous solution exists on some interval $[0, h]$, and that this solution is unique. Under these assumptions we begin by recalling some results from [37]. A first result in this connection is as follows.

Theorem 6.25. *Under the above hypotheses we have that* $y \in C^{m-1}[0, h]$.

Proof. In view of Lemma 6.2, the function y satisfies the Volterra equation

$$y(x) = p(x) + J_0^n[f(\cdot, y(\cdot))](x)$$

with p being some polynomial whose precise form is not of interest at the moment. Let $k \in \{0, 1, 2 \ldots, m-1\}$ (this implies $k < n$) and differentiate this equation k times:

$$\begin{aligned}
D^k y(x) &= D^k p(x) + D^k J_0^n[f(\cdot, y(\cdot))](x) \\
&= D^k p(x) + D^k J_0^k J_0^{n-k}[f(\cdot, y(\cdot))](x) \\
&= D^k p(x) + J_0^{n-k}[f(\cdot, y(\cdot))](x)
\end{aligned}$$

in view of the semigroup property of fractional integration and (1.1). Now recall that y is continuous; thus the argument of the integral operator J_0^{n-k} is a continuous function. Hence, in view of the polynomial structure of p and Theorem 2.5, the function on the right-hand side of the equation is continuous, and so the function on the left, viz. $D^k y$, must be continuous too. $\qquad \square$

Theorem 6.26. *Assume the hypotheses of Theorem 6.25. Moreover let* $n > 1$, $n \notin \mathbb{N}$ *and* $f \in C^1(G)$. *Then* $y \in C^m(0, h]$. *Furthermore,* $y \in C^m[0, h]$ *if and only if* $f(0, y_0^{(0)}) = 0$.

Remark 6.12. Since the function f and the initial value $y_0^{(0)}$ are given, it is easy to check whether the condition $f(0, y_0^{(0)}) = 0$ is fulfilled or not.

Remark 6.13. A comparison with Theorem 6.B reveals a significant difference between the classical and the fractional setting: In the former, smoothness of the given function f implies smoothness of the solution y; in the latter this holds only under certain additional conditions. This is not unexpected because of what we found in Example 6.4.

Proof. We introduce the abbreviation $z(t) := f(t, y(t))$ and write out the identity stated in the previous proof for $k = m - 1$:

$$D^{m-1}y(x) = D^{m-1}p(x) + J_0^{n-m+1}z(x)$$

$$= D^{m-1}p(x) + \frac{1}{\Gamma(n-m+1)} \int_0^x (x-t)^{n-m}z(t)\,dt.$$

We differentiate once again, recall that p is a polynomial of degree $m - 1$ and find

$$D^m y(x) = D^m p(x) + \frac{1}{\Gamma(n-m+1)} \frac{d}{dx} \int_0^x (x-t)^{n-m}z(t)\,dt$$

$$= \frac{1}{\Gamma(n-m+1)} \frac{d}{dx} \int_0^x u^{n-m}z(x-u)\,du$$

$$= \frac{1}{\Gamma(n-m+1)} \left(x^{n-m}z(0) + \int_0^x u^{n-m}z'(x-u)\,du \right)$$

$$= \frac{1}{\Gamma(n-m+1)} x^{n-m}f(0, y_0^{(0)}) + J_0^{n-m+1}z'(x). \tag{6.30}$$

Since $n > 1$ we deduce that $m \geq 2$, and thus Theorem 6.25 asserts that $y \in C^1[0, h]$. An explicit calculation gives that

$$z'(t) = \frac{\partial}{\partial t}f(t, y(t)) + \frac{\partial}{\partial y}f(t, y(t))y'(t).$$

Consequently, by our differentiability assumption on f, the function z' is continuous, and so $J_0^{n-m+1}z' \in C[0, h]$ too (cf. Theorem 2.5). The fact that $m > n$ then finally yields that the right-hand side of (6.30), and therefore also the left-hand side of this equation, i.e. the function $D^m y$, is always continuous on the half-open interval $(0, h]$ whereas it is continuous on the closed interval $[0, h]$ if and only if $f(0, y_0^{(0)}) = 0$. $\square$

It is possible to generalize this idea and to keep Remarks 6.12 and 6.13 valid:

Theorem 6.27. *Assume the hypotheses of Theorem 6.25. Moreover let $k \in \mathbb{N}$, $n > k$, $n \notin \mathbb{N}$ and $f \in C^k(G)$. Let $z(t) := f(t, y(t))$. Then $y \in C^{m+k-1}(0, h]$. Furthermore, $y \in C^{m+k-1}[0, h]$ if and only if z has a k-fold zero at the origin.*

The proof is based on a repeated differentiation of (6.30); we leave the details as an exercise to the reader.

A common feature of the results above is that they always require a relatively high order n of the differential operator in order to prove that the solution y possesses a large number of derivatives in the half-open interval $(0, h]$ or even in the closed interval $[0, h]$. However, it is also possible to obtain similar results if n is small if we impose stronger smoothness conditions on f. This follows from our next statement.

Theorem 6.28. *Assume the hypotheses of Theorem 6.25. Moreover let $f \in C^k(G)$. Then $y \in C^k(0, h] \cap C^{m-1}[0, h]$, and for $\ell = m, m+1, \ldots, k$ we have $y^{(\ell)}(x) = O(x^{n-\ell})$ as $x \to 0$.*

This theorem is a special case of [21, Theorem 2.1]. The proof given there is based on methods from the theory of Fredholm integral equations that are beyond the scope of this text. Therefore we do not give any details here and refer the reader to the original paper of Brunner et al. [21] instead.

We must note one specific point here. From our theorems above (and those that will follow in the remainder of this section) the reader may be lead into the belief that a smooth (differentiable or even analytic) function f on the right-hand side of the differential equation will necessarily lead to a non-smooth behaviour of the solution, at least in the neighbourhood of the starting point. Indeed, in the classical literature one often finds statements like "it is easily seen that y cannot be smooth [if f is analytic]" [120, p. 89]. The fact that the analytic function $y(x) = 1$ solves the initial value problem $D_{*0}^n y(x) = y(x) - 1$, $y(0) = 1$, which has an analytic function $f(x,y) = y - 1$ on its right-hand side, gives an easy counterexample to these statements. Fortunately we can give an extension of Theorem 6.27 that fully characterizes the situations where we may have smoothness for the given function f and the solution y simultaneously. We note that the statement below is a special case of a more general result for a larger class of equations [37, Theorem 3.1].

Theorem 6.29. *Consider the initial value problem (6.29), and assume that f is analytic on a suitable set G. Define $T(x) := \sum_{j=0}^{m-1} y_0^{(j)} x^j / j!$. Then, y is analytic if and only if $f(x, T(x)) = 0$ for all x.*

Remark 6.14. Once again we note that f is the given function from the right-hand side of the differential equation under consideration, and T is also known because it is a polynomial whose coefficients can be computed from the given initial conditions. Thus, in practice it is always possible to check whether the condition $f(x, T(x)) = 0$ for all x is satisfied or not.

Proof. The direction "$\Leftarrow$" is a simple consequence of the fact that a solution (and hence, by uniqueness, *the* solution) of the initial value problem is $y = T$ because $D_{*0}^n y = D_{*0}^n T = 0 = f(\cdot, T(\cdot))$. In other words, the solution is a classical polynomial and hence analytic.

For the other direction, we assume y to be analytic. Then, since f is analytic, the right-hand side of the differential equation is analytic too. Therefore, since y is assumed to be the solution of the equation, the left-hand side $D_{*0}^n y$ must be analytic as well. But Theorem 3.15 asserts that y and $D_{*0}^n y$ can only be analytic simultaneously if $D_{*0}^n y = 0$. This implies that y must be a polynomial of degree $m - 1$, and since y satisfies our initial conditions we find that $y = T$. We thus conclude

$$0 = D_{*0}^n y(x) = f(x, y(x)) = f(x, T(x))$$

for all x. □

The proof of Theorem 6.29 actually gives us some additional information.

Corollary 6.30. *Consider the initial value problem (6.29), and assume that f is analytic on a suitable set G. If the solution y of this initial value problem is analytic, then*

$$y(x) = \sum_{j=0}^{m-1} \frac{y_0^{(j)}}{j!} x^j.$$

In other words, the only class of analytic functions that can arise as the solution of a Caputo-type fractional differential equation of order n with an analytic given function on the right-hand side consists of the polynomials of order $m-1$. We can express this fact in a different but also quite instructive manner.

Corollary 6.31. *Assume that the solution of the initial value problem (6.29) is analytic, but not a polynomial of degree $m-1$. Then, the function f is not analytic.*

It thus turns out that, in general (i.e. if we do not deal with the special case of Theorem 6.29), we must expect some derivative of the solution to have a discontinuity at the origin. The nature of this discontinuity can be analyzed to obtain some quite useful information on the precise behaviour of the solution near the origin. In this context we have the following results that are very similar to those of Lubich [120, §2]. It must be noted however that the original paper of Lubich contains a number of small errors. We shall give the correct formulations here, with the basic elements of the proofs being influenced by ideas of Hennecke [Hennecke, T., 2006, private communication]. As a matter of fact it is possible to show that these results actually hold for a large class of Volterra equations that contains our fractional differential equations as a special case. However we shall not go into details on this aspect here and refer the interested reader to [37, §3] instead.

The precise nature of the results that we can present depends on whether or not the order n of the differential equation is rational. We begin by looking at the case of rational values of n.

Theorem 6.32. *Let $n = p/q$ where $p \geq 1$ and $q \geq 2$ are two relatively prime integers. Consider the initial value problem (6.29) and assume that f can be written in the form $f(x,y) = \bar{f}(x^{1/q}, y)$ where $\bar{f}$ is analytic in a neighbourhood of $(0, y_0^{(0)})$. Then, there exists a uniquely determined analytic function $\bar{y} : (-r, r) \to \mathbb{R}$ with some $r > 0$ such that $y(x) = \bar{y}(x^{1/q})$ for $x \in [0, r)$.*

Notice that our basic assumptions assert the existence of a solution on some interval $[0, h)$, whereas Theorem 6.32 gives us a representation valid in $[0, r)$ with some r. This parameter r will be the radius of convergence of the power series representation of $\bar{y}$. Our method of proof will allow us to conclude that $r > 0$ but it will not tell us anything about the relation between r and h. The same observation applies to the theorems and corollaries following below.

Before we come to the proof of Theorem 6.32 we note two immediate consequences of this result.

Corollary 6.33. *Let $n = p/q$ where $p \geq 1$ and $q \geq 2$ are two relatively prime integers. Consider the initial value problem (6.29) and assume that f is analytic in a neighbourhood of $(0, y_0^{(0)})$. Then, there exists a uniquely determined analytic function $\bar{y} : (-r, r) \to \mathbb{R}$ with some $r > 0$ such that $y(x) = \bar{y}(x^{1/q})$ for $x \in [0, r)$.*

Proof. Since f itself is analytic, it can be written in the form of a convergent power series

$$f(x,y) = \sum_{j,k=0}^{\infty} f_{jk} x^j y^k.$$

It is then evident that the function $\bar{f}$ with

$$\bar{f}(x,y) := \sum_{j,k=0}^{\infty} f_{jk} x^{qj} y^k$$

is analytic too and satisfies the relation $f(x,y) = \bar{f}(x^{1/q}, y)$. Thus, the claim follows directly from Theorem 6.32.   □

Our second corollary is almost immediately evident:

Corollary 6.34. *Under the assumptions of Theorem 6.32 or Corollary 6.33, the solution y of the initial value problem (6.29) can be written in a neighbourhood of the origin in the form of the convergent series*

$$y(x) = \sum_{i=0}^{\infty} \bar{y}_i x^{i/q}$$

with certain coefficients $\bar{y}_i$. In addition, $\bar{y}_i = 0$ if $i < p$ and $i/q \notin \mathbb{N}$.

Proof. The fact that y can be represented in the indicated way is a direct consequence of Theorem 6.32. The fact that $\bar{y}_i = 0$ if $i < p$ and $i/q \notin \mathbb{N}$ will be shown in the proof of that theorem; see (6.35) below.   □

Proof (of Theorem 6.32). As noted above, our proof uses methods similar to those used by Lubich [120, §2]. We proceed by constructing a formal solution in the form of a so-called Psi series or Puiseux series [96, Chapter 7]. The structure of this series will match our requirements, i.e. it will be a power series in $x^{1/q}$. The coefficients of the series are determined recursively in such a way that the series can be seen to be a formal solution of the initial value problem at hand. Finally we will show that this formal series is actually convergent. This part of the proof will be based on a suitable modification of Lindelöf's majorant method (see [95, §2.5] for this method applied to ordinary differential equations of integer order). It then follows that the function defined by this convergent series is a solution of the initial value problem and that it can be represented in the form that we have claimed.

Let us thus consider a function

$$y(x) = \sum_{i=0}^{\infty} \bar{y}_i x^{i/q} \tag{6.31}$$

where $y(0) = y_0^{(0)}$. We need to show that we can choose the coefficients $\bar{y}_i$ such that the series converges and that the function represented by the series solves our initial value problem.

Substituting (6.31) into (6.2) (that we know to be equivalent to the initial value problem) and using the series expansion of $\bar{f}$ around the point $(0, y_0^{(0)})$, viz.

$$f(x, y(x)) = \bar{f}(x^{1/q}, y(x)) = \sum_{\rho, \sigma = 0}^{\infty} f_{\rho\sigma} x^{\rho/q} (y - y_0^{(0)})^{\sigma}$$

$$= \sum_{\rho, \sigma = 0}^{\infty} f_{\rho\sigma} x^{\rho/q} \left(\sum_{i=1}^{\infty} \bar{y}_i x^{i/q} \right)^{\sigma},$$

we find

$$\sum_{i=0}^{\infty} \bar{y}_i x^{i/q} = \sum_{k=0}^{m-1} \frac{x^k}{k!} y_0^{(k)} + \frac{1}{\Gamma(n)} \int_0^x (x-t)^{n-1} \sum_{\rho, \sigma = 0}^{\infty} f_{\rho\sigma} t^{\rho/q} \left[\sum_{i=1}^{\infty} \bar{y}_i t^{i/q} \right]^{\sigma} dt. \quad (6.32)$$

We rearrange the term in brackets as

$$\left[\sum_{i=1}^{\infty} \bar{y}_i t^{i/q} \right]^{\sigma} = \sum_{j=0}^{\infty} \left(\sum_{i_1 + \ldots + i_\sigma = j} \bar{y}_{i_1} \cdot \ldots \cdot \bar{y}_{i_\sigma} \right) t^{j/q}. \quad (6.33)$$

As $i \geq 1$, the case $j = 0$ in the first sum on the right hand side only occurs for $\sigma = 0$. In this case we set the coefficient to the correct value 1, in accordance with the usual convention for empty products. In all other cases the second sum on the right hand extends over all products $\bar{y}_{i_1} \cdot \ldots \cdot \bar{y}_{i_\sigma}$ with σ factors, where the sum of the indices i_κ equals j, because exactly these terms contribute to the coefficient of $t^{j/q}$ in the rearranged series. The indices are labelled, so no multinomials arise, but formally different products can be equal. This rearrangement is allowed if we can prove the required convergence properties. Assuming uniform convergence, we can also exchange the order of summation and integration and integrate term by term. If we do so, we get

$$\sum_{i=0}^{\infty} \bar{y}_i x^{i/q} = \sum_{k=0}^{m-1} \frac{x^k}{k!} y_0^{(k)}$$

$$+ \sum_{\rho, \sigma, j = 0}^{\infty} f_{\rho\sigma} \left(\frac{\Gamma(\frac{\rho+j}{q}+1)}{\Gamma(\frac{\rho+j+p}{q}+1)} \sum_{i_1 + \ldots + i_\sigma = j} \bar{y}_{i_1} \cdot \ldots \cdot \bar{y}_{i_\sigma} \right) x^{(\rho+j+p)/q}. \quad (6.34)$$

We now compare the coefficients of $x^{i/q}$ on both sides of this equation. The result of this comparison depends on i. For $i < p$ we obtain

$$\bar{y}_i = \begin{cases} \dfrac{y_0^{(i/q)}}{(i/q)!} & \text{if } i = 0, q, 2q, \ldots, (m-1)q, \\ 0 & \text{else,} \end{cases} \quad (6.35)$$

whereas for $i \geq p$ we find

$$\bar{y}_i = \sum_{p+j+\rho=i} \sum_{\sigma=0}^{\infty} f_{\rho\sigma} \left(\frac{\Gamma(\frac{p+j}{q}+1)}{\Gamma(\frac{p+j+p}{q}+1)} \sum_{i_1+\ldots+i_\sigma=j} y_{i_1} \cdot \ldots \cdot y_{i_\sigma} \right). \qquad (6.36)$$

Thus, the coefficients $\bar{y}_i$ are determined uniquely for $0 \leq i < p$ because of (6.35). For $i \geq p$ we see that (6.36) is a recurrence relation: It states that its left-hand side, viz. $\bar{y}_i$, only depends on coefficients $\bar{y}_{i_\ell}$ where

$$i_\ell \leq i_1 + \cdots + i_\sigma = j = i - p - \rho < i$$

because $p > 0$. In other words, $\bar{y}_i$ can be expressed in terms of coefficients with lower indices. This means we have a unique formal solution, and we need to show that the series obtained in this way converges locally absolutely and uniformly.

It remains to prove that this formal generalized power series is absolutely and uniformly convergent in a neighbourhood of the origin. To this end we will use a generalization of Lindelöf's majorant method. The majorant for our formal solution is based on taking absolute values of the coefficients of f and of the initial values. Thus we set

$$F(x^{1/q}, y(x)) := \sum_{\rho,\sigma=0}^{\infty} |f_{\rho\sigma}| x^{\rho/q} (y - |y_0^{(0)}|)^\sigma.$$

This series is known to converge because $\bar{f}$ is analytic and hence has an absolutely convergent series expansion. We then look at the Volterra equation

$$Y(x) = \sum_{k=0}^{m-1} \frac{x^k}{k!} |y_0^{(k)}| + \frac{1}{\Gamma(n)} \int_0^x (x-t)^{n-1} F(t, Y(t)) \, dt. \qquad (6.37)$$

The formal solution Y of this equation may be computed in exactly the same way as above. It is immediately clear that Y is a majorant for y, and that all coefficients of Y are positive. Thus, we now need to prove that the series expansion of $Y(r)$ converges for some $r > 0$. The positivity of the coefficients of Y then implies that the series expansion of $Y(x)$ converges uniformly for all $x \in [0, r]$, and the majorant criterion tells us that the expansion of $y(x)$ converges absolutely and uniformly for these x too.

The idea in the convergence proof of Lindelöf is that the finite partial sum

$$P_{\ell+1}(x) = \sum_{i=0}^{\ell+1} Y_i x^{i/q}, \qquad (6.38)$$

of the formal solution of (6.37) can be bounded in terms of P_ℓ, F and the initial values. The key observation is the inequality

$$P_{\ell+1}(x) \leq \sum_{k=0}^{m-1} \frac{x^k}{k!} |y_0^{(k)}| + \frac{1}{\Gamma(n)} \int_0^x (x-t)^{n-1} F(t, P_\ell(t)) \, dt \qquad (6.39)$$

that follows from the recursive computation of the coefficients. If we expand the right-hand side in a series as above, the low order terms up to $x^{(\ell+1)/q}$ exactly cancel the left-hand side by construction, while there will in general be additional terms of higher order. These are all nonnegative because all coefficients involved are nonnegative.

Let us choose some $b > 0$ and define $C_1 := \sum_{k=0}^{m-1} b^k |y_0^{(k)}|/k!$. Moreover let $C_2 :=$ $\max_{(x,w)\in[0,b]\times[0,2C_1]} |F(x,w)|/\Gamma(n+1)$. Finally, define $r := \min\{b, (C_1/C_2)^{1/n}\}$. Our goal is now to prove the inequality

$$|P_\ell(x)| \leq 2C_1 \quad \text{for all } \ell = 0, 1, 2, \ldots \text{ and all } x \in [0, r]. \tag{6.40}$$

The proof is based on mathematical induction over ℓ. The induction basis $\ell = 0$ is evident since $P_0(x) = Y_0 = |y_0^{(0)}| \leq C_1$ by definition of C_1. For the induction step from ℓ to $\ell+1$ we recall that $|P_{\ell+1}(x)| = P_{\ell+1}(x)$ and write, using (6.39),

$$
\begin{aligned}
|P_{\ell+1}(x)| &\leq \sum_{k=0}^{m-1} \frac{x^k}{k!} |y_0^{(k)}| + \frac{1}{\Gamma(n)} \int_0^x (x-t)^{n-1} F(t, P_\ell(t)) \, dt \\
&\leq \sum_{k=0}^{m-1} \frac{r^k}{k!} |y_0^{(k)}| + \frac{1}{\Gamma(n)} \max_{t\in[0,x]} |F(t, P_\ell(t))| \int_0^x (x-t)^{n-1} \, dt \\
&\leq C_1 + x^n \frac{1}{\Gamma(n+1)} \max_{t\in[0,x]} |F(t, P_\ell(t))| \\
&\leq C_1 + r^n \frac{1}{\Gamma(n+1)} \max_{(t,w)\in[0,r]\times[0,2C_1]} |F(t, w)| \\
&\leq C_1 + r^n C_2 \leq 2C_1.
\end{aligned}
$$

Thus the sequence of partial sums of the majorant is uniformly bounded on $[0, r]$. In view of the positivity of all its coefficients it is also monotone. Therefore the majorant is absolutely convergent, and since it has the structure of a power series, it is also uniformly convergent on compact subsets of $[0, r)$. Arguing with the usual majorant criterion, we conclude the same properties for the series expansion (6.31) which finally tells us that our above interchange of summation and integration was legal. Thus the proof is complete. $\square$

If n is irrational then we find a slightly different result.

Theorem 6.35. *Let n be a positive irrational number. Consider the initial value problem (6.29) and assume that f can be written in the form $f(x,y) = \bar{f}(x, x^n, y)$ where $\bar{f}$ is analytic in a neighbourhood of $(0, 0, y_0^{(0)})$. Then, there exists a uniquely determined analytic function $\bar{y} : (-r, r) \times (-r^n, r^n) \to \mathbb{R}$ with some $r > 0$ such that $y(x) = \bar{y}(x, x^n)$ for $x \in [0, r)$.*

Once again we note two corollaries before coming to the proof of Theorem 6.35.

Corollary 6.36. *Let n be a positive irrational number. Consider the initial value problem (6.29) and assume that f is analytic in a neighbourhood of $(0, y_0^{(0)})$. Then, there exists a uniquely determined analytic function $\bar{y}: (-r, r) \times (-r^n, r^n) \to \mathbb{R}$ with some $r > 0$ such that $y(x) = \bar{y}(x, x^n)$ for $x \in [0, r)$.*

This result can be deduced from Theorem 6.35 in the same way as Corollary 6.33 has been shown to follow from Theorem 6.32.

The second corollary is an expansion of the solution y that immediately follows from Theorem 6.35.

Corollary 6.37. *Under the assumptions of Theorem 6.35, y is of the form*

$$y(x) = \sum_{\mu,\nu=0}^{\infty} \bar{y}_{\mu\nu} x^{\mu+n\nu}.$$

Proof (of Theorem 6.35). As in the proof of Theorem 6.32, the assumptions imply the local existence of a unique continuous solution y of (6.2) where $y(0) = y_0^{(0)}$. This time we substitute $\sum_{i=0}^{\infty} \bar{y}_{i_1 i_2} x^{i_1 + i_2 n}$ into (6.2), and by computations as in the proof of Theorem 6.32 we obtain relations similar to (6.35) and (6.36) for the coefficients, namely

$$\bar{y}_{i_1,0} = \frac{y_0^{(i_1)}}{(i_1)!} \qquad \text{for } 0 \le i_1 \le m-1 \tag{6.41}$$

and

$$\bar{y}_{i_1 i_2} - \sum_{\substack{m_1+j_1=i_1 \\ m_2+j_2+1=i_2}} \gamma_{m_1 m_2 j_1 j_2} f_{m_1 m_2 \sigma} \sum_{\substack{u_1+\ldots+u_\sigma=j_1 \\ v_1+\ldots+v_\sigma=j_2}} \bar{y}_{u_1 v_1} \cdot \ldots \cdot \bar{y}_{u_\sigma v_\sigma} \tag{6.42}$$

for the remaining cases where the coefficients $\gamma_{m_1 m_2 j_1 j_2}$ are given by

$$\gamma_{m_1 m_2 j_1 j_2} - \frac{\Gamma(m_1 + j_1 + (m_2 + j_2)n + 1)}{\Gamma(m_1 + j_1 + (m_2 + j_2 + 1)n + 1)}. \tag{6.43}$$

The coefficient $\bar{y}_{i_1 i_2}$ is uniquely determined in terms of the coefficients corresponding to smaller exponents. Hence we can arrange the exponents with respect to increasing magnitude and compute each coefficient uniquely from a (sometimes empty) subset of the coefficients of smaller exponents that have been computed beforehand.

The rest of the proof is the analogous to the rational case; we therefore omit the details. □

Theorem 6.35 gives us a representation of the solution in terms of an analytic function $\bar{y}$ of two variables, whereas the analytic function of Theorem 6.32 (dealing with rational n) was a function of just one variable. Simple examples show that we cannot expect the latter to hold in general in the irrational case. We can, however, construct a representation like the one of Theorem 6.35 in the rational case. The problem here is that this representation is not unique any more because many of the exponents $\mu + n\nu$ appearing in the form of the expansion shown in Corollary 6.37 may be represented as a linear combination of 1 and n in more than one way (for example, if $n = p/q$ in lowest terms, then $\mu + n\nu = (\mu - p) + n(\nu + q)$). To illustrate this phenomenon, we look at a simple example.

Example 6.5. Consider the initial value problem

$$D_*^{1/2}y(x) = y(x), \qquad y(0) = 0. \tag{6.44}$$

The unique continuous solution is $y(x) \equiv 0$. Thus, one possible representation for y of the form indicated in Corollary 6.37 is obtained by setting $\bar{y}_{\mu\nu} = 0$ for all μ and ν. Alternatively we may choose $\bar{y}_{00} = 0$, $\bar{y}_{\mu,0} = s_\mu$ and $\bar{y}_{0,2\mu} = -s_\mu$ for $\mu \geq 1$ where $(s_\mu)_{\mu=1}^\infty$ is an arbitrary sequence that decays sufficiently fast. Then we have

$$\bar{y}(x, x^{1/2}) = \sum_{\mu=0}^\infty s_\mu x^\mu - \sum_{\mu=0}^\infty s_\mu (x^{1/2})^{2\mu} = 0.$$

Therefore we can represent the solution in an infinite number of ways. This type of non-uniqueness obviously applies to all equations of the form (6.29) that obey the conditions imposed in Theorem 6.32.

A close inspection of the proofs reveals the possibility to obtain comparable results under the weaker assumption that f only is a C^k function for some $k \in \mathbb{N}$. Specifically, in the rational case we have:

Theorem 6.38. *Let $n = p/q$ where $p \geq 1$ and $q \geq 2$ are two relatively prime integers. Consider the initial value problem (6.29) and assume that f can be written in the form $f(x, y) = \bar{f}(x^{1/q}, y)$ where $\bar{f} \in C^j([0, h^*] \times [y_0^{(0)} - K, y_0^{(0)} + K])$ with some $K > 0$, $h^* > 0$ and $j \in \mathbb{N}$. Then, the solution y of (6.29) has an asymptotic expansion in powers of $x^{1/q}$ as $x \to 0$. In particular, the smallest noninteger exponent in this expansion is n.*

If n is irrational then we can prove the following statement.

Theorem 6.39. *Let n be a positive irrational number. Consider the initial value problem (6.29) and assume that f can be written in the form $f(x, y) = \bar{f}(x, x^n, y)$ where $\bar{f} \in C^j([0, h^*] \times [0, (h^*)^n] \times [y_0^{(0)} - K, y_0^{(0)} + K])$ with some $K > 0$, $h^* > 0$ and $j \in \mathbb{N}$. Then, the solution y of (6.29) has an asymptotic expansion in mixed powers of x and x^n as $x \to 0$.*

Proof (of Theorems 6.38 and 6.39). The proof is based on a combination of the proofs of Theorems 6.32 and 6.35, respectively, and ideas of Lubich [120, Corollary 3]. In particular, instead of building up an infinite series expansion for $\bar{y}$ as in (6.31) or its counterpart in the irrational case, we only construct a truncated power series $\bar{y}_N(x)$, say, plus the remainder term $R(x)$, where the degree N of the truncated series is to be computed later. In a similar way we expand the function $\bar{f}$ according to

Taylor's theorem in the form of a polynomial in $x^{1/q}$ and y around the point $(0, y_0^{(0)})$ (in the rational case) or in x, x^n and y around $(0, 0, y_0^{(0)})$ (in the irrational case), plus the corresponding remainder term. We can then proceed as above and come up with a relation similar to (6.34), only that the infinite series are now truncated and that the remainder terms come into the equation. Thus we can find the direct definition for the first coefficients of the truncated expansion as in (6.35) and the recurrence relation for the following coefficients as in (6.36) where the latter now does not hold for all $i > p$ any more but only up to the point that the truncation of the expansion for $\bar{f}$ permits. This determines the value N where our truncated series $\bar{y}_N$ ends. It is then evident that $\bar{y}_N(x)$ has an asymptotic expansion of the required form, and a straightforward estimation of the remainder term $R(x)$ based on the generalization of (6.34) shows that the remainder only contains higher order terms that do not destroy the asymptotics of $\bar{y} = \bar{y}_N + R$.   □

The theorems above have given us a large amount of information about the smoothness properties of the solutions of fractional differential equations, and in particular about the exact behaviour of the solution as $x \to 0$, most notably the formal asymptotic expansion. In a concrete application however it may be possible to say even more. An aspect of special significance, for example in view of the development of numerical methods, is the question for the precise values of the constants in this expansion, and most importantly the question whether certain coefficients vanish. A suitable generalization of the Taylor expansion technique for ordinary differential equations described in [88, Chapter I.8] could be useful in this context. Precise results in this connection seem to be unknown at the moment though.

6.5 Boundary Value Problems

In this final section of this chapter we move away from the initial value problems discussed so far and turn our attention towards boundary value problems. Many different types of boundary conditions are conceivable; we restrict ourselves to those types that appear to be most significant. For further reading, in particular with respect to other classes of problems, we refer to the survey of Agarwal et al. [4].

As always in this chapter, the differential equation under consideration is

$$D_{*0}^n y(x) = f(x, y(x)), \qquad x \in [0, T], \tag{6.45}$$

with some $n > 0$.

We know from the theory of initial value problems that the number of conditions that we must impose in order to obtain reasonable existence and uniqueness results is $\lceil n \rceil$. Thus, we first look at the case $0 < n < 1$ where it is appropriate to impose exactly one boundary condition. The most natural form for such a condition is

$$ay(0) + by(T) = c \tag{6.46}$$

with certain constants $a, b, c \in \mathbb{R}$. In this case we can reformulate the boundary value problem as a Fredholm integral equation in the following way.

Lemma 6.40. *Let $0 < n < 1$, $a, b, c \in \mathbb{R}$ and $a + b \neq 0$. Moreover assume that $f : [0, T] \times \mathbb{R} \to \mathbb{R}$ is continuous. Then, the function $y \in C[0, T]$ is a solution of the boundary value problem (6.45), (6.46) if and only if it is a solution of the integral equation*

$$y(x) = \frac{c}{a+b} + \frac{1}{\Gamma(n)} \int_0^x (x-t)^{n-1} f(t, y(t)) \, dt$$
$$- \frac{1}{\Gamma(n)} \frac{b}{a+b} \int_0^T (T-t)^{n-1} f(t, y(t)) \, dt. \tag{6.47}$$

Two special cases of this result need to be mentioned: If $b = 0$ then (6.46) reduces to an initial condition, and thus we recover Lemma 6.2. And if $a = 0$ then (6.46) becomes a terminal condition and we are in the situation of Theorem 6.18.

Proof. Equation (6.47) implies

$$y(0) = \frac{c}{a+b} - \frac{1}{\Gamma(n)} \frac{b}{a+b} \int_0^T (T-t)^{n-1} f(t, y(t)) \, dt$$

and

$$y(T) = \frac{c}{a+b} + \frac{1}{\Gamma(n)} \frac{a}{a+b} \int_0^T (T-t)^{n-1} f(t, y(t)) \, dt$$

from which (6.46) follows. Moreover, an application of the differential operator D_{*0}^n to both sides of (6.47) yields (6.45). Thus, y solves the boundary value problem if it solves the integral equation.

On the other hand, if y solves the differential equation (6.45) then we know from Lemma 6.2 that it also satisfies the Volterra equation

$$y(x) = y(0) + \frac{1}{\Gamma(n)} \int_0^x (x-t)^{n-1} f(t, y(t)) \, dt \tag{6.48}$$

with some (presently unknown) quantity $y(0)$. Taken at $x = T$ this reads

$$y(T) = y(0) + \frac{1}{\Gamma(n)} \int_0^T (T-t)^{n-1} f(t, y(t)) \, dt.$$

We can plug this into the boundary condition (6.46) and derive

$$(a+b)y(0) + \frac{1}{\Gamma(n)} b \int_0^T (T-t)^{n-1} f(t, y(t)) \, dt = c. \tag{6.49}$$

Now we solve (6.48) for $y(0)$ and insert the result into (6.49) which yields

$$(a+b)y(x) - \frac{a+b}{\Gamma(n)} \int_0^x (x-t)^{n-1} f(t, y(t)) \, dt + \frac{b}{\Gamma(n)} \int_0^T (T-t)^{n-1} f(t, y(t)) \, dt = c$$

which is equivalent to the Fredholm integral equation (6.47).   □

Remark 6.15. In the case $a + b = 0$ we can only obtain the weaker result that any solution of the boundary value problem (6.45), (6.46) satisfies the integral equation

$$c = \frac{1}{\Gamma(n)} b \int_0^T (T-t)^{n-1} f(t, y(t)) \, dt.$$

This follows by proceeding as in the second part of the proof of Lemma 6.40 up to (6.49) which is just what we have claimed. However, we cannot deduce that any solution of this integral equation solves the initial value problem.

The substantial difference between (6.47) and the integral equation of Remark 6.15 is that the former is an integral equation of the second kind whereas the latter is an equation of the first kind. It is well known that Fredholm equations of the second kind are much more well behaved than those of the first kind with respect to questions of existence and uniqueness of solutions and the continuity of the solutions with respect to the given data [109].

Lemma 6.40 immediately allows us to deduce a simple existence and uniqueness theorem (see also [16]).

Theorem 6.41. *Assume the hypotheses of Lemma 6.40. If additionally f satisfies a Lipschitz condition with Lipschitz constant L with respect to its second variable, and if*

$$LT^n \left(1 + \frac{|b|}{|a+b|}\right) < \Gamma(n+1) \tag{6.50}$$

then the boundary value problem (6.45), (6.46) has a unique solution $y \in C[0, T]$.

Proof. From Lemma 6.40 we can see that we need to prove that the operator A, defined by

$$Ay(x) := \frac{c}{a+b} + \frac{1}{\Gamma(n)} \int_0^x (x-t)^{n-1} f(t, y(t)) \, dt$$
$$- \frac{1}{\Gamma(n)} \frac{b}{a+b} \int_0^T (T-t)^{n-1} f(t, y(t)) \, dt.$$

has a unique fixed point. It is clear that the operator maps $C[0, T]$ into itself and that

$$|Ay(x) - A\tilde{y}(x)| \le \frac{1}{\Gamma(n)} \int_0^x (x-t)^{n-1} |f(t, y(t)) - f(t, \tilde{y}(t))| \, dt$$
$$+ \frac{1}{\Gamma(n)} \frac{|b|}{|a+b|} \int_0^T (T-t)^{n-1} |f(t, y(t)) - f(t, \tilde{y}(t))| \, dt$$

$$\leq \frac{L}{\Gamma(n)} \|y - \tilde{y}\|_\infty \left(\int_0^x (x-t)^{n-1} \, dt + \frac{|b|}{|a+b|} \int_0^T (T-t)^{n-1} \, dt \right)$$

$$\leq \frac{L}{\Gamma(n+1)} T^n \left(1 + \frac{|b|}{|a+b|} \right) \|y - \tilde{y}\|_\infty$$

which implies, under our assumption, that A is a contraction. Thus, by Banach's fixed point theorem, we obtain that A indeed has a unique fixed point. $\square$

Under slightly different conditions we can also derive an existence theorem.

Theorem 6.42. *Assume the hypotheses of Lemma 6.40. If additionally f is uniformly bounded by an absolute constant then the boundary value problem (6.45), (6.46) has a solution $y \in C[0,T]$.*

Proof. We use the same operator A as in the previous proof. Our goal is now to show that it has at least one fixed point. To this end we want to invoke Schauder's fixed point theorem. Thus all we need to show is that $X := \{Ay : y \in C[0,T]\}$ is a relatively compact set. A necessary and sufficient condition for this to hold is contained in the Arzelà–Ascoli Theorem: We need to show that X is uniformly bounded and equicontinuous. But the uniform boundedness of X is a trivial consequence of the definition of A and the boundedness of f, and the equicontinuity can be shown just as in the proof of Theorem 6.1. $\square$

In the theory of differential equations of integer order, the most important class of boundary value problems is the one with second order differential equations, i.e. equations where one imposes two boundary conditions and not just one as we have done so far. Transferred to Caputo-type fractional differential equations, this corresponds to equations of order $n \in (1,2)$. Thus we shall now take a look at this type of equations. Once again we begin by converting the given boundary value problem into an equivalent Fredholm equation. We shall impose two boundary conditions similar to those of (6.46); however, in line with classical theory, we now allow first derivatives of the solution at the end points 0 and T to appear in addition to the function values themselves. Thus, our boundary value problem now has the form

$$D_{*0}^n y(x) = f(x, y(x)), \qquad (6.51a)$$

$$a_{j0}y(0) + a_{j1}y'(0) + b_{j0}y(T) + b_{j1}y'(T) = c_j \qquad (j = 1, 2), \qquad (6.51b)$$

with some $n \in (1,2)$. The equivalence result then reads as follows.

Lemma 6.43. *Let $1 < n < 2$, $a_{jk}, b_{jk}, c_k \in \mathbb{R}$ ($j = 1, 2$, $k = 0, 1$) and assume that $f : [0,T] \times \mathbb{R} \to \mathbb{R}$ is continuous. If the matrix*

$$M := \begin{pmatrix} a_{10} + b_{10} & a_{11} + Tb_{10} + b_{11} \\ a_{20} + b_{20} & a_{21} + Tb_{20} + b_{21} \end{pmatrix}$$

is regular, then we have that $y \in C^1[0,T]$ is a solution of the boundary value problem (6.51) if and only if it is a solution of the Fredholm integral equation

$$y(x) = \alpha_1 + \alpha_2 x + \frac{1}{\Gamma(n)} \int_0^x (x-t)^{n-1} f(t, y(t)) \, dt \qquad (6.52)$$

where α_1, α_2 are determined by the linear system

$$M \begin{pmatrix} \alpha_1 \\ \alpha_2 \end{pmatrix} = \begin{pmatrix} c_1 - \int_0^T \left[b_{10} \dfrac{(T-t)^{n-1}}{\Gamma(n)} + b_{11} \dfrac{(T-t)^{n-2}}{\Gamma(n-1)} \right] f(t, y(t)) \, dt \\ c_2 - \int_0^T \left[b_{20} \dfrac{(T-t)^{n-1}}{\Gamma(n)} + b_{21} \dfrac{(T-t)^{n-2}}{\Gamma(n-1)} \right] f(t, y(t)) \, dt \end{pmatrix}. \qquad (6.53)$$

At first sight (6.52) looks like a Volterra equation. However, this is not the case as is evident from the definition of the quantities α_1 and α_2: The Fredholm operators on the right-hand side of (6.53) enter (6.52) in this way.

Proof. The proof is very similar to the proof of Lemma 6.40. We apply Lemma 6.2 to find the equivalence of the differential equation with the integral equation (6.52) with certain constants α_1 and α_2. The differentiability of the solution follows from Theorem 6.25. The two boundary conditions provide two linear equations that allow to determine the constants α_1 and α_2; the resulting equation system is just (6.53). We leave the details to the reader. $\qquad \square$

Based on this result we can now formulate analoga of Theorems 6.41 and 6.42. We begin with the latter.

Theorem 6.44. *Assume the hypotheses of Lemma 6.43. Moreover let f be uniformly bounded by an absolute constant. Then, the boundary value problem (6.51) has a solution $y \in C^1[0, T]$.*

Proof. We proceed much as in the proof of Theorem 6.42: We use Lemma 6.43 to rewrite the boundary value problem in the form of the integral equation (6.52), note that the existence of a solution to this equation can be interpreted as the existence of a fixed point of a properly chosen operator A with $Ay(x)$ being defined as the right-hand side of (6.52), and use the boundedness of f to deduce the existence of such a fixed point via Schauder's theorem. $\qquad \square$

Finally, the uniqueness theorem reads as follows.

Theorem 6.45. *Assume the hypotheses of Lemma 6.43. Moreover let f satisfy a Lipschitz condition with respect to the second variable with a sufficiently small Lipschitz constant. Then, the boundary value problem (6.51) has a unique solution $y \in C^1[0, T]$.*

Remark 6.16. In the case of boundary value problems of order $n \in (0, 1)$, we also have existence and uniqueness only if the Lipschitz constant of f is sufficiently small. It is evident from (6.50) that the maximal allowed value L of the Lipschitz constant in this connection depends on the parameters of the boundary condition, i.e. on the values a, b and T. Similarly, in the case $1 < n < 2$ under consideration now, the allowed value of the Lipschitz constant will depend on a_{jk}, b_{jk} and T.

Proof. The proof goes along the same lines as the proof of Theorem 6.41. We omit the details.   □

We conclude the treatment of boundary value problems for Caputo differential equations at this point and only note, for further reference, that other types of boundary conditions are occasionally used as well; see, e.g., the discussion in [4]. Moreover, in principle, boundary value problems of higher order can be treated in an analogous manner. However, in view of their minor significance in practice, we shall not discuss this topic here explicitly.

Exercises

Exercise 6.1. Fill in the details of the first part of the proof of Lemma 6.2.

Exercise 6.2. Give explicit proofs for Corollaries 6.6 and 6.7.

Exercise 6.3. Show that, as claimed in the proof of Lemma 6.19, the function $\Phi : \mathbb{R}_+ \to \mathbb{R}$ defined by $\Phi(x) := (\varepsilon_1 + \varepsilon_3)E_n(\varepsilon_2 x^n)$, is the solution of the integral equation

$$\Phi(x) = \varepsilon_1 + \varepsilon_3 + \frac{\varepsilon_2}{\Gamma(n)} \int_0^x (x-t)^{n-1} \Phi(t) \, dt.$$

Exercise 6.4. Consider the fractional differential equation

$$D_{*0}^{1/3} y(x) = x^2 + xy(x)$$

with initial condition $y(0) = -1$. For this initial value problem, construct the operator A from the proof of Theorem 6.5 and determine the first five elements of the corresponding Picard iteration sequence.

Exercise 6.5. Consider the fractional differential equation

$$D_{*0}^{3/2} y(x) = x + (xy(x))^2$$

with initial conditions $y(0) = 3$ and $y'(0) = 1$. For this initial value problem, construct the operator A from the proof of Theorem 6.5 and determine the first three elements of the corresponding Picard iteration sequence.

Exercise 6.6. Give a proof for Theorem 6.27.

Exercise 6.7. Fill in the details of the proof of Lemma 6.43.

Exercise 6.8. Provide the details of the proof of Theorem 6.45.

Exercise 6.9. Compute the maximal allowed value of the Lipschitz constant of f mentioned in Remark 6.16.

Chapter 7
Single-Term Caputo Fractional Differential Equations: Advanced Results for Special Cases

With the fundamentals of a theory for fractional differential equations with Caputo derivatives being in place now, we next attempt to give some additional information on certain particularly important special cases of equations. Specifically this includes a more precise analysis of the properties of solutions to linear equations in Sects. 7.1 (where we will look at initial value problems) and 7.2 (which is focused on boundary value problems), the investigation of the long-term behaviour of solutions defined on sufficiently large (and, in particular, unbounded) intervals in Sect. 7.3, and a brief look in Sect. 7.4 at how singular equations can lead to solutions with properties that differ substantially from those that we have seen for regular problems in Chap. 6.

7.1 Initial Value Problems for Linear Equations

In the first two sections of this chapter we restrict our attention to a special class of equations that nevertheless is very important in many applications: linear fractional differential equations. It is a common observation in many areas of mathematics that the linearity assumption allows to derive more precise statements. The same holds true for fractional differential equations. In particular, explicit expressions for the solutions of such equations can often be obtained. We have already found a special case of this situation in Theorem 6.11. In general it turns out that the Laplace transform is an extremely useful tool for the analysis of linear fractional differential equations; we shall use it here too. The fundamental definitions and properties of this transform are repeated in Appendix D.3. We begin the investigations in this section by providing generalizations of the integration theorem and the differentiation theorem for Laplace transforms (parts (c) and (d) of Theorem D.11) to Riemann–Liouville fractional integrals and Caputo fractional derivatives. It is rather obvious that for $n \in \mathbb{N}$ we recover the classical statements. Here and throughout the rest of the text we denote the Laplace transform operator by $\mathscr{L}$.

K. Diethelm, *The Analysis of Fractional Differential Equations*,
Lecture Notes in Mathematics 2004, DOI 10.1007/978-3-642-14574-2_7,
© Springer-Verlag Berlin Heidelberg 2010

Theorem 7.1. *Assume that* $f : [0, \infty) \to \mathbb{R}$ *is such that* $\mathscr{L}f$ *exists on* $[s_0, \infty)$ *with some* $s_0 \in \mathbb{R}$. *Let* $n > 0$ *and* $m := \lceil n \rceil$. *Then, for* $s > \max\{0, s_0\}$ *we have*

$$\mathscr{L}J_0^n f(s) = \frac{1}{s^n}\mathscr{L}f(s)$$

and

$$\mathscr{L}D_{*0}^n f(s) = s^n \mathscr{L}f(s) - \sum_{k=1}^{m} s^{n-k} f^{(k-1)}(0).$$

Proof. Let $g(x) = x^{n-1}/\Gamma(n)$. Then, by Example D.1 (b) and the linearity of the Laplace transform, $\mathscr{L}g(s) = 1/s^n$ for $s > 0$. Moreover, by definition of the Riemann–Liouville integral operator, $J_0^n f$ is the convolution of f and g, and therefore the convolution theorem (Theorem D.11 (b)) implies that

$$\mathscr{L}J_0^n f(s) = \mathscr{L}g(s) \cdot \mathscr{L}f(s) = \frac{1}{s^n}\mathscr{L}f(s)$$

for $s > \max\{0, s_0\}$.

Furthermore, we recall that $D_{*0}^n f = J_0^{m-n} D^m f$, and thus – in view of what we have just shown and the differentiation theorem –

$$\mathscr{L}D_{*0}^n f(s) = \mathscr{L}J_0^{m-n} D^m f(s) = \frac{1}{s^{m-n}}\mathscr{L}D^m f(s)$$

$$= s^{n-m}\left(s^m \mathscr{L}f(s) - \sum_{k=1}^{m} s^{m-k} f^{(k-1)}(0)\right)$$

$$= s^n \mathscr{L}f(s) - \sum_{k=1}^{m} s^{n-k} f^{(k-1)}(0). \qquad \square$$

Using these results we are in a position to solve another example equation that is slightly more advanced than the equations considered in Theorem 6.11.

Example 7.1. We are looking for the solution of the initial value problem

$$D_{*0}^n y(x) = -y(x) - q(x), \qquad y(0) = 2,$$

with some $n \in (0, 1)$ and some function q.

The approach is based on the Laplace transform method. Applying the Laplace transform to the initial value problem, we derive

$$s^n \mathscr{L}y(s) - 2s^{n-1} = -\mathscr{L}y(s) - \mathscr{L}q(s)$$

in view of Theorem 7.1 and Example D.1. We solve this equation for $\mathscr{L}y(s)$ and obtain

$$\mathscr{L}y(s) = \frac{2s^{n-1}}{s^n + 1} - \frac{\mathscr{L}q(s)}{s^n + 1}.$$

Now we need to find the inverse Laplace transforms of the functions on the right-hand side. From Theorem 4.5 we know that

$$z(x) = E_n(-x^n) \Rightarrow \mathscr{L}z(s) = \frac{s^{n-1}}{s^n+1}.$$

Moreover, we have for this function z that

$$\mathscr{L}z'(s) = s\mathscr{L}z(s) - z(0) = \frac{s^n}{s^n+1} - 1 = -\frac{1}{s^n+1}.$$

by the differentiation theorem for the Laplace transform. Combining this with the convolution theorem for the Laplace transform, we find

$$y(x) = 2E_n(-x^n) + \int_0^x q(x-t)\frac{d}{dt}E_n(-t^n)\,dt.$$

We can actually generalize the observations of this example and develop an explicit formula for the solution of a simple class of equations, the linear equations with constant coefficients. We shall once again discover the significance of the Mittag-Leffler functions E_n and E_{n_1,n_2}.

Theorem 7.2. *Let $n > 0$, $m = \lceil n \rceil$ and $\lambda \in \mathbb{R}$. The solution of the initial value problem*

$$D_{*0}^n y(x) = \lambda y(x) + q(x), \qquad y^{(k)}(0) = y_0^{(k)} \quad (k = 0, 1, \ldots, m-1), \qquad (7.1)$$

where $q \in C[0,h]$ is a given function, can be expressed in the form

$$y(x) = \sum_{k=0}^{m-1} y_0^{(k)} u_k(x) + \tilde{y}(x) \qquad (7.2a)$$

with

$$\tilde{y}(x) = \begin{cases} J_0^n q(x) & \text{if } \lambda = 0, \\ \dfrac{1}{\lambda} \displaystyle\int_0^x q(x-t)u_0'(t)\,dt & \text{if } \lambda \neq 0, \end{cases} \qquad (7.2b)$$

where $u_k(x) := J_0^k e_n(x)$, $k = 0, 1, \ldots, m-1$, and $e_n(x) := E_n(\lambda x^n)$.

Remark 7.1. In the case $0 < n < 1$ we may use the above mentioned definitions of u_0 and e_n to rewrite the representation (7.2) of the solution of the initial value problem in the form

$$y(x) = y_0^{(0)} E_n(\lambda x^n) + n \int_0^x q(x-t)t^{n-1} E_n'(\lambda t^n)\,dt \qquad (7.3)$$

which holds both for $\lambda = 0$ and $\lambda \neq 0$. Using the power series expansion of the Mittag-Leffler functions one can easily see that this is equivalent to

$$y(x) = y_0^{(0)} E_n(\lambda x^n) + \int_0^x q(x-t)t^{n-1} E_{n,n}(\lambda t^n)\, dt.$$

Alternatively we may rewrite (7.3) as

$$y(x) = y_0^{(0)} E_n(\lambda x^n) + n \int_0^x (x-t)^{n-1} E_n'(\lambda(x-t)^n)q(t)\, dt. \tag{7.4}$$

In the limit case $n \to 1-$ this reduces to the well known formula

$$y(x) = y_0^{(0)} e^{\lambda x} + \int_0^x e^{\lambda(x-t)} q(t)\, dt$$

that is usually derived by the variation-of-constants method. For future reference we note that (7.4) remains valid in a vector-valued setting, i.e. if y and q are functions mapping to $\mathbb{R}^N$ for some N, $y_0^{(0)} \in \mathbb{R}^N$, and λ is an $N \times N$ matrix. More information about such multi-dimensional problems will be given at the end of this section.

Proof (of Theorem 7.2). In the case $\lambda = 0$ we have that $e_n(x) = E_n(0) = 1$ and therefore $u_k(x) = x^k/k!$ for every k. Thus the claim can be verified by a direct application of the relevant differential operators to the given representation of y.

For $\lambda \neq 0$ we will prove the following facts:

(a) The functions u_k satisfy the homogeneous differential equation $D_{*0}^n u_k = \lambda u_k$ ($k = 0, 1, \ldots, m-1$), and they fulfil the initial conditions $u_k^{(j)}(0) = \delta_{kj}$ (Kronecker's delta) for $j, k = 0, 1, \ldots, m-1$
(b) The function $\tilde{y}$ is a solution of the inhomogeneous differential equation with homogeneous initial conditions

Then, the superposition principle gives the claim.

Concerning (a), we know that

$$e_n(x) = \sum_{j=0}^{\infty} \frac{\lambda^j x^{nj}}{\Gamma(1+jn)}.$$

Thus,

$$u_k(x) = J_0^k e_n(x) = \sum_{j=0}^{\infty} \frac{\lambda^j x^{nj+k}}{\Gamma(1+jn+k)}.$$

Using this representation, we can see that u_k solves the homogeneous differential equation (cf. the proof of Theorem 4.3). Moreover we see that $u_k^{(k)}(0) = D^k J_0^k e_n(0) = e_n(0) = 1$. For $j < k$ we have $u_k^{(j)}(0) = D^j J_0^k e_n(0) = J_0^{k-j} e_n(0) = 0$ because of the continuity of e_n, and for $j > k$ we find

$$u_k^{(j)}(0) = D^j J_0^k e_n(0) = D^{j-k} e_n(0) = 0$$

because

$$e_n(x) = 1 + \frac{\lambda x^n}{\Gamma(1+n)} + \frac{\lambda^2 x^{2n}}{\Gamma(1+2n)} + \cdots$$

which implies

$$D^{j-k} e_n(x) = \frac{\lambda x^{n+k-j}}{\Gamma(1+n+k-j)} + \frac{\lambda^2 x^{2n+k-j}}{\Gamma(1+2n+k-j)} + \cdots \to 0$$

for $x \to 0$ (recall that $1 \le j - k \le m - 1 < n$). This completes the proof of (a).

Concerning statement (b), we have

$$\tilde{y}(x) = \frac{1}{\lambda} \int_0^x q(x-t) u_0'(t)\, dt = \frac{1}{\lambda} \int_0^x q(x-t) e_n'(t)\, dt$$

$$= \frac{1}{\lambda} \int_0^x q(t) e_n'(x-t)\, dt.$$

The first factor in the integral is continuous by assumption and the second factor is (at least improperly) integrable. Thus the integral exists everywhere and is a continuous function of x, and $\tilde{y}(0) = 0$. Moreover, for $n > 1$ (i.e. $m \ge 2$) we have, by the standard rules for the differentiation of parameter integrals with respect to the parameter,

$$D\tilde{y}(x) = \frac{1}{\lambda} \int_0^x q(t) e_n''(x-t)\, dt + \frac{1}{\lambda} q(x) \underbrace{e_n'(0)}_{=0}.$$

By the same argument as above (continuity of q and at worst weak singularity of e_n'') we see that $\tilde{y}'(0) = 0$. Proceeding in this manner, we find that

$$D^k \tilde{y}(x) = \frac{1}{\lambda} \int_0^x q(t) e_n^{(k+1)}(x-t)\, dt$$

for $k = 0, 1, \ldots, m-1$, and in all these cases the argumentation gives $D^k \tilde{y}(0) = 0$. Thus $\tilde{y}$ satisfies the required homogeneous initial conditions, and it remains to prove that it solves the inhomogeneous differential equation. To this end we write

$$e_n'(u) = \frac{d}{du} e_n(u) = \frac{d}{du} E_n(\lambda u^n) = \lambda n u^{n-1} E_n'(\lambda u^n)$$

$$= \lambda n u^{n-1} \sum_{j=1}^{\infty} \frac{j(\lambda u^n)^{j-1}}{\Gamma(1+jn)} = \lambda u^{n-1} \sum_{j=1}^{\infty} \frac{(\lambda u^n)^{j-1}}{\Gamma(jn)} = \sum_{j=1}^{\infty} \frac{\lambda^j u^{nj-1}}{\Gamma(jn)}$$

and thus

$$\tilde{y}(x) = \frac{1}{\lambda} \int_0^x q(t) e_n'(x-t)\, dt = \frac{1}{\lambda} \int_0^x q(t) \sum_{j=1}^{\infty} \frac{\lambda^j (x-t)^{jn-1}}{\Gamma(jn)}\, dt$$

$$= \sum_{j=1}^{\infty} \lambda^{j-1} \frac{1}{\Gamma(jn)} \int_0^x q(t)(x-t)^{jn-1}\, dt = \sum_{j=1}^{\infty} \lambda^{j-1} J_0^{jn} q(x).$$

Hence

$$D_{*0}^n \tilde{y}(x) = \sum_{j=1}^{\infty} \lambda^{j-1} D_{*0}^n J_0^{jn} q(x) = \sum_{j=1}^{\infty} \lambda^{j-1} J_0^{(j-1)n} q(x)$$

$$= \sum_{j=0}^{\infty} \lambda^j J_0^{jn} q(x) = q(x) + \sum_{j=1}^{\infty} \lambda^j J_0^{jn} q(x) = q(x) + \lambda \tilde{y}(x).$$

The interchange of summation and integration or, respectively, summation and differentiation is possible here in view of the convergence properties of the series expansion for e_n' and the continuity of q. $\square$

In practice two special cases of linear fractional differential equations are of particular importance because they can be interpreted as fractional generalizations of two fundamental integer-order differential equations of physics. These two classical equations are the *relaxation equation*

$$Dy(x) = -\mu y(x) + q(x), \qquad y(0) = y_0, \tag{7.5}$$

and the *oscillation equation*

$$D^2 y(x) = -\mu y(x) + q(x), \qquad y(0) = y_0^{(0)}, \quad Dy(0) = y_0^{(1)}, \tag{7.6}$$

where in both cases we assume $\mu > 0$ for physical reasons.

We begin our considerations that will lead to the desired fractional generalization with the classical relaxation equation (7.5). As is well known, its solution is

$$y(x) = e^{-\mu x} \left(y_0 + \int_0^x q(t) e^{\mu t} \, dt \right) = y_0 e^{-\mu x} + \int_0^x q(t) e^{\mu(t-x)} \, dt$$

which is just the case $n = 1$ and $\lambda = -\mu$ of (7.2). The solution decays exponentially as $x \to \infty$. Incidentally, this decay behaviour is our reason for choosing $\mu > 0$; the choice $\mu < 0$ would lead to a blow-up of the solution and hence to instabilities that are physically impossible in the systems usually described by such an equation.

A natural generalization to our fractional setting would consist of replacing the first-order differential operator by D_{*0}^n with some suitable n while keeping the initial condition unchanged. Thus, since our initial condition only relates to y itself, we may choose an arbitrary $n \in (0, 1)$. In this way we derive the so-called (simple) *fractional relaxation equation*

$$D_{*0}^n y(x) = -\mu y(x) + q(x), \qquad y(0) = y_0, \tag{7.7}$$

with $0 < n < 1$ and $\mu > 0$. The solution of the homogeneous equation is given by

$$y_{\text{hom}}(x) = y_0 E_n(-\mu x^n)$$

according to Theorem 7.2, and it turns out that this solution behaves in a way that differs somehow from the behaviour mentioned above in the case $n = 1$:

Theorem 7.3. *Let* $\mu > 0$, $0 < n < 1$ *and* $u_0(x) = E_n(-\mu x^n)$.

(a) *The function* u_0 *is completely monotonic on* $(0, \infty)$, *i.e.* $(-1)^k D^k u_0(x) \geq 0$ *for every* $x > 0$ *and every* $k \in \mathbb{N}_0$.
(b) *For* $x \to \infty$ *we have*

$$u_0(x) = \frac{x^{-n}}{\mu \Gamma(1-n)} (1 + o(1)).$$

In other words, the solution decays algebraically, i.e. much slower than the exponential decay known for the classical first-order relaxation equation. (Statement (a) implies that the solution is monotonically decreasing since Du_0 is nonpositive; statement (b) shows the speed of the decay.) A process showing this very slow decay behaviour is sometimes called *ultraslow* [82]. The classical proof of this theorem is based on an analysis of the properties of the Laplace transform of u_0 in the complex plane. It requires methods beyond the scope of this text, and therefore we shall not include it here. The details of this proof may be found in the original works of Gorenflo and Mainardi [80, 81]. However, it is possible to provide an alternative proof that avoids the use of Laplace transform techniques. We shall provide the necessary machinery in Sect. 7.3 and return to this question there (cf. pp. 162ff.). For the moment we only present the plots of Fig. 7.1 showing the case $\mu = 1$ for the purpose of illustration. Notice that the dash-dotted line corresponding to $n = 1$ shows the classical solution e^{-x}; it is apparent that this function decays much faster than the three others.

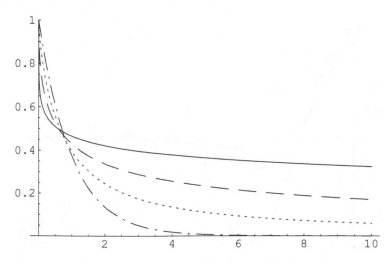

Fig. 7.1 Plots of $E_n(-x^n)$ for $n = 1/4$ (*continuous line*), $n = 1/2$ (*dashed line*), $n = 3/4$ (*dotted line*), and $n = 1$ (*dash–dotted line*)

In a similar way, we can see that the (simple) *fractional oscillation equation*

$$D_{*0}^n y(x) = -\mu y(x) + q(x), \qquad y(0) = y_0^{(0)}, \quad Dy(0) = y_0^{(1)} \qquad (7.8)$$

with $1 < n < 2$ and $\mu > 0$ generalizes the classical oscillation equation (7.6). The corresponding homogeneous problem has the two linearly independent solutions

$$u_0(x) = E_n(-\mu x^n) \qquad \text{and} \qquad u_1(x) = J_0 u_0(x)$$

according to Theorem 7.2. We can say something about the behaviour of these two functions as well.

Theorem 7.4. *Let $\mu > 0$, $1 < n < 2$ and $u_0(x) = E_n(-\mu x^n)$ and $u_1(x) = J_0 u_0(x)$. Then, for $x \to \infty$ we have*

$$u_0(x) = \frac{x^{-n}}{\mu \Gamma(1-n)}(1 + o(1)) \qquad \text{and} \qquad u_1(x) = \frac{x^{1-n}}{\mu \Gamma(2-n)}(1 + o(1)).$$

Thus the decay is once again algebraic, but (as is evident from Figs. 7.2 and 7.3, again displaying the case $\mu = 1$) not monotonic. In particular, we rediscover the classical solutions $u_0(x) = \cos x$ and $u_1(x) = \sin x$ in the graphs for $n = 2$; these functions of course do not decay at all. A proof of the theorem is given in [81, §3].

Theorems 7.3 and 7.4 give us some valuable information about the speed of the decay of the function $u_0(x) = E_n(-\mu x^n)$ and its first primitive u_1. A related important question is the question for the changes of sign of u_0. Some simple properties can already be read off from Figs. 7.1 and 7.2. A more precise statement is contained in the following result.

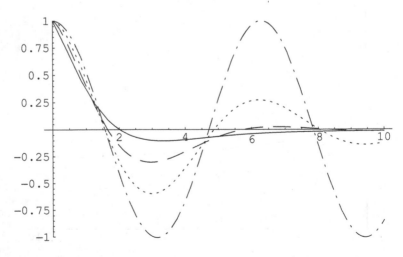

Fig. 7.2 Plots of $u_0(x)$ for $n = 5/4$ (*continuous line*), $n = 3/2$ (*dashed line*), $n = 7/4$ (*dotted line*), and $n = 2$ (*dash–dotted line*). $\mu = 1$ was chosen in all cases

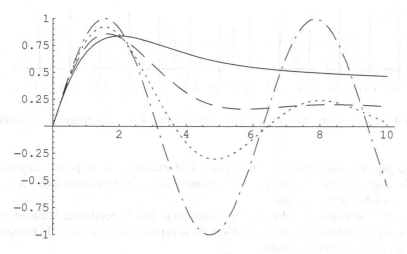

Fig. 7.3 Plots of $u_1(x)$ for $n = 5/4$ (*continuous line*), $n = 3/2$ (*dashed line*), $n - 7/4$ (*dotted line*), and $n = 2$ (*dash–dotted line*). $\mu = 1$ was chosen in all cases

Theorem 7.5. *Consider the function* $u_0(x) = E_n(-\mu x^n)$ *with some* $\mu > 0$.

(a) In the case $0 < n < 1$, u_0 *does not have any zeros on* $[0, \infty)$
(b) In the case $1 < n < 2$, *the number of zeros of* u_0 *on* $[0, \infty)$ *(counting multiplicities) is finite and odd*

Remember that in the case $n - 2$ we had $u_0(x) = \cos x$ which has infinitely many zeros on $[0, \infty)$.

Proof. Theorem 7.3 tells us that, for $0 < n < 1$, u_0 decays monotonically on $[0, \infty)$ and that $\lim_{x \to \infty} u_0(x) = 0$. Hence it cannot have any zeros which proves (a).

For statement (b) we first note that the property $u_0(0) = 1$ is a direct consequence of the definition of u_0 and the power series expansion of the Mittag-Leffler function E_n. In addition we note that the asymptotic result of Theorem 7.4 tells us that, for large x, $u_0(x)$ behaves as $x^{-n}/\Gamma(1 - n)$. Since we now have $1 < n < 2$ we find that $\Gamma(1 - n) < 0$, and hence we see that $u_0(x)$ approaches 0 from below. Thus $u_0(x) < 0$ for sufficiently large x. It follows that the number of zeros of u_0 must be finite and odd.   □

It would certainly be of interest to find out the precise location of the zeros of u_0 in the case $1 < n < 2$ or the values of n for which the number of zeros changes. The author is presently only aware of some first steps towards a conclusive answer to these questions that can be found in the work of Gorenflo and Mainardi [80]. Their results deal with the normalized case $\mu = 1$. Obvious substitutions can be used to transfer them to the general case. In fact they have computed the Mittag-Leffler functions numerically and found the relations between n and the number of zeros of u_0 indicated in Fig. 7.4. The transition from k to $k + 2$ zeros, say, takes place

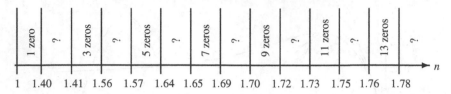

Fig. 7.4 Number of zeros of u_0 for $\mu = 1$ and various values of n (horizontal axis not to scale)

in the gaps marked with a question mark. Unfortunately, more precise analytical results concerning the location of the transition points in the general case seems to be unavailable at the moment.

Apart from these data related to the transition points, Gorenflo and Mainardi also give some information about the location of the zeros of u_0 themselves. In particular they prove the following results.

Theorem 7.6. *If $\varepsilon > 0$ is sufficiently small then the function $u_0(x) = E_n(-x^n)$ with $n = 1 + \varepsilon$ has exactly one zero $x^* > 0$. This zero satisfies the relation*

$$x^* = (1 + o(1))\ln(2/\varepsilon) \quad as \ \varepsilon \to 0.$$

It thus turns out that x^* grows towards ∞ at a very slow rate as $\varepsilon \to 0$. The proof is again based on estimating the Laplace transform of u_0 in the complex domain; we refer to [80, §4] for the details. Notice however that the approximation $x^* = (1 + o(1))\ln(2/\varepsilon)$ is actually quite accurate already for moderately large values of ε. Take, for example, $\varepsilon = 1/4$, i.e. $n = 5/4$. Then, the estimated value for the location of the only zero of u_0 is $\ln(2/\varepsilon) = \ln 8 \approx 2.08$, which matches the exact value very nicely as can be seen by a look at the plot of u_0 in Fig. 7.2 (the continuous line).

In a similar way one can investigate the behaviour as $n \to 2$:

Theorem 7.7. *The function $u_0(x) = E_n(-x^n)$ with $n = 2 - \varepsilon$ has $Z(\varepsilon)$ zeros in $[0, \infty)$, where $Z(\varepsilon)$ is an odd number for all $\varepsilon \in (0, 1)$ that satisfies the relation $Z(\varepsilon) = 12\pi^{-2}\varepsilon^{-1}(1 + o(1))\ln \varepsilon^{-1}$ as $\varepsilon \to 0$. Denoting the largest of these zeros by x^*, we have that $x^* = 12\pi^{-1}\varepsilon^{-1}(1 + o(1))\ln \varepsilon^{-1}$ as $\varepsilon \to 0$.*

Once again we refer to [80, §4] for a proof of this result and restrict our comments to the observation that a comparison of numerically computed values for the largest zero and the asymptotic value of Theorem 7.7 shows that the asymptotic approximation is less accurate than the one of Theorem 7.6 in the sense that now ε must be much closer to 0 in order for the approximation to be good in terms of the absolute error.

An asymptotic formula for the location of all zeros of Mittag-Leffler functions (i.e. not only the zeros located on the negative real axis) has been provided by Sedletskij [173]. His result can be applied to two-parameter Mittag-Leffler functions with arbitrary parameters. In the special case that we have dealt with in the two theorems above, it reads as follows.

Theorem 7.8. *Let* $0 < n_1 < 2$ *and* $n_2 > 0$. *Then, for each* $m \in \mathbb{Z}$ *we have* $E_{n_1,n_2}(z_m) = 0$ *where*

$$z_m = \left(2\pi i m - n_1 \tau_{n_2}(\ln 2\pi |m| + i\frac{\pi}{2}\operatorname{sgn} m) + \gamma_{n_2} + O(|m|^{-n_1}) + O(|m|^{-1}\ln|m|)\right)^{n_1}$$

with some constants τ_{n_2} *and* γ_{n_2} *that only depend on* n_2.

We refer to the original article of Sedletskij [173] for a proof of this theorem and for some additional information, in particular for further results concerning Mittag-Leffler functions with differently chosen parameters.

In this context we can also note that a discussion of the location of the non-real zeros of one-parameter Mittag-Leffler functions may already be found in the early paper by Wiman [192]. Moreover, the recent work of Seybold and Hilfer [174] contains some numerical data on the location of the zeros of the two-parameter Mittag-Leffler function $E_{0.8,0.9}$. Their methods might be applicable to obtain corresponding data for other Mittag-Leffler functions too. Finally we mention the work of Gorenflo et al. [79] who have provided a numerical algorithm for the computation of function values of Mittag-Leffler functions and their derivatives. Of course, a combination of this method with a standard rootfinding algorithm like the Newton-Raphson scheme can yield numerical values for the zeros of Mittag-Leffler functions with given parameters.

The considerations in this section have so far only dealt with linear equations with constant coefficients. If we allow equations with nonconstant coefficients then the theory becomes much more cumbersome. Nevertheless there are still some results that we can prove. Thus the object of interest is now the fractional differential equation

$$D_{*0}^n y(x) = f(x)y(x) + g(x) \tag{7.9a}$$

subject to the initial conditions

$$y^{(k)}(0) = y_0^{(k)} \qquad (k = 0, 1, \dots, \lceil n \rceil - 1). \tag{7.9b}$$

We begin with a very simple theorem.

Theorem 7.9. *Let* $n > 0$, $m = \lceil n \rceil$ *and* $f, g \in C[0,h]$ *with some* $h > 0$. *Then, the initial value problem (7.9) has a unique solution* $y \in C^{m-1}[0,h]$.

Proof. The fact that the initial value problem has a unique continuous solution on the complete interval $[0,h]$ follows from the basic existence and uniqueness result given in Theorem 6.8. The differentiability property is a consequence of Theorem 6.25. $\square$

A theoretical method that sometimes provides useful information on the solution of the initial value problem (7.9) is based on a concept whose foundations are similar to a concept used in the proof of Theorem 6.8. Specifically we rewrite the initial value problem in the Volterra form and rearrange some of the terms, thus obtaining

$$y(x) = \sum_{k=0}^{m-1} y_0^{(k)} \frac{x^k}{k!} + \frac{1}{\Gamma(n)} \int_0^x (x-t)^{n-1} (f(t)y(t) + g(t)) \, dt$$

$$= \sum_{k=0}^{m-1} y_0^{(k)} \frac{x^k}{k!} + J_0^n g(x) + \frac{1}{\Gamma(n)} \int_0^x (x-t)^{n-1} f(t) y(t) \, dt. \qquad (7.10)$$

Here we now define

$$T^*(x) := \sum_{k=0}^{m-1} y_0^{(k)} \frac{x^k}{k!} + J_0^n g(x) \qquad (7.11)$$

and see that T^* is continuous on $[0, h]$ if g is continuous in the same interval. Introducing the notation

$$k(x,t) := \frac{1}{\Gamma(n)} (x-t)^{n-1} f(t),$$

we then write

$$\phi_0(x) := T^*(x)$$

and

$$\phi_j(x) := J_0^n (f \cdot \phi_{j-1})(x) = \int_0^x k(x,t) \phi_{j-1}(t) \, dt \quad (j = 1, 2, \ldots).$$

Finally we define the *jth iterated kernel* k_j for $j = 1, 2, \ldots$ via the recurrence relation

$$k_1(x,t) := k(x,t) \quad \text{and} \quad k_j(x,t) := \int_t^x k(x,\tau) k_{j-1}(\tau,t) \, d\tau \quad (j = 2, 3, \ldots).$$

Armed with these tools we can then prove an explicit representation for the solution of (7.9):

Theorem 7.10. *Let $f, g \in C[0, h]$ and $n > 0$. Then, the unique solution y of the initial value problem (7.9) in the interval $[0, h]$ can be expressed in the form*

$$y(x) = T^*(x) + \int_0^x R(x,t) T^*(t) \, dt$$

where T^* is defined in (7.11) and the function R is given by

$$R(x,t) = \sum_{j=1}^{\infty} k_j(x,t).$$

In particular, this series is uniformly convergent.

Definition 7.1. The function R introduced in Theorem 7.10 is called the *resolvent kernel* of the Volterra equation (7.10).

For the proof we need some auxiliary statements.

Lemma 7.11. *Let $f, g \in C[0,h]$ and $n > 0$. For $j = 1, 2, \ldots$, we have:*

(a) k_j is continuous on $\{(x,t) : 0 \le t < x \le h\}$.

(b)

$$|k_j(x,t)| \le \frac{1}{\Gamma(jn)} \|f\|_\infty^j (x-t)^{jn-1},$$

and k_j is continuous on the closed triangle $\{(x,t) : 0 \le t \le x \le h\}$ for $j \ge 1/n$.

(c)

$$\phi_j(x) = \int_0^x k_j(x,t) T^*(t) \, dt.$$

Proof. Part (a) follows directly from the definition of the k_j.

The proof of (b) is based on the induction principle. The induction basis ($j = 1$) is an immediate consequence of the definition of k_1. For the induction step ($j - 1 \mapsto j$) we have, in view of the definition of k and the induction hypothesis,

$$
\begin{aligned}
|k_j(x,t)| &= \left| \int_t^x k(x,\tau) k_{j-1}(\tau,t) \, d\tau \right| \\
&\le \|f\|_\infty \frac{1}{\Gamma(n)} \frac{\|f\|_\infty^{j-1}}{\Gamma((j-1)n)} \int_t^x (x-\tau)^{n-1} (\tau-t)^{(j-1)n-1} \, d\tau \\
&\le \frac{1}{\Gamma(n)} \frac{\|f\|_\infty^j}{\Gamma((j-1)n)} \int_t^x (x-\tau)^{n-1} (\tau-t)^{(j-1)n-1} \, d\tau \\
&= \frac{1}{\Gamma(n)} \frac{\|f\|_\infty^j}{\Gamma((j-1)n)} (x-t)^{jn} \frac{\Gamma(n)\Gamma((j-1)n)}{\Gamma(jn)}
\end{aligned}
$$

which proves the desired inequality. The continuity is a direct consequence of this.

For statement (c), we also proceed by mathematical induction. The induction basis ($j = 1$) is an immediate consequence of the definitions of T^*, ϕ_1 and k_1. For the induction step ($j - 1 \mapsto j$), we use the definition of ϕ_j and the induction hypothesis and find

$$\phi_j(x) = \int_0^x k(x,t) \phi_{j-1}(t) \, dt = \int_0^x k(x,t) \int_0^t k_{j-1}(t,\tau) T^*(\tau) \, d\tau \, dt.$$

In view of statements (a) and (b), the integrability of all the functions involved here is not a problem, and in particular we may interchange the order of integration. This yields

$$
\begin{aligned}
\phi_j(x) &= \int_0^x \int_\tau^x k(x,t) k_{j-1}(t,\tau) T^*(\tau) \, dt \, d\tau \\
&= \int_0^x T^*(\tau) \int_\tau^x k(x,t) k_{j-1}(t,\tau) \, dt \, d\tau = \int_0^x T^*(\tau) k_j(x,\tau) \, d\tau
\end{aligned}
$$

as required. $\qquad\square$

Proof (of Theorem 7.10). We first show the uniform convergence of the series that defines the resolvent kernel. In this context the first $\lceil 1/n \rceil$ summands do not play a role; it is sufficient to look at the series $\sum_{j=\lceil 1/n \rceil+1}^{\infty} k_j(x,t)$. For this series, Lemma 7.11 (b) provides the majorant

$$\sum_{j=\lceil 1/n \rceil+1}^{\infty} \frac{\|f\|_\infty^j}{\Gamma(jn)}(x-t)^{jn-1} \leq \sum_{j=\lceil 1/n \rceil+1}^{\infty} \frac{\|f\|_\infty^j}{\Gamma(jn)}h^{jn-1} = \frac{1}{h}\sum_{j=\lceil 1/n \rceil+1}^{\infty} \frac{(\|f\|_\infty h^n)^j}{\Gamma(jn)}$$

which, using the root test, is easily seen to converge. This implies the uniform convergence. We stress at this point that this uniform convergence result does not imply the continuity of R since the first few summands of the series may be unbounded and hence discontinuous. Nevertheless, in view of Lemma 7.11 (a), R is at least continuous on the set $\{(x,t) : 0 \leq t < x \leq h\}$. Moreover, we may combine this uniform convergence result with the statement of Lemma 7.11 (c) to conclude that the series $\sum_{j=1}^{\infty} \phi_j$ is uniformly convergent too.

To show that the solution y can be expressed in the indicated manner, we note that (7.10) and (7.11) reveal that

$$y(x) = T^*(x) + \int_0^x k(x,t)y(t)\,dt.$$

On the other hand,

$$\int_0^x R(x,t)T^*(t)\,dt = \sum_{j=1}^{\infty} \int_0^x k_j(x,t)T^*(t)\,dt = \sum_{j=1}^{\infty} \phi_j(x)$$

$$= \sum_{j=0}^{\infty} \int_0^x k(x,t)\phi_j(t)\,dt = \int_0^x k(x,t)\sum_{j=0}^{\infty} \phi_j(t)\,dt$$

in view of the definition of R, Lemma 7.11 (c) and the definition of ϕ_j. Thus, in order to complete the proof it suffices to show that

$$\sum_{j=0}^{\infty} \phi_j = y.$$

We had already noted that the series on the left-hand side of this equation is uniformly convergent. Since all its summands are continuous on $[0,h]$, the limit function is continuous too. Our proof of the identity is based on our knowledge that y is the unique solution of the integral equation

$$y(x) = \sum_{k=0}^{m-1} y_0^{(k)}\frac{x^k}{k!} + \frac{1}{\Gamma(n)}\int_0^x (x-t)^{n-1}(f(t)y(t)+g(t))\,dt.$$

If we can show that the series $\sum_{j=0}^{\infty} \phi_j$ solves this integral equation too, we will have accomplished our goal. To this end, we use the definitions of T^*, k, ϕ_0 and ϕ_j, $j = 1, 2, \ldots$, and note that the uniform convergence allows us to interchange summation and integration to write

$$\sum_{j=0}^{\infty} \phi_j(x) = T^*(x) + \frac{1}{\Gamma(n)} \int_0^x (x-t)^{n-1} f(t) \sum_{j=1}^{\infty} \phi_{j-1}(t) \, dt$$

$$= \sum_{k=0}^{m-1} y_0^{(k)} \frac{x^k}{k!} + J_0^n g(x) + \frac{1}{\Gamma(n)} \int_0^x (x-t)^{n-1} f(t) \sum_{j=0}^{\infty} \phi_j(t) \, dt$$

$$= \sum_{k=0}^{m-1} y_0^{(k)} \frac{x^k}{k!} + \frac{1}{\Gamma(n)} \int_0^x (x-t)^{n-1} \left(f(t) \sum_{j=0}^{\infty} \phi_j(t) + g(t) \right) dt. \qquad \square$$

Occasionally it is of interest to augment the explicit series representation of the resolvent kernel above by the following result stating that the resolvent kernel itself is the solution of an integral equation which is closely related to our original one.

Theorem 7.12. *Under the assumptions of Theorem 7.10, the resolvent kernel R satisfies the* resolvent equation

$$R(x,t) = k(x,t) + \int_t^x k(x,\tau) R(\tau,t) \, d\tau$$

for $0 \leq t < x \leq h$.

Proof. From the series representation of Theorem 7.10 and the fact that the uniform convergence permits us to interchange summation and integration, we deduce

$$\int_t^x k(x,\tau) R(\tau,t) \, d\tau = \int_t^x k(x,\tau) \sum_{j=1}^{\infty} k_j(\tau,t) \, d\tau = \sum_{j=1}^{\infty} \int_t^x k(x,\tau) k_j(\tau,t) \, d\tau$$

$$= \sum_{j=1}^{\infty} k_{j+1}(x,t) = R(x,t) - k(x,t). \qquad \square$$

In Theorem 7.2 and Remark 7.1 we had developed a number of representations for the solution of initial value problems for linear equations with constant coefficients. In the multi-dimensional case, i.e. in the case where the unknown solution y is a mapping to $\mathbb{R}^N$ for some N and λ is an $N \times N$ matrix (from now on denoted by Λ), these representations require the evaluation of Mittag-Leffler functions of matrix-valued arguments. While this is no problem in theory since the relevant series are known to be convergent, it is a most cumbersome method in practice. We thus now present an alternative approach that is usually much simpler to handle. Our path follows the lines laid out by Odibat [144] and Bonilla et al. [18]. In order to retain a close relation to the classical technique for first-order equations [30, §4.6], we restrict our attention to differential equations of order $n \in (0, 1)$.

Let us thus consider the fractional differential equation

$$D_{*0}^n y(x) = \Lambda y(x) + q(x), \tag{7.12}$$

with $0 < n < 1$, an $N \times N$ matrix Λ, a given function $q : [0,h] \to \mathbb{C}^N$ and an unknown solution $y : [0,h] \to \mathbb{C}^N$. We shall first discuss the construction of the general solution for this equation; the initial condition that then leads to a special solution will be added later.

As usual we start with the homogeneous problem corresponding to (7.12), i.e. with the case $q \equiv 0$. In the classical situation $n = 1$ we know that we can use an approach of the form $y(x) = u\exp(\Lambda x)$ with a suitable vector u. Since we have already found that the Mittag-Leffler function $E_n(\lambda x^n)$ takes the role of $\exp(\lambda x)$ in the one-dimensional case, it is natural to seek a solution that is a linear combination of expressions of the form

$$y(x) = uE_n(\lambda x^n) \tag{7.13}$$

with suitable vectors $u \in \mathbb{C}^N$ and scalars $\lambda \in \mathbb{C}$ that need to be determined. Inserting (7.13) into the given homogeneous differential equation

$$D_{*0}^n y(x) = \Lambda y(x), \tag{7.14}$$

we obtain, in view of Theorem 4.3,

$$u\lambda E_n(\lambda x^n) = \Lambda u E_n(\lambda x^n).$$

Since $E_n(\lambda x^n) \neq 0$ (this follows from Theorem 7.5 for $\lambda < 0$ and directly from the power series representation of E_n for $\lambda \geq 0$; for complex λ it can be proved too), this implies

$$\lambda u = \Lambda u,$$

i.e. λ must be an eigenvalue of the matrix Λ, and u must be a corresponding eigenvector. Now, if all k-fold eigenvalues of Λ have k eigenvectors, then we know that the set of all these eigenvectors is linearly independent and hence it forms a basis of $\mathbb{C}^N$. We have thus shown:

Theorem 7.13. *Let $\lambda_1, \ldots, \lambda_N$ be the eigenvalues of Λ and $u^{(1)}, \ldots, u^{(N)}$ be the corresponding eigenvectors. Then, the general solution of the differential equation (7.14) has the form*

$$y(x) = \sum_{\ell=1}^N c_\ell u^{(\ell)} E_n(\lambda_\ell x^n)$$

with certain constants $c_\ell \in \mathbb{C}$. The unique solution of this differential equation subject to the initial condition $y(0) = y_0$ is characterized by the linear system

$$y_0 = (u^{(1)}, \ldots, u^{(N)})(c_1, \ldots, c_N)^{\mathrm{T}}.$$

The linear system for the computation of the coefficients c_ℓ of the solution of the initial value problem is obtained by setting $x = 0$ in the general solution. The fact that the linear system has a unique solution follows because of the above mentioned linear independence of the eigenvectors that form the coefficient matrix of this system.

Example 7.2. The eigenvalues of

$$\Lambda = \begin{pmatrix} 2 & -1 \\ 4 & -3 \end{pmatrix}$$

are 1 (with eigenvector $(1,1)^{\mathrm{T}}$) and -2 (with eigenvector $(1,4)^{\mathrm{T}}$); thus the general solution for the differential equation $D_{*0}^n y(x) = \Lambda y(x)$ with $0 < n < 1$ is

$$y(x) = c_1 \begin{pmatrix} 1 \\ 1 \end{pmatrix} E_n(x^n) + c_2 \begin{pmatrix} 1 \\ 4 \end{pmatrix} E_n(-2x^n)$$

with arbitrary constants c_1 and c_2. The unique solution that satisfies the initial condition $y(0) = (4, -3)^{\mathrm{T}}$ is obtained by choosing $c_1 = -7/3$ and $c_2 = 19/3$.

Of course, it is well known that there exist matrices that do not have a full set of eigenvectors, i.e. matrices that have a k-fold eigenvalue with some $k > 1$ to which less than k eigenvectors are available. In this case, the theory of Theorem 7.13 is not applicable, and we must resort to a different representation of the general solution.

Specifically we know from elementary Linear Algebra that, given any square (real or complex) matrix Λ, there exists a nonsingular matrix B such that $\Phi = B^{-1}\Lambda B$ has the so-called Jordan form

$$\Phi = \begin{pmatrix} \Phi_1 & 0 & \cdots & 0 \\ 0 & \Phi_2 & \ddots & 0 \\ \vdots & & \ddots & 0 \\ 0 & \cdots & 0 & \Phi_k \end{pmatrix} \tag{7.15}$$

where the Jordan boxes are square matrices of the form

$$\Phi_\mu = \begin{pmatrix} \lambda_\mu & 1 & 0 & 0 & \cdots & 0 \\ 0 & \lambda_\mu & 1 & 0 & \cdots & 0 \\ 0 & 0 & \lambda_\mu & 1 & \ddots & 0 \\ \vdots & & \ddots & \ddots & \ddots & \vdots \\ 0 & & \cdots & 0 & \lambda_\mu & 1 \\ 0 & & \cdots & & 0 & \lambda_\mu \end{pmatrix} \tag{7.16}$$

with r_μ rows and columns. In particular we have $r_1 + r_2 + \ldots + r_k = N$, and the eigenvalues of Λ can be found in the main diagonal of Φ in their respective multiplicities. We note that all diagonal entries of each Jordan block Φ_k coincide with each other. Different Jordan blocks may have the same diagonal entries though. (For example, in the Jordan representation of the N-dimensional unit matrix we have $k = N, r_\mu = 1$

for all μ and $\lambda_\mu = 1$ for all μ.) In view of the simple structure of a Jordan block, the system

$$D_{*0}^n z(x) = \Phi_k z(x), \quad \text{i.e.} \quad \begin{cases} D_{*0}^n z_1(x) & = \lambda_k z_1(x) + z_2(x), \\ \vdots & \vdots \\ D_{*0}^n z_{r_k-1}(x) & = \lambda_k z_{r_k-1}(x) + z_{r_k}(x), \\ D_{*0}^n z_{r_k}(x) & = \lambda_k z_{r_k}(x) \end{cases}$$

is easily solved by backward substitution and an application of Theorem 7.2 for each component of z. Evaluating the integrals arising in Theorem 7.2 explicitly, we find that the columns of the matrix $Z = (z_{ij})_{i,j=1}^{r_k}$ with

$$z_{ij}(x) = \begin{cases} 0 & \text{if } i > j, \\ x^{(j-i)n} D^{j-i} E_n(\lambda_k x^n) & \text{else} \end{cases}$$

form r_k linearly independent solutions of $D_{*0}^n z(x) = \Phi_k z(x)$. Observing that

$$E_n(\Phi x^n) = \begin{pmatrix} E_n(\Phi_1 x^n) & 0 & \cdots & 0 \\ 0 & E_n(\Phi_2 x^n) & \ddots & 0 \\ \vdots & & \ddots & 0 \\ 0 & \cdots & 0 & E_n(\Phi_k x^n) \end{pmatrix},$$

it follows that we may combine the results for the individual Jordan blocks into one system of linearly independent solutions (and hence a basis for the solution space) of the full system $D_{*0}^n z(x) = \Phi z(x)$ in a straightforward way. Let us illustrate this procedure by the example

$$\Phi = \begin{pmatrix} \lambda_1 & 1 & 0 & 0 & 0 & 0 \\ 0 & \lambda_1 & 1 & 0 & 0 & 0 \\ 0 & 0 & \lambda_1 & 0 & 0 & 0 \\ 0 & 0 & 0 & \lambda_2 & 0 & 0 \\ 0 & 0 & 0 & 0 & \lambda_3 & 1 \\ 0 & 0 & 0 & 0 & 0 & \lambda_3 \end{pmatrix}.$$

Here we have $k = 3$, $r_1 = 3$, $r_2 = 1$ and $r_3 = 2$. This gives rise to the solutions being the columns of the matrix

$$\begin{pmatrix} E_n(\lambda_1 x^n) & x^n E_n'(\lambda_1 x^n) & x^{2n} E_n''(\lambda_1 x^n) & 0 & 0 & 0 \\ 0 & E_n(\lambda_1 x^n) & x^n E_n'(\lambda_1 x^n) & 0 & 0 & 0 \\ 0 & 0 & E_n(\lambda_1 x^n) & 0 & 0 & 0 \\ 0 & 0 & 0 & E_n(\lambda_2 x^n) & 0 & 0 \\ 0 & 0 & 0 & 0 & E_n(\lambda_3 x^n) & x^n E_n'(\lambda_3 x^n) \\ 0 & 0 & 0 & 0 & 0 & E_n(\lambda_3 x^n) \end{pmatrix}.$$

But now, in view of the relation $\Phi = B^{-1} \Lambda B$, the solution y of the system (7.14) is related to the solution z created above in a very simple way: We have $\Lambda = B \Phi B^{-1}$

and hence (7.14) becomes $D_{*0}^n y = B \Phi B^{-1} y$, i.e. $D_{*0}^n B^{-1} y = \Phi B^{-1} y$. Introducing the substitution $z = B^{-1} y$ or, equivalently, $y = Bz$, we arrive at the equation $D_{*0}^n z(x) = \Phi z(x)$ that we have just solved. Thus, we obtain the desired solution y of (7.14) in the form

$$y(x) = Bz(x)$$

where $z(x)$ has the indicated form. We have thus proved the following result.

Theorem 7.14. *For each k-fold eigenvalue λ of the matrix Λ we have k linearly independent solutions of the homogeneous linear differential equation (7.14) that can be represented in the form*

$$y_\ell(x) = \pi^{(\ell)}(x), \qquad \ell = 1, 2, \ldots, k,$$

where the $\pi^{(\ell)}(x)$ are N-dimensional vectors whose component functions $\pi_j^{(\ell)}$, $j = 1, 2, \ldots, N$, are of the form

$$\pi_j^{(\ell)}(x) = \sum_{\mu=0}^{\ell-1} c_j^{(\ell,\mu)} x^{\mu n} D^\mu E_n(\lambda x^n). \tag{7.17}$$

The combination of these solutions for all eigenvalues leads to N linearly independent solutions of the system (7.14), i.e. to a basis of the space of all solutions of this system.

Note that $\pi^{(1)}(x) = c^{(1,0)} E_n(\lambda x^n)$, and so $c^{(1,0)}$ must be an eigenvector of Λ associated to the eigenvalue λ.

Remark 7.2. In the limit case $n \to 1$, Theorem 7.14 reduces to the well known classical observation that the k solutions of $Dy = \Lambda y$ that correspond to the k-fold eigenvalue λ of the matrix Λ have the form $y(x) = \pi_{k-1}(x) \exp(\lambda x)$ where π_{k-1} is a vector-valued polynomial of degree at most $k - 1$ with suitably chosen coefficients.

Example 7.3. The matrix

$$\Lambda = \begin{pmatrix} 1 & 1 & 1 \\ 2 & 1 & -1 \\ 0 & -1 & 1 \end{pmatrix}$$

has the single eigenvalue -1 with eigenvector $(-3, 4, 2)^T$ and the double eigenvalue 2 with eigenvector $(0, 1, -1)^T$ but no second eigenvector. Thus the general solution for the differential equation $D_{*0}^n y(x) = \Lambda y(x)$ with $0 < n < 1$ takes the form

$$y(x) = c_1 \begin{pmatrix} -3 \\ 4 \\ 2 \end{pmatrix} E_n(-x^n) + c_2 \begin{pmatrix} 0 \\ 1 \\ -1 \end{pmatrix} E_n(2x^n)$$

$$+ c_3 \left[c^{(2,0)} E_n(2x^n) + c^{(2,1)} x^n E_n'(2x^n) \right]$$

with arbitrary constants c_1, c_2 and c_3 and suitably chosen vectors $c^{(2,0)}$ and $c^{(2,1)}$. By inserting this representation into the differential equation, rewriting the Mittag-Leffler functions in power series form and comparing the coefficients of the resulting power series on both sides of the resulting equation, we find that $c^{(2,0)} = (1,0,1)^{\mathrm{T}}$ and $c^{(2,1)} = (0,1,-1)^{\mathrm{T}}$.

Remark 7.3. We already know that the vector-valued coefficients of the functions $E_n(\lambda x^n)$ in the solution representation are just the eigenvectors of Λ with respect to the eigenvalue λ. Obviously, these vectors are independent of n. The example above indicates that the coefficients $c^{(\ell,\mu)}$ that arise in the representation of Theorem 7.14 do not depend on n either. Thus we can compute these coefficients by looking at the special case $n = 1$ of the problem and use the values obtained in this way for $n \in (0,1)$ too. The advantage of this approach is that the cumbersome calculation via the power series representation of the Mittag-Leffler function and the corresponding comparison of coefficients can be completely avoided. Rather, it is sufficient to invoke the classical theory [30, §4.6] that tells us that the vectors $c^{(\ell,\mu)}$ can be obtained in a simple way as the eigenvectors and suitable multiples of the generalized eigenvectors of Λ.

Theorem 7.14 holds for any matrix Λ with complex coefficients. If the given problem is real, i.e. if the coefficients of Λ are real, then one is typically interested in a real solution, i.e. in a solution with real vectors $y^{(\ell)}$ and real-valued basis functions instead of the complex-valued functions $E_n(\lambda_\ell \cdot)$ and their derivatives. The approach via Theorem 7.2 and Remark 7.1 never leads us out of the real numbers in such a case and thus automatically provides such a real solution, but it involves the cumbersome computation of Mittag-Leffler functions of matrix arguments. The method of Theorems 7.13 and 7.14 avoids this complication but it does not always produce a real solution directly since the eigenvalues and eigenvectors of a real matrix need not be real. However, we may construct a real solution from the solution provided by Theorem 7.13 in the usual way that is commonly employed for first-order equations [30, pp. 80–81], namely by considering the real and imaginary parts of the complex solutions, respectively:

Theorem 7.15. *If Λ is a real matrix and the set of solutions of the system (7.14) constructed in Theorem 7.14 contains complex functions, then a purely real basis of the set of solutions can be obtained by the following manipulation: For each k-fold nonreal eigenvalue $\lambda = a + \mathrm{i}b$, remove the solution vectors $\pi^{(\ell)}$, $\ell = 1,\ldots,k$, associated to this eigenvalue and those corresponding to the eigenvalue $\bar{\lambda} = a - \mathrm{i}b$ from the set of solutions constructed in Theorem 7.13, and insert the vectors $\widehat{\pi}^{(\ell)}$ and $\widetilde{\pi}^{(\ell)}$ ($\ell = 1,\ldots,k$) instead, where*

$$\widehat{\pi}_j^{(\ell)}(x) = \sum_{\mu=0}^{\ell-1} \frac{1}{2} \mathrm{Re}\, c_j^{(\ell,\mu)} x^{\mu n} (D^\mu E_n((a+\mathrm{i}b)x^n) + D^\mu E_n((a-\mathrm{i}b)x^n))$$

$$- \sum_{\mu=0}^{\ell-1} \frac{1}{2\mathrm{i}} \mathrm{Im}\, c_j^{(\ell,\mu)} x^{\mu n} (D^\mu E_n((a+\mathrm{i}b)x^n) - D^\mu E_n((a-\mathrm{i}b)x^n))$$

and

$$\tilde{\pi}_j^{(\ell)}(x) = \sum_{\mu=0}^{\ell-1} \frac{1}{2} \operatorname{Im} c_j^{(\ell,\mu)} x^{\mu n} (D^\mu E_n((a+ib)x^n) + D^\mu E_n((a-ib)x^n))$$

$$+ \sum_{\mu=0}^{\ell-1} \frac{1}{2i} \operatorname{Re} c_j^{(\ell,\mu)} x^{\mu n} (D^\mu E_n((a+ib)x^n) - D^\mu E_n((a-ib)x^n)).$$

We only illustrate Theorem 7.15 by means of an example and leave the proof as an exercise. Here we merely note that the power series representation of the Mittag-Leffler functions implies that expressions of the form $D^\mu E_n(z) + D^\mu E_n(\bar{z})$ and $(D^\mu E_n(z) - D^\mu E_n(\bar{z}))/i$ that arise in this theorem are indeed real.

Example 7.4. The eigenvalues of the matrix

$$\Lambda = \begin{pmatrix} -1 & 0 & 0 \\ 2 & 1 & -9 \\ 3 & 6 & 1 \end{pmatrix}$$

are -1 and $1 \pm \sqrt{54}i$ with eigenvectors $(58, -31, 6)^T$ and $(0, \pm\sqrt{6}i, 2)^T$, respectively. Thus, Theorem 7.13 is applicable, and we derive that the general complex solution for the differential equation $D_{*0}^n y(x) = \Lambda y(x)$ with $0 < n < 1$ is

$$y(x) = c_1 \begin{pmatrix} 58 \\ -31 \\ 6 \end{pmatrix} E_n(-x^n) + c_2 \begin{pmatrix} 0 \\ \sqrt{6}i \\ 2 \end{pmatrix} F_n((1+\sqrt{54}i)x^n)$$

$$+ c_3 \begin{pmatrix} 0 \\ -\sqrt{6}i \\ 2 \end{pmatrix} E_n((1-\sqrt{54}i)x^n)$$

with arbitrary $c_1, c_2, c_3 \in \mathbb{C}$. A solution using real-valued functions can be given as

$$y(x) = c_1 \begin{pmatrix} 58 \\ -31 \\ 6 \end{pmatrix} E_n(-x^n)$$

$$+ \frac{1}{2} \left[c_2 \begin{pmatrix} 0 \\ 0 \\ 2 \end{pmatrix} + c_3 \begin{pmatrix} 0 \\ \sqrt{6} \\ 0 \end{pmatrix} \right] \left[E_n((1+\sqrt{54}i)x^n) + E_n((1-\sqrt{54}i)x^n) \right]$$

$$+ \frac{1}{2i} \left[c_3 \begin{pmatrix} 0 \\ 0 \\ 2 \end{pmatrix} - c_2 \begin{pmatrix} 0 \\ \sqrt{6} \\ 0 \end{pmatrix} \right] \left[E_n((1+\sqrt{54}i)x^n) - E_n((1-\sqrt{54}i)x^n) \right]$$

with arbitrary $c_1, c_2, c_3 \in \mathbb{R}$.

We have thus completed the fundamental theory of homogeneous linear systems of fractional differential equations with constant coefficients. In particular we have found methods to compute all their solutions. Hence we are in a position to handle the corresponding inhomogeneous problem (7.12) in a very simple way by invoking the variation-by-constants approach described in Theorem 7.2 and Remark 7.1.

7.2 Boundary Value Problems for Linear Equations

Let us briefly leave the area of initial value problems and turn our attention towards boundary value problems for a short time. We keep, however, our restriction on the class of differential equations under consideration to the linear equations.

Specifically, we consider the problem

$$D_{*0}^n y(x) = f(x)y(x) + g(x), \tag{7.18a}$$

$$U_j[y] = c_j \qquad (j = 1, 2, \ldots, \sigma) \tag{7.18b}$$

where
$$U_j[y] := a_{j0}y(0) + a_{j1}y'(0) + b_{j0}y(T) + b_{j1}y'(T) \tag{7.18c}$$

with some $\sigma \in \mathbb{N}$, $n > 0$, given functions $f, g \in C[0, T]$ and given parameters a_{jk}, b_{jk} and c_j (see also (6.51)). Evidently we shall be looking for a solution on the interval $[0, T]$. It is clear that the set of solutions of the corresponding homogeneous problem, i.e. the special case of (7.18) where $g \equiv 0$ and $c_j = 0$ for all j, is a linear space that we shall denote by D_{hom}. Our first fundamental result then looks as follows. Here and in the following, we write $m = \lceil n \rceil$ as usual.

Theorem 7.16. *Let* $y_1, \ldots, y_m$ *be linearly independent solutions of the homogeneous differential equation associated to (7.18a), and let r be the rank of the matrix*

$$M_{\text{hom}} := \begin{pmatrix} U_1[y_1] & \cdots & U_1[y_m] \\ \vdots & & \vdots \\ U_\sigma[y_1] & \cdots & U_\sigma[y_m] \end{pmatrix}.$$

Then, the solution space D_{hom} *of the homogeneous boundary value problem has the property*
$$\dim D_{\text{hom}} = m - r.$$

Proof. The general solution y of the homogeneous differential equation has the form $y = \sum_{k=1}^m \alpha_k y_k$ ($\alpha_k \in \mathbb{R}$). The homogeneous boundary conditions lead to the system

$$M_{\text{hom}} \begin{pmatrix} \alpha_1 \\ \vdots \\ \alpha_m \end{pmatrix} = 0. \tag{7.19}$$

By assumption, the rank of M_{hom} is r, and hence the system has $m - r$ linearly independent solution vectors $\alpha^{(j)} = (\alpha_1^{(j)}, \ldots, \alpha_m^{(j)})^\mathsf{T}$, $j = 1, \ldots, m - r$. Now consider the functions $\tilde{y}_j = \sum_{k=1}^m \alpha_k^{(j)} y_k$, $j = 1, \ldots, m - r$. These are $m - r$ solutions of the homogeneous boundary value problem. To see that they are linearly independent, we proceed as follows. Assume that

$$0 = \sum_{j=1}^{m-r} \beta_j \tilde{y}_j = \sum_{k=1}^{m} \left(\sum_{j=1}^{m-r} \alpha_k^{(j)} \beta_j \right) y_k.$$

The linear independence of the y_k implies that all the coefficients in the parentheses must vanish. We can combine these m scalar conditions into one condition in vector notation,

$$\sum_{j=1}^{m-r} \alpha^{(j)} \beta_j = 0.$$

But we already know that the $\alpha^{(j)}$ are linearly independent, and thus we conclude $\beta_j = 0$ for all j. This completes the proof of the linear independence of the functions $\tilde{y}_j$, $j = 1, \ldots, m - r$.

It remains to show that every solution y of the homogeneous boundary value problem can be expressed as a linear combination of the functions $\tilde{y}_1, \ldots, \tilde{y}_{m-r}$. To this end, let y be such a solution of the homogeneous boundary value problem. Since it must in particular be a solution of the homogeneous differential equation, it must have a representation in the form $y = \sum_{k=1}^m \alpha_k y_k$, and since it also satisfies the boundary conditions, the coefficients α_k have the property (7.19), and we had seen above that the vector $\alpha = (\alpha_1, \ldots, \alpha_m)^\mathsf{T}$ can be written as a linear combination of the $\alpha^{(j)}$, $j = 1, 2, \ldots, m - r$, with coefficients β_j, say. But this implies

$$y(x) = \sum_{k=1}^{m} \alpha_k y_k = \sum_{k=1}^{m} \sum_{j=1}^{m-r} \alpha_k^{(j)} \beta_j y_k = \sum_{j=1}^{m-r} \beta_j \sum_{k=1}^{m} \alpha_k^{(j)} y_k = \sum_{j=1}^{m-r} \beta_j \tilde{y}_j$$

which is the required property. $\qquad \square$

For the inhomogeneous boundary value problem we can state the following result.

Theorem 7.17. *The general solution of the boundary value problem (7.18) has the form $y = y_{\text{hom}} + y_{\text{inhom}}$ where y_{hom} is the general solution of the associated homogeneous problem and y_{inhom} is a particular solution of the inhomogeneous problem.*

Proof. In view of the linearity of the problem, this statement can be proved by the usual elementary methods from Linear Algebra. $\qquad \square$

Note that we do not claim that a solution of the inhomogeneous problem exists. It only describes the structure of the set of solutions if a particular solution exists. The next theorem provides a criterion that allows us to determine whether this is the case or not.

Theorem 7.18. *Let* $y_1, \ldots, y_m$ *be linearly independent solutions of the homogeneous differential equation associated to (7.18a), and let* $\tilde{y}$ *be a particular solution of the inhomogeneous differential equation (7.18a) itself. Moreover denote*

$$M_{\text{inhom}} := \begin{pmatrix} U_1[y_1] & \cdots & U_1[y_m] & c_1 - U_1[\tilde{y}] \\ \vdots & & \vdots & \vdots \\ U_\sigma[y_1] & \cdots & U_\sigma[y_m] & c_\sigma - U_\sigma[\tilde{y}] \end{pmatrix}.$$

Then, the inhomogeneous boundary value problem (7.18) has a solution if and only if the matrix M_{inhom} *has the same rank as the matrix* M_{hom} *introduced in Theorem 7.16.*

Proof. Any solution y of the inhomogeneous differential equation (7.18a) can be expressed in the form $y = \tilde{y} + \sum_{k=1}^{m} \alpha_k y_k$. A solution to the inhomogeneous boundary value problem (7.18) thus exists if and only if we can find constants $\alpha_1, \ldots, \alpha_m$ such that

$$c_j = U_j[y] = U_j[\tilde{y}] + \sum_{k=1}^{m} \alpha_k U_j[y_k] \qquad (j = 1, 2, \ldots, \sigma).$$

Clearly, this is the case if and only if

$$M_{\text{hom}} \begin{pmatrix} \alpha_1 \\ \vdots \\ \alpha_m \end{pmatrix} = \begin{pmatrix} c_1 - U_1[\tilde{y}] \\ \vdots \\ c_\sigma - U_\sigma[\tilde{y}] \end{pmatrix}.$$

But it is well known from Linear Algebra that this is equivalent to the requirement that the coefficient matrix M_{hom} of this system have the same rank as the extended matrix obtained by adding the column vector on the right-hand side of the equation to the matrix, i.e. the matrix M_{inhom}. □

Obviously, the case that the matrix M_{hom} is a square matrix, i.e. the case $\sigma = m$, is particularly important. In this situation we can give the following information.

Theorem 7.19. *Consider the boundary value problem (7.18) subject to the condition* $\sigma = m$. *Let* M_{hom} *be defined as in Theorem 7.16.*

(a) *If* $\det M_{\text{hom}} \neq 0$ *then the homogeneous boundary value problem associated to (7.18) only has the trivial solution* $y_{\text{hom}} \equiv 0$, *and the inhomogeneous problem (7.18) itself has a unique solution*

(b) *If* $\det M_{\text{hom}} = 0$ *then the homogeneous boundary value problem associated to (7.18) has nontrivial solutions*

Proof. Using elementary Linear Algebra, this immediately follows from the preceding theorem. □

7.3 Stability of Fractional Differential Equations

We shall now leave the class of boundary value problems and return to focus our attention on initial value problems. Having gained some insight into the behaviour of linear equations, we now consider the general case and turn our attention towards the question for the stability of a given fractional differential equation (or, more frequently, for a system of fractional differential equations). In the classical case of integer-order equations, this is a well known and important area of research. Various notions of stability are commonly discussed; we refer to [30, Chapter 5] and the references therein for a nicely readable introduction. As indicated in this reference, stability issues are usually investigated for first-order equations, i.e. for equations requiring only one initial condition to guarantee the uniqueness of the solution. We restrict our attention here to a class of problems that is as close to this case as possible. Thus our object of study in this section is the differential equation

$$D_{*0}^n y(x) = f(x, y(x)) \qquad \text{with } n \in (0,1). \tag{7.20}$$

Here y may be a function mapping to $\mathbb{R}^N$ for an arbitrary $N \in \mathbb{N}$; of course f must then be defined on a suitable subset of $\mathbb{R}^{N+1}$. In the classical case $n = 1$ one would allow arbitrary initial points [30, Chapter 5]; Theorem 6.17 allows us to do the same in the fractional case.

When talking about stability, one is interested in the behaviour of the solutions of (7.20) for $x \to \infty$. Therefore we will only consider problems whose solutions y exist on $[0, \infty)$. Moreover, a few additional assumptions are required that we will impose throughout this section. The first of these assumptions is that f is defined on a set $G := [0, \infty) \times \{w \in \mathbb{R}^N : \|w\| < W\}$ with some $0 < W \le \infty$. The norm in this definition of G may be an arbitrary norm on $\mathbb{R}^N$. Our second assumption is that f is continuous on its domain of definition and that it satisfies a Lipschitz condition there. This asserts that the initial value problem consisting of (7.20) and the initial condition $y(0) = y_0$ has a unique solution on the interval $[0, b)$ with some $b \le \infty$ if $\|y_0\| \le W$. And finally we assume that

$$f(x, 0) = 0 \qquad \text{for all } x \ge 0. \tag{7.21}$$

This condition implies that the function $y(x) = 0$ is a solution of (7.20). Under these hypotheses we may formulate our main concepts.

Definition 7.2. (a) The solution $y(x) = 0$ of the differential equation (7.20), subject to the assumptions mentioned above, is called *stable* if, for any $\varepsilon > 0$ there exists some $\delta > 0$ such that the solution of the initial value problem consisting of the differential equation (7.20) and the initial condition $y(0) = y_0$ satisfies $\|y(x)\| < \varepsilon$ for all $x \ge 0$ whenever $\|y_0\| < \delta$.

(b) The solution $y(x) = 0$ of the differential equation (7.20), subject to the assumptions mentioned above, is called *asymptotically stable* if it is stable and there exists some $\gamma > 0$ such that $\lim_{x \to \infty} \|y(x)\| = 0$ whenever $\|y_0\| < \gamma$.

In Sect. 6.3 we had seen that, under the usual continuity and Lipschitz assumptions on f, the solution of a fractional differential equation does not change much over some finite interval if we perturb the initial values by a small magnitude. The essence of the notion of stability is the extension of this idea to unbounded intervals: The trivial solution is stable if a small change in the initial value leads to a small change of the solution over the complete positive half-line. Obviously this is much stronger than the continuous dependence on the given data discussed in Sect. 6.3. Asymptotic stability is even stronger since it requires the solution of the perturbed problem not only to remain close to the original solution but actually to converge to the latter.

Remark 7.4. In Definition 7.2 we have only discussed properties of the identically vanishing solution of a fractional differential equation satisfying the condition of (7.21). We may transfer these concepts and the results below to the behaviour of arbitrary solutions of equations that may or may not satisfy (7.21) by a procedure similar to that used in the proof of Theorem 6.15: A solution y of the differential equation $D_{*0}^n y(x) = g(x,y(x))$ is said to be (asymptotically) stable if and only if the zero solution of $D_{*0}^n z(x) = f(x,z(x))$ with $f(x,z) := g(x,z+y(x)) - g(x,y(x))$ is (asymptotically) stable.

We begin our analysis by looking at a very simple special case, the homogeneous linear differential equation with constant coefficients (see also [132]).

Theorem 7.20. *Consider the N-dimensional fractional differential equation system $D_{*0}^n y(x) = \Lambda y(x)$, where Λ is an arbitrary constant $N \times N$ matrix.*

(a) *The solution $y(x) = 0$ of the system is asymptotically stable if and only if all eigenvalues λ_j $(j = 1, 2, \ldots, N)$ of Λ satisfy $|\arg \lambda_j| > n\pi/2$.*

(b) *The solution $y(x) = 0$ of the system is stable if and only if the eigenvalues satisfy $|\arg \lambda_j| \geq n\pi/2$ and all eigenvalues with $|\arg \lambda_j| = n\pi/2$ have a geometric multiplicity that coincides with their algebraic multiplicity (i.e. an eigenvalue that is an ℓ-fold zero of the characteristic polynomial has ℓ linearly independent eigenvectors).*

Notice that in the limit case $n \to 1$ we recover the well known classical result [30, p. 94] that the eigenvalues must have negative real parts in case (a) and nonpositive real parts and a full set of eigenvectors if the real parts are zero for case (b).

Proof. Theorems 7.13 and 7.14 give us the information about the precise form of all solutions of the differential equation under consideration. The fact that these solutions behave in the required way then follows from Theorems 4.4 and 4.6. □

This result enables us to investigate the stability properties of the problems discussed in Examples 7.2, 7.3 and 7.4.

Example 7.5. (a) The solution $y \equiv 0$ of Example 7.2 is unstable because the coefficient matrix has the eigenvalue $\lambda_1 = 1$ that is real and strictly positive (i.e. it has $\arg \lambda_1 = 0$). This corresponds to the fact that the associated component of the general solution, viz. the function $(1,1)^T E_n(x^n)$, grows without bound as $x \to \infty$.

(b) Similarly, the coefficient matrix of Example 7.3 has a real and positive eigenvalue 2, and so we observe instability in this example too.

(c) The coefficient matrix of Example 7.2 has the three eigenvalues $\lambda_1 = -1$ (with $\arg \lambda_1 = \pi$), $\lambda_2 = 1 + \sqrt{54}i$ (with $\arg \lambda_2 = \arccos(1/\sqrt{55}) \approx 1.4355$) and $\lambda_3 = \bar{\lambda}_2 = 1 - \sqrt{54}i$ (with $\arg \lambda_3 = -\arg \lambda_2$). Thus we have asymptotic stability if and only if $n < 2|\arg \lambda_2|/\pi = 2\arccos(1/\sqrt{55})/\pi \approx 0.9139$ and stability (since all eigenvalues are simple) if and only if $n \leq 2|\arg \lambda_2|/\pi$.

Remark 7.5. From case (c) of this example we conclude that the stability properties of the zero solution of a fractional differential equation may depend on the order n of the equation. Specifically, as is clear from Theorem 7.20, in the case of a homogeneous linear system with constant coefficients we may say that there exists a threshold value n^*, say, such that the system is asymptotically stable if $n < n^*$ and unstable if $n > n^*$. In other words, the stability properties can be improved by reducing the order n of the differential operator. An analogous statement applies to other (nonlinear) types of differential equations. This is a common observation in the theory of fractional dynamical systems [119] where systems tend to exhibit chaotic behaviour if the order of the differential operators is larger than the threshold value n^* and remain stable if the order is less than n^*.

In the non-fractional case, i.e. for $n = 1$, quite deep results are known in the case of homogeneous linear equations with non-constant coefficients. Most of the proofs of these results rely on the fact that explicit expressions for the solutions of the corresponding differential equations are known. In the fractional case, such explicit expressions do not seem to be available. Therefore a transfer of these results remains an open problem.

There are, however, some other methods that one can use to obtain results on the long-term behaviour of solutions of fractional differential equations. A possible approach is indicated in [115, Theorem 4.6]. Transferred to our setting, the result reads as follows.

Theorem 7.21. *Consider the initial value problem*

$$D^n_{*0}y(x) = f(x, y(x)), \qquad y(0) = y_0 > 0,$$

where $0 < n < 1$ and $f : [0, \infty) \times [0, y_0] \to (-\infty, 0]$ is continuous and satisfies a Lipschitz condition with respect to the second variable. Moreover assume that $f(x, 0) = 0$ for all x. Then, the unique solution y of this initial value problem exists on $[0, \infty)$ and satisfies

$$0 < y(x) \leq y_0 \quad \text{for all } x \geq 0.$$

Proof. We rewrite the initial value problem in the corresponding Volterra form,

$$y(x) = y_0 + \frac{1}{\Gamma(n)} \int_0^x (x-t)^{n-1} f(t, y(t)) \, dt,$$

and define

$$f(t,z) := \begin{cases} f(t,y_0) & \text{for } z > y_0, \\ 0 & \text{for } z < 0, \end{cases}$$

thus extending the domain of definition of f to $[0,\infty) \times \mathbb{R}$. Obviously, this extended function f is still continuous and fulfils a Lipschitz condition with respect to its second variable. Thus, by Corollary 6.9, the initial value problem with this extended function f has a unique solution y. We now need to show that this solution actually satisfies the inequality $0 \le y(x) \le y_0$ for all x. This then implies that $(x, y(x))$ is always in the original (non-extended) domain of definition of f from which we conclude that y is also the unique solution of the original initial value problem.

To prove the required inequalities we proceed as follows. Our first observation is that, since $y(0) = y_0 > 0$ and y is continuous, there exists some $x_0 > 0$ such that $y(x) > 0$ for $x \in [0, x_0]$. Now assume that y has a change of sign. Then, we may find some x_1 and x_2 such that $0 < x_1 < x_2$ and

$$y(x) \begin{cases} > 0 & \text{for } 0 \le x < x_1, \\ = 0 & \text{for } x = x_1, \\ < 0 & \text{for } x_1 < x \le x_2. \end{cases}$$

Thus, by definition of f and its extension,

$$f(x, y(x)) \begin{cases} \le 0 & \text{for } 0 \le x < x_1, \\ = 0 & \text{for } x_1 \le x \le x_2. \end{cases}$$

Introducing this observation into the Volterra equation above, we find that

$$0 > y(x_2) = y_0 + \frac{1}{\Gamma(n)} \int_0^{x_2} (x_2 - t)^{n-1} f(t, y(t)) \, dt$$

$$= y_0 + \frac{1}{\Gamma(n)} \int_0^{x_1} (x_2 - t)^{n-1} f(t, y(t)) \, dt + \frac{1}{\Gamma(n)} \int_{x_1}^{x_2} (x_2 - t)^{n-1} f(t, y(t)) \, dt$$

$$= y_0 + \frac{1}{\Gamma(n)} \int_0^{x_1} (x_2 - t)^{n-1} f(t, y(t)) \, dt$$

$$\ge y_0 + \frac{1}{\Gamma(n)} \int_0^{x_1} (x_1 - t)^{n-1} f(t, y(t)) \, dt = y(x_1) = 0$$

which is the required contradiction that shows that y cannot have a change of sign. Thus, $y(x) \ge 0$ for all $x \ge 0$.

Using this observation, we then conclude from the Volterra form of our initial value problem, using the fact that $f(t, y) \le 0$ for all t and y by definition, that $y(x) \le y_0$.

Finally, the strict inequality $y(x) > 0$ then follows from Corollary 6.16.   □

Under additional assumptions we may also say something about the limit of $y(x)$ as $x \to \infty$.

Theorem 7.22. *Assume the hypotheses of Theorem 7.21. Moreover, assume that for all $a > 0$ and all continuous functions $Y : [0, \infty) \to [a, y_0]$ there holds*

$$\lim_{x \to \infty} J_0^n[f(\cdot, Y(\cdot))](x) = -\infty.$$

Then,

$$\lim_{x \to \infty} y(x) = 0$$

if the limit exists.

Proof. In view of Theorem 7.21 it is clear that $0 \le y(x) \le y_0$ for all x, and hence $\lim_{x \to \infty} y(x) \ge 0$ if the limit exists. We shall prove indirectly that the limit must be zero. To this end let us assume that $\lim_{x \to \infty} y(x) = \lambda > 0$. We may then find an x_0 such that $y(x) \ge \lambda/2$ for all $x \ge x_0$. Now we define

$$Y(x) := \begin{cases} y(x) & \text{for } x > x_0, \\ y(x_0) & \text{for } x \le x_0. \end{cases}$$

Obviously, Y is a continuous function satisfying $Y(x) \ge \lambda/2$ for all $x \ge 0$, and from the Volterra form of the initial value problem under consideration we see that

$$y(x) = y_0 + J_0^n[f(\cdot, y(\cdot))](x) = y_0 + J_0^n[f(\cdot, y(\cdot)) - f(\cdot, Y(\cdot))](x)$$
$$+ J_0^n[f(\cdot, Y(\cdot))](x).$$

On the right-hand side of this equation, the first term is a constant and the last term tends to $-\infty$ as $x \to \infty$ by assumption. We shall show below that the second term remains bounded as x grows. Form this observation we then conclude that $y(x) \to -\infty$ as $x \to \infty$ which contradicts the fact that $y(x) \ge 0$ that follows from Theorem 7.21. Thus, our assumption $\lim_{x \to \infty} y(x) > 0$ must be false.

To prove the boundedness of the second term in the equation above for $x \to \infty$, we write

$$J_0^n[f(\cdot, y(\cdot)) - f(\cdot, Y(\cdot))](x) = \frac{1}{\Gamma(n)} \int_0^x (x - t)^{n-1}[f(t, y(t)) - f(t, Y(t))]\, dt$$

and observe that the term in brackets vanishes for $t > x_0$ by definition of Y. Thus,

$$|J_0^n[f(\cdot, y(\cdot)) - f(\cdot, Y(\cdot))](x)| = \frac{1}{\Gamma(n)} \left| \int_0^{x_0} (x - t)^{n-1}[f(t, y(t)) - f(t, Y(t))]\, dt \right|$$
$$\le \frac{2}{\Gamma(n)} \sup_{t \in [0, x_0], z \in [0, y_0]} |f(t, z)| \int_0^{x_0} (x - t)^{n-1}\, dt$$
$$= \frac{2}{\Gamma(n+1)} \sup_{t \in [0, x_0], z \in [0, y_0]} |f(t, z)|(x^n - (x - x_0)^n).$$

The expression on the right-hand side is easily seen to be a positive and monotonically decreasing function of x and hence bounded as required. $\qquad \square$

Remark 7.6. A particular situation where the assumptions of Theorem 7.21 are satisfied is the case of a homogeneous linear fractional differential equation, i.e. the case $f(x,z) = -\mu(x)z$, with some nonnegative continuous and bounded function μ. If, in addition, $\mu(x) \geq \mu_0 > 0$ for all $x \geq 0$ with a suitable constant μ_0, then Theorem 7.22 is applicable too. A sufficient condition for this to hold is that $\mu(x) = \mu_0$ is a positive constant.

These observations now place us, as indicated in Sect. 7.1, in a position to prove Theorem 7.3 without using Laplace transform techniques.

Proof (of Theorem 7.3). The function u_0 under consideration in Theorem 7.3 is the unique solution of the homogeneous fractional relaxation equation $D_{*0}^n u_0(x) = -\mu u_0(x)$ with some $\mu > 0$ subject to the initial condition $u_0(x) = 1$. This corresponds to the case $f(t,z) = -\mu z$ of Theorems 7.21 and 7.22. It is easily seen that the assumptions of both these theorems are fulfilled. We thus obtain $0 < u_0(x) \leq 1$ for all $x \in [0,\infty)$ by Theorem 7.21 and $\lim_{x\to\infty} u_0(x) = 0$ by Theorem 7.22.

From the explicit representation of u_0 in terms of the Mittag-Leffler function and the power series expansion of the latter we find that u_0 has infinitely many continuous derivatives on the open interval $(0,\infty)$ and that $\lim_{x\to 0+} u_0'(x) = -\infty$. We can then differentiate the Volterra form of the initial value problem for $x > 0$ which yields

$$u_0'(x) = -\frac{\mu}{\Gamma(n)} x^{n-1} - \frac{\mu}{\Gamma(n)} \int_0^x (x-t)^{n-1} u_0'(t)\,dt.$$

This is a Volterra equation for u_0'. We have seen above that $u_0'(x) < 0$ for $x \in (0,\varepsilon)$ with some $\varepsilon > 0$. With this knowledge we can then handle the Volterra equation for u_0' in the same way as we had done with the equation for y in the proof of Theorem 7.21 to show that u_0' does not have a change of sign. A repeated application of this step (differentiation of the Volterra equation and showing that the solution does not change its sign) then gives the properties of the derivatives of u_0 stated in part (a) of Theorem 7.3.

A particular consequence of the results that we have just proved is the fact that u_0 is a monotonically decreasing function that converges to zero as its argument grows to ∞. It is therefore worth trying to model the asymptotic behaviour of $u_0(x)$ in this limit case via the approach

$$u_0(x) = cx^{-\beta} + o(x^{-\beta}) \quad \text{as } x \to \infty$$

with certain constants $\beta > 0$ and $c \in \mathbb{R}$. We may insert this relation into the Volterra form of the initial value problem to obtain

$$
\begin{aligned}
cx^{-\beta} + o(x^{-\beta}) = u_0(x) &= 1 - \frac{\mu}{\Gamma(n)} \int_0^x (x-t)^{n-1} u_0(t)\,dt \\
&= 1 - \frac{c\mu}{\Gamma(n)} \int_0^x (x-t)^{n-1} t^{-\beta}\,dt + \int_0^x (x-t)^{n-1} o(t^{-\beta})\,dt \\
&= 1 - \frac{c\mu}{\Gamma(n)} x^{n-\beta} \frac{\Gamma(n)\Gamma(1-\beta)}{\Gamma(n+1-\beta)} + o(x^{n-\beta}).
\end{aligned}
$$

Since we have no term of order x^0 on the left-hand side, we must not have such a term on the right-hand side either. Thus the constant 1 and the next highest term on the right-hand side, i.e. the term involving $x^{n-\beta}$, must cancel each other. This yields

$$\beta = n \quad \text{and} \quad c = \frac{1}{\mu \Gamma(1-n)}$$

which proves part (b) of the theorem. $\square$

7.4 Singular Equations

The problems considered so far have all been regular in the sense that the function f on the right-hand side of the differential equation (6.1a) has been at least continuous (and, in most cases, even differentiable a certain number of times). In some applications however one encounters equations where this is not the case. Therefore we will now conclude this chapter with a section devoted to some results concerning such singular problems. We will not provide a completely general analysis; rather we will restrict our attention to some particularly important special cases which will nevertheless demonstrate a rich variety of phenomena that may be encountered in the investigation of singular fractional differential equations.

An important and frequently used result in the theory of equations without singularities was the equivalence between the initial value problem (6.1) and the corresponding Volterra integral equation formulation (6.2) that we had established in Lemma 6.2. Our first observation in the context of this section is that in the presence of singularities we may lose this equivalence. This can be seen from the following example.

Example 7.6. For $1 < n < 2$, consider the integral equation

$$y(x) = 1 + \frac{1}{\Gamma(n)} \int_0^x (x-t)^{n-1} \frac{1}{y(t)-1} \, dt \tag{7.22}$$

that formally corresponds to the initial value problem

$$D_{*0}^n y(x) = \frac{1}{y(x)-1}, \quad y(0) = 1, \quad y'(0) = 0. \tag{7.23}$$

An explicit calculation shows that (7.22) is solved by the functions

$$y(x) = 1 \pm \frac{\sqrt{\Gamma(1-n/2)}}{\sqrt{\Gamma(1+n/2)}} x^{n/2}.$$

Thus we observe that the integral equation formulation has more than one continuous solution. However we also see that neither of these solutions is differentiable at $x = 0$, and therefore these solutions do not satisfy the second initial condition. We have thus found functions that solve the integral equation (7.22) but not the corresponding initial value problem (7.23).

A different prototypical example of a singular fractional differential equation that has important applications in practice has been introduced by Joulin [99] who uses it to model the propagation of a flame in the context of a thermo-diffusive model with high activation energies using a gaseous mixture with simple chemistry $A \to B$. In these circumstances, he shows that the radius of the flame at time x is given by $y(x)$, where the function y is the solution of the initial value problem

$$y(x)D_{*0}^{1/2}y(x) = y(x)\ln y(x) + Eq(x), \qquad y(0) = 0. \qquad (7.24)$$

Here the function q describes a time-dependent point source energy, and therefore it is assumed to be nonnegative, continuous and integrable on $\mathbb{R}_+$, and E represents the intensity of this heat source such that $E \cdot \|q\|_{L_1(0,\infty)}$ is the total amount of energy introduced into the system. It is sometimes assumed that q is normalized such that $\|q\|_{L_1(0,\infty)} = 1$; we found it convenient not to impose this requirement. For our purposes it will be useful to state the differential equation in an explicit form, i.e. we solve it for $D_{*0}^{1/2}y$ and obtain the initial value problem in the representation

$$D_{*0}^{1/2}y(x) = f(x,y(x)) \quad \text{with} \quad f(x,w) := \ln w + E\frac{q(x)}{w}, \qquad y(0) = 0, \quad (7.25)$$

which is equivalent to the original form (7.24). It is immediately evident that the function f is not continuous in any neighbourhood of the initial point $(0,0)$.

The model described by (7.24) which can be justified in a mathematically rigorous way [110] has some rather natural important questions associated with it. Apart from the most obvious one for an (exact or approximate) solution for a specific choice of the parameters E and q, one is often strongly interested in the qualitative behaviour of the solution. Analytically, it is possible to prove the following result that indicates that we have to deal with a bifurcation phenomenon [8, Theorem 0.2]:

Theorem 7.23. *Assume that there exists some $x_0 > 0$ such that $q(x) > 0$ for $x \in (0,x_0)$ and $q(x) = 0$ else. Then, the initial value problem (7.24) has a unique continuous solution y. Moreover, there exists a critical value $E_{\mathrm{crit}}(q)$ such that*

- *if $E > E_{\mathrm{crit}}(q)$ then y is defined on $[0,\infty)$ and $\lim_{x\to\infty} y(x) = \infty$*
- *if $E = E_{\mathrm{crit}}(q)$ then y is defined on $[0,\infty)$ and $\lim_{x\to\infty} y(x) = 1$*
- *if $E < E_{\mathrm{crit}}(q)$ then there exists some finite $x_{\max} > x_0$ such that y is defined on $[0,x_{\max}]$ and $\lim_{x\to x_{\max}} y(x) = 0$*

The proof of this theorem may be found in [8]. It is based on the observation that the solution of the fractional initial value problem can be written as the solution of a parabolic partial differential equation that has the classical form for such equations and that, in particular, does not contain any fractional derivatives. The required properties then follow from the standard theory of parabolic partial differential equations.

Apart from the properties described in Theorem 7.23 above, a number of other results on analytical aspects of (7.25) are of interest.

The first observation deals with the asymptotic behaviour of the solution y of our problem (7.25) as $x \to 0$. It is taken from [8, Proposition 1.2].

Theorem 7.24. *Assume that $q(x) \geq 0$ for $x \geq 0$ and that $q(x) = q_0 x^\beta (1 + o(1))$ as $x \to 0$ with some $\beta \in [0, 1/2)$. Then, as $x \to 0$, the solution y of the initial value problem (7.24) behaves as*

$$y(x) = \rho_\beta x^{1/4 + \beta/2} (1 + o(1))$$

where

$$\rho_\beta = \left(E q_0 \frac{\Gamma\left(\dfrac{3}{4} - \dfrac{\beta}{2}\right)}{\Gamma\left(\dfrac{5}{4} - \dfrac{\beta}{2}\right)} \right)^{1/2}.$$

The simplest special case of this result, $\beta = 0$, already reveals that the asymptotic behaviour of the solution is significantly different from the behaviour that we would expect from the solution of a nonsingular equation as described by Theorems 6.38 and 6.39. Notice that this result is cited in [7, Proposition 2.1] with a factor $\sqrt{\pi}$ accidentally omitted. The significance of the condition $\beta < 1/2$ in Theorem 7.24 is explained by the following result taken from [8, Proposition 1.4].

Theorem 7.25. *Assume that $q(x) = q_0 x^\beta (1 + o(1))$ as $x \to 0$ with some $\beta \geq 1/2$. Then, as $x \to 0$, the solution y of the initial value problem (7.24) behaves as*

$$y(x) = \frac{1}{\sqrt{\pi}} x^{1/2} |\ln x| \cdot (1 + o(1)).$$

So the asymptotic behaviour of the exact solution near the origin changes as the parameter β crosses the value $1/2$.

Similar results can be derived in the case that the support of q is unbounded [8, Theorem 0.1]. More information on related questions may be found in [161]. From the point of view of applications however the situation discussed in Theorem 7.23 is by far the most relevant. Stated explicitly, it says that the flame will quench in finite time if the energy added to the system is smaller than the critical level E_{crit}, and it will burn persistently if the energy is above E_{crit}. For safety considerations it is therefore very important to find out the value of E_{crit} (or at least lower bounds for it) if one is interested in keeping the fire under control. On the other hand, sometimes one is interested in constructing a permanently burning flame, and then one needs to know E_{crit} (or at least upper bounds for it) in order to find an efficient process that uses as little energy as possible. Thus it is certainly justified to investigate the initial value problem (7.24) thoroughly, using both analytical and numerical approaches. This is even more emphasized by the observation [110, p. 570] that Joulin's ideas can be carried over to a much larger class of experiments, and so one should expect that a successful procedure for (7.24) will also be able to handle models from this large class too. Such a study of numerical methods for this problem has been described, e.g., in [54]. From the analytical side however, the knowledge at the moment seems to be very limited.

Exercises

Exercise 7.1. Let $u_0(x) = E_n(-x^n)$ for $1 < n < 2$ be the function investigated in Theorems 7.6 and 7.7. Compute this function numerically and plot the resulting graph of the function for

$$n \in \{1.001, 1.01, 1.1, 1.5, 1.9, 1.99, 1.999\}$$

on a sufficiently large interval. Based on the numerical results, determine the number of zeros of these functions and give approximate values for the largest and smallest zeros.
Hint: Efficient algorithms for the numerical computation of Mittag-Leffler functions are described in [79] and [175].

Exercise 7.2. Give a proof for Theorem 7.15.

Exercise 7.3. Give a proof for Theorem 7.24 under the assumption that $q(x) = \tilde{q}(x)x^\beta$ where the function $\tilde{q}$ is analytic in a neighbourhood of the origin and satisfies $\tilde{q}(0) = q_0$.
Hint: Assume that the solution y of (7.25) can be written as a generalized power series, viz. $y(x) = \sum_{j=0}^\infty c_j x^{\lambda_j}$. Insert this into the differential equation and compare left- and right-hand side to determine the values of c_j and λ_j, and prove that the series converges.

Exercise 7.4. Give a proof for Theorem 7.25 under the assumption that $q(x) = \tilde{q}(x)x^\beta$ where the function $\tilde{q}$ is analytic in a neighbourhood of the origin and satisfies $\tilde{q}(0) = q_0$.
Hint: Proceed as in Exercise 7.3, except that now the assumption on y reads $y(x) = \sum_{j=0}^\infty c_j x^{\lambda_j} \ln x$.

Chapter 8
Multi-Term Caputo Fractional Differential Equations

Up to this point, we have only considered so-called *single-term equations*, i.e. equations where only one differential operator is involved. In certain cases though we need to solve equations containing more than one differential operator. A classical example is the so-called *Bagley–Torvik equation*

$$AD_{*0}^2 y(x) + BD_{*0}^{3/2} y(x) + Cy(x) = f(x)$$

where A, B and C are certain constants and f is a given function. This equation arises, for example, in the modelling of the motion of a rigid plate immersed in a Newtonian fluid. It was originally proposed in [184] and is thoroughly discussed, e.g., in [153, §8.3.2] and, from a numerical point of view, in [44]. Another example for an application of equations with more than one fractional derivative is the *Basset equation*

$$D^1 y(x) + bD_{*}^n y(x) + cy(x) = f(x), \qquad y(0) = y_0,$$

where $0 < n < 1$. This equation is most frequently, but not exclusively, used with $n = 1/2$. It describes the forces that occur when a spherical object sinks in a (relatively less dense) incompressible viscous fluid; see, e.g., [14, 15, 125].

A most general type of equations that includes the examples mentioned above would be

$$g(x, y(x), D_{*0}^{n_1} y(x), D_{*0}^{n_2} y(x), \dots, D_{*0}^{n_k} y(x)) = 0$$

with $0 < n_1 < n_2 < \cdots < n_k$ and a certain function g. An equation of this type will be called a *multi-term* (or, more precisely, *k-term*) *fractional differential equation*. Since for such a general equation almost no results seem to be known, we restrict our attention to certain special cases (for the most part we will look at explicit equations, i.e. equations that can be solved for the highest order derivative $D_{*0}^{n_k} y(x)$) and we investigate the most fundamental properties.

We first introduce a concept that will be useful throughout the investigation of multi-term equations and that is applicable to the fully general class of multi-term equations mentioned above, so there is no need to specialize on a subset of these equations yet.

K. Diethelm, *The Analysis of Fractional Differential Equations*,
Lecture Notes in Mathematics 2004, DOI 10.1007/978-3-642-14574-2_8,
© Springer-Verlag Berlin Heidelberg 2010

Definition 8.1. The fractional differential equation

$$g(x, y(x), D_{*0}^{n_1}y(x), D_{*0}^{n_2}y(x), \ldots, D_{*0}^{n_k}y(x)) = 0$$

with $0 < n_1 < n_2 < \cdots < n_k$ and a certain function g is called *commensurate* if the numbers $n_1, n_2, \ldots, n_k$ are commensurate, i.e. if the quotients n_μ/n_ν are rational numbers for all $\mu, \nu \in \{1, 2, \ldots k\}$.

Remark 8.1. Some authors (see, e.g., [144]) use this terminology in a different sense; they apply it to systems of fractional differential equations of the form

$$D_{*0}^{n_k}y_k(x) = f_k(x, y_1(x), \ldots, y_N(x)), \qquad k = 1, 2, \ldots, N$$

and say that such a system is commensurate if $n_1 = n_2 = \ldots = n_N$. We refrain from this use of the notion and stick to our Definition 8.1 because the latter is more in keeping with the traditional use of the concept that is common in number theory.

The Bagley–Torvik equation is an example for a commensurate fractional differential equation in our sense; a counterexample is

$$D_{*0}^1 y(x) + D_{*0}^{1/\pi} y(x) = 0.$$

Whether the Basset equation is commensurate or not depends on the precise value of n: The equation is commensurate if and only if n is a rational number.

Commensuracy is important because it allows us to follow an approach that is well known from the theory of integer-order differential equations: We may transform the given equation into a system of differential equations involving only one differential operator. In this way we can invoke the theory of one-term equations as described in Chap. 6 in order to investigate questions of existence and uniqueness or to derive other properties of the solutions. Moreover we can use numerical methods for one-term equations as described, e.g., in the survey article [39] to approximate the solutions.

It is of course true that the assumption of commensuracy significantly restricts the class of differential equations that may be considered, but we can say that many equations derived in physical or engineering applications do have this property. Apart from the Bagley–Torvik equation, we refer to Koeller's model of viscoelastic materials (e.g., copolymers) described in [107, eqs. (3.9) and (3.10)]. Specifically we draw the reader's attention to the physical motivation of the commensuracy assumption given by Bagley and Calico in the second section of their paper [9]. Incidentally, Rossikhin and Shitikova [164] have recently pointed out that Koeller's equation is just a special case of a more general model for viscoelastic behaviour introduced in a formally different but equivalent way by Rabotnov [157].

To make the considerations slightly simpler, from now on we introduce the hypothesis mentioned above on the given differential equation: We assume that it can be solved explicitly for the highest order derivative, i.e. that the differential equation may be rewritten in the form

$$D_{*0}^{n_k}y(x) = f(x, y(x), D_{*0}^{n_1}y(x), D_{*0}^{n_2}y(x), \ldots, D_{*0}^{n_{k-1}}y(x))$$

with a suitable function f. Our approach is essentially based on the following equivalence theorem taken from [45] that we shall extend later. The key statement of this theorem (and Theorem 8.2 below) is that the situation with commensurate multi-term fractional differential equations subject to appropriately chosen initial values is essentially the same as for differential equations of higher integer order: Such initial value problems may be rewritten in the form of a single-order system of equations.

Theorem 8.1. *Consider the equation*

$$D_{*0}^{n_k}y(x) = f(x, y(x), D_{*0}^{n_1}y(x), D_{*0}^{n_2}y(x), \dots, D_{*0}^{n_{k-1}}y(x)) \tag{8.1a}$$

subject to the initial conditions

$$y^{(j)}(0) = y_0^{(j)}, \qquad j = 0, 1, \dots, \lceil n_k \rceil - 1 \tag{8.1b}$$

where $n_k > n_{k-1} > \dots > n_1 > 0$, $n_j - n_{j-1} \le 1$ for all $j = 2, 3, \dots, k$ and $0 < n_1 \le 1$. Assume that $n_j \in \mathbb{Q}$ for all $j = 1, 2, \dots, k$, define M to be the least common multiple of the denominators of $n_1, n_2, \dots, n_k$ and set $\gamma := 1/M$ and $N := Mn_k$. Then this initial value problem is equivalent to the system of equations

$$
\begin{aligned}
D_{*0}^{\gamma}y_0(x) &= y_1(x), \\
D_{*0}^{\gamma}y_1(x) &= y_2(x), \\
&\vdots \\
D_{*0}^{\gamma}y_{N-2}(x) &= y_{N-1}(x), \\
D_{*0}^{\gamma}y_{N-1}(x) &= f(x, y_0(x), y_{n_1/\gamma}(x), \dots, y_{n_{k-1}/\gamma}(x)),
\end{aligned} \tag{8.2a}
$$

together with the initial conditions

$$y_j(0) = \begin{cases} y_0^{(j/M)} & \text{if } j/M \in \mathbb{N}_0, \\ 0 & \text{else}, \end{cases} \tag{8.2b}$$

in the following sense.

1. *Whenever $Y := (y_0, \dots, y_{N-1})^{\mathrm{T}}$ with $y_0 \in C^{\lceil n_k \rceil}[0, b]$ for some $b > 0$ is the solution of the system (8.2), the function $y := y_0$ solves the multi-term equation initial value problem (8.1).*
2. *Whenever $y \in C^{\lceil n_k \rceil}[0, b]$ is a solution of the multi-term initial value problem (8.1), the vector function $Y := (y_0, \dots y_{N-1})^{\mathrm{T}} := (y, D_{*0}^{\gamma}y, D_{*0}^{2\gamma}y, \dots, D_{*0}^{(N-1)\gamma}y)^{\mathrm{T}}$ solves the multidimensional initial value problem (8.2).*

Proof. In the case $n_j \in \mathbb{N}$ for all j we have $M = 1$ and $\gamma = 1$, and therefore we recover a standard result from the theory of ordinary differential equations (of integer order). Thus, from now on we assume that at least one of the n_j is not an integer. As a consequence we have $M \ge 2$ and hence $0 < \gamma \le 1/2$.

In order to prove the first claim, we have to assume that $(y_0, \ldots, y_{N-1})^T$ is a solution of the system, and we define $y := y_0$. Then, by a repeated application of Lemma 3.13 (which is legal since $\gamma < 1$) in combination with the system (8.2a) we have

$$D_{*0}^{\gamma} y(x) = D_{*0}^{\gamma} y_0(x) = y_1(x),$$
$$D_{*0}^{2\gamma} y(x) = D_{*0}^{\gamma} D_{*0}^{\gamma} y(x) = D_{*0}^{\gamma} y_1(x) = y_2(x),$$
$$D_{*0}^{3\gamma} y(x) = D_{*0}^{\gamma} D_{*0}^{2\gamma} y(x) = D_{*0}^{\gamma} y_2(x) = y_3(x),$$

$$\vdots \tag{8.3}$$

$$D_{*0}^{(N-1)\gamma} y(x) = D_{*0}^{\gamma} D_{*0}^{(N-2)\gamma} y(x) = D_{*0}^{\gamma} y_{N-2}(x) = y_{N-1}(x),$$
$$D_{*0}^{N\gamma} y(x) = D_{*0}^{\gamma} D_{*0}^{(N-1)\gamma} y(x) = D_{*0}^{\gamma} y_{N-1}(x)$$
$$= f(x, y_0(x), y_{n_1/\gamma}(x), \ldots, y_{n_k/\gamma}(x))$$
$$= f(x, y_0(x), D_{*0}^{n_1} y(x), \ldots, D_{*0}^{n_{k-1}} y(x)).$$

By definition, $N\gamma = M n_k / M = n_k$, and hence the left-hand side of the last equation is simply $D_{*0}^{n_k} y(x)$. Thus the function y satisfies the differential equation (8.1a). Moreover it is evident from the equation system (8.3) that, for $\ell = 0, 1, \ldots, \lceil n_k \rceil - 1$ we have

$$y^{(\ell)}(0) = D_{*0}^{\ell} y(0) = y_{\ell\gamma}(0) = y_{\ell/M}(0) = y_0^{(\ell)},$$

and so the function y also satisfies the initial conditions (8.1b).

For the second claim we have to assume that y satisfies the multi-term equation (8.1a) and the initial conditions (8.1b). The equation system (8.3) is valid in this case too, and therefore it follows that the vector-valued function Y indeed satisfies (8.2a) and the initial conditions $y_j(0) = y_0^{(j/M)}$ whenever $j/M \in \mathbb{N}_0$. Finally, an application of Lemma 3.11 reveals that $y_j(0) = 0$ in the other cases. $\square$

In many practical applications one has $n_k < 1$. A close inspection of the proof reveals that we can relax the rationality condition of Theorem 8.1 in such a case. We can then state the following modification.

Theorem 8.2. *Consider the equation (8.1a) subject to the initial conditions (8.1b) as in Theorem 8.1. Let now $1 \geq n_k > n_{k-1} > \ldots > n_1 > 0$, and assume the equation to be commensurate. Define $\tilde{n}_j := n_j / n_1$ for $j = 1, \ldots, k$, let $\tilde{M}$ be the least common multiple of the denominators of the values $\tilde{n}_1, \ldots, \tilde{n}_k$ and set $\gamma := n_1 / \tilde{M}$ and $N := \tilde{M} n_k / n_1$. Then this initial value problem is equivalent to the system of equations (8.2a) together with the initial conditions (8.2b) in the same sense as in Theorem 8.1.*

Remark 8.2. In the case of Theorem 8.2, due to the restriction $n_k \leq 1$ the initial conditions (8.2b) of the system can be simplified to

$$y_j(0) = \begin{cases} y_0^{(0)} & \text{if } j = 0, \\ 0 & \text{else.} \end{cases}$$

Proof. The proof is identical to the proof of the previous theorem. In that result we had to impose the rationality assumption only because we needed to make sure that γ is a unit fraction (i.e. $1/\gamma$ is an integer). The reason for this was that we had to combine the given initial values (corresponding to integer order derivatives) with the system (8.2a) in such a way that we had equations corresponding to integer order derivatives too. In the present case this is not necessary because the only given initial condition is related to the function y itself. $\qquad\square$

As an immediate consequences of this result, we can deduce an existence theorem and a uniqueness theorem.

Theorem 8.3. *Assume the hypotheses of Theorem 8.1 or Theorem 8.2. Moreover let $K > 0$, $h^* > 0$, and $G := [0, h^*] \times [y_0^{(0)} - K, y_0^{(0)} + K] \times \prod_{j=1}^k T_j$ where $T_j = [-K, K]$ if $n_j \notin \mathbb{N}_0$ and $T_j = [y_0^{(n_j)} - K, y_0^{(n_j)} + K]$ else. If $f : G \to \mathbb{R}$ is continuous, then the multi-term initial value problem (8.1) has a solution on the interval $[0, h]$ with some $h > 0$.*

Proof. By the equivalence statement of Theorem 8.1 or 8.2, respectively, we may reduce the existence question for the multi-term equation to the existence question for the vector-valued equation (8.2a) with initial conditions (8.2b). It is then easily seen that for this equation we may use the existence result from Theorem 6.1 (see also Remark 6.1) to deduce that a solution exists on the interval $[0, h]$ with a suitable $h > 0$. $\qquad\square$

Theorem 8.4. *Assume the hypotheses of Theorem 8.1 or Theorem 8.2. Moreover define the set G as in Theorem 8.3. If $f : G \to \mathbb{R}$ is continuous and satisfies a Lipschitz condition with respect to all variables except for the first, then there exists some $h > 0$ such that the multi-term initial value problem (8.1) has a unique solution on the interval $[0, h]$.*

Proof. As in the proof of Theorem 8.3, we use the equivalence statement of Theorems 8.1 or 8.2, respectively, to reduce the uniqueness question for the multi-term equation to the same question for the vector-valued equation (8.2a) subject to the initial conditions (8.2b). This time we invoke the uniqueness result of the Picard–Lindelöf Theorem 6.5 (see also Remark 6.1) to deduce that the problem (8.2) indeed has a unique solution. $\qquad\square$

Our next goal is to derive some Gronwall-type results. Specifically we show that, under small variations in the orders n_j in the multi-term differential equation (8.1a) but subject to the assumption that all other given data remain unchanged, we can give a uniform bound on the change in the solution on any closed bounded interval $[0, h]$. We state and prove the result for an equation with two terms, but the generalisation to multi-term equations is straightforward. Note that a closely related discussion of similar results that apply to integral equations may be found, for example, in [20].

Lemma 8.5 (First Gronwall inequality for two-term equations). *Let $n_2 > 0$ and $n_1, \tilde{n}_1 \in (0, n_2)$ be chosen so that the equations*

$$D_{*0}^{n_2} y(x) = f(x, y(x), D_{*0}^{n_1} y(x)) \tag{8.4a}$$

subject to the initial conditions

$$y(0) = y_0, y'(0) = y_0', \ldots, y^{(\lceil n_2 \rceil - 1)}(0) = y_0^{(\lceil n_2 \rceil - 1)} \tag{8.4b}$$

and

$$D_{*0}^{n_2} z(x) = f(x, z(x), D_{*0}^{\tilde{n}_1} z(x)) \tag{8.5a}$$

subject to the same initial conditions

$$z(0) = y_0, z'(0) = y_0', \ldots, z^{(\lceil n_2 \rceil - 1)}(0) = y_0^{(\lceil n_2 \rceil - 1)} \tag{8.5b}$$

(where f satisfies a Lipschitz condition in its second and third arguments on a suitable domain) have unique continuous solutions $y, z : [0, h] \to \mathbb{R}$. We assume further that $\lfloor n_1 \rfloor = \lfloor \tilde{n}_1 \rfloor$. Then there exist constants K_1 and K_2 such that

$$|y(x) - z(x)| \leq K_1 |n_1 - \tilde{n}_1| E_{n_2}(K_2 h^{n_2}) \tag{8.6}$$

for all $x \in [0, h]$.

Remark 8.3. Note that if $n_1, \tilde{n}_1, n_2$ are all rational then the equations (8.4) and (8.5) may be rewritten as systems of equations as described in Theorem 8.1, and both equations have unique continuous solutions in view of Theorem 8.4.

Proof. The essential steps of the proof are very similar to those encountered in the theorems in Sect. 6.3: We write the solutions y and z in the form of the equivalent Volterra integral equations

$$y(x) = \sum_{j=0}^{\lceil n_2 \rceil - 1} \frac{y_j}{j!} x^j + \frac{1}{\Gamma(n_2)} \int_0^x (x - t)^{n_2 - 1} f(t, y(t), D_{*0}^{n_1} y(t)) \, dt \tag{8.7}$$

and

$$z(x) = \sum_{j=0}^{\lceil n_2 \rceil - 1} \frac{y_j}{j!} x^j + \frac{1}{\Gamma(n_2)} \int_0^x (x - t)^{n_2 - 1} f(t, z(t), D_{*0}^{\tilde{n}_1} z(t)) \, dt \tag{8.8}$$

and we fix $h > 0$. Subtracting we obtain the relation

$$y(x) - z(x) = \frac{1}{\Gamma(n_2)} \int_0^x (x - t)^{n_2 - 1} \left(f(t, y(t), D_{*0}^{n_1} y(t)) - f(t, y(t), D_{*0}^{\tilde{n}_1} y(t)) \right) dt$$

$$+ \frac{1}{\Gamma(n_2)} \int_0^x (x - t)^{n_2 - 1} \left(f(t, y(t), D_{*0}^{\tilde{n}_1} y(t)) - f(t, z(t), D_{*0}^{\tilde{n}_1} z(t)) \right) dt.$$

Now, with $m \in \mathbb{N}$ chosen so that $m-1 < n_1, \tilde{n}_1 < m$, and bearing in mind that y is the unique solution to (8.4) on $[0,h]$ we can estimate (using the Lipschitz condition on f and the definition of the Caputo derivative) the first term on the right-hand side:

$$\left| \frac{1}{\Gamma(n_2)} \int_0^x (x-t)^{n_2-1} \left(f(t,y(t),D_{*0}^{n_1}y(t)) - f(t,y(t),D_{*0}^{\tilde{n}_1}y(t)) \right) dt \right| \leq K_1 |n_1 - \tilde{n}_1|$$

uniformly for $x \in [0,h]$. Moreover we can use the Lipschitz conditions on f in the second term on the right-hand side to give

$$\left| \frac{1}{\Gamma(n_2)} \int_0^x (x-t)^{n_2-1} \left(f(t,y(t),D_{*0}^{\tilde{n}_1}y(t)) - f(t,z(t),D_{*0}^{\tilde{n}_1}z(t)) \right) dt \right| \leq K_2 J_0^{n_2} |y-z|(x)$$

by evaluating the integral representations of the fractional derivatives of y and z. If we put $\delta(x) = y(x) - z(x)$ it follows that

$$|\delta(x)| \leq K_1 |n_1 - \tilde{n}_1| + K_2 J_0^{n_2} |\delta|(x). \tag{8.9}$$

Equation (8.9) now allows us to conclude by means of Lemma 6.19 that

$$|\delta(x)| \leq K_1 |n_1 - \tilde{n}_1| E_{n_2}(K_2 h^{n_2})$$

uniformly for $x \in [0,h]$ and the proof is complete. $\qquad\square$

An analogous statement is true if we vary the order of the other differential operator.

Lemma 8.6 (Second Gronwall inequality for two-term equations). *Let $n_1 > 0$ and $n_2, \tilde{n}_2 \in (n_1, \infty)$ be chosen such that $\lceil \tilde{n}_2 \rceil = \lceil n_2 \rceil$ and so that the equations (8.4) and*

$$D_{*0}^{\tilde{n}_2}z(x) = f(x,z(x),D_{*0}^{n_1}z(x)) \tag{8.10a}$$

subject to the same initial conditions

$$z(0) = y_0, z'(0) = y_0', \ldots, z^{(\lceil n_2 \rceil - 1)}(0) = y_0^{(\lceil n_2 \rceil - 1)} \tag{8.10b}$$

(where f satisfies a Lipschitz condition in its second and third arguments on a suitable domain) have unique continuous solutions $y, z : [0,h] \to \mathbb{R}$. Then

$$\sup_{x \in [0,h]} |y(x) - z(x)| = O(n_2 - \tilde{n}_2). \tag{8.11}$$

Proof. The proof is essentially the same as the proof of Theorem 6.22; we omit the details. $\qquad\square$

We use the conclusions of Lemmas 8.5 and 8.6 several times. First we derive an existence and uniqueness theorem for more general multi-term nonlinear equations (with non-commensurate multiple derivatives). As before, we give a proof for the two-term equation; the result can be easily generalised for equations with more terms.

Theorem 8.7. *Let the bounded function* $f : [0,h] \times \mathbb{R}^2 \to \mathbb{R}$ *satisfy a uniform Lipschitz condition in its second and third arguments and be continuous in its first argument. It follows that the equation*

$$D_{*0}^{n_2}y(x) = f(x, y(x), D_{*0}^{n_1}y(x)) \qquad (8.12)$$

where $n_2 > n_1 > 0$, *subject to the initial conditions*

$$D^k y(0) = y_0^{(k)}, \qquad k = 0, 1, \ldots, \lceil n_2 \rceil - 1,$$

has a unique continuous solution on $[0,h]$.

Proof. Case 1: $n_1, n_2 \in \mathbb{Q}$. We observe that the result is already established when n_1 and n_2 are rational because of Theorem 8.4.

Case 2: $n_2 \in \mathbb{Q}$, $n_1 \notin \mathbb{Q}$. We construct a sequence $(n_{1,j})_{j=1}^{\infty}$ of rational numbers whose limit is n_1. Without loss of generality, we may assume that all members of the sequence are contained in the interval $(\lfloor n_1 \rfloor, \lfloor n_1 + 1 \rfloor)$. Clearly by case 1 the equation (8.12) with n_1 replaced in turn by each $n_{1,j}$, subject to the given initial conditions, has a unique continuous solution $y^{[j]}$ on $[0,h]$ whose Caputo derivative of order n_2 is also continuous. We now use equation (8.9) together with the fact that $n_{1,j} - n_1 \to 0$ as $j \to \infty$ to conclude that the sequence of solutions converges uniformly on $[0,h]$ to a continuous function y with $D_{*0}^{n_2}y$ also being continuous. It remains to prove that this function y is the solution of (8.12).

To this end, let us define $r_j(z) := \|D_{*0}^{n_2}z - f(\cdot, z(\cdot), D_{*0}^{n_{1,j}}z(\cdot))\|_{L_\infty[0,h]}$ for any z such that $D_{*0}^{n_2}z$ is continuous. By definition, we immediately obtain $r_j(y^{[j]}) = 0$ for all j. Since $y^{[j]} \to y$ uniformly as $j \to \infty$ and r is continuous on the space we consider, we find that

$$r_j(y^{[j]}) - r_j(y) \to 0 \qquad \text{as } j \to \infty.$$

Therefore y is a solution of the given initial value problem.

Because of the Lipschitz condition on f, we can prove the uniqueness of the solution by the usual Picard iteration techniques, i.e. we proceed as in the proof of Theorem 6.5.

Case 3: $n_2 \notin \mathbb{Q}$. Here we use a sequence $(n_{2,j})$ of rational numbers satisfying $n_2 < n_{2,j} < \lceil n_2 \rceil$ for all j and $\lim_{j \to \infty} n_{2,j} = n_2$. For each j we can use either case 1 or case 2 (depending on whether or not $n_1 \in \mathbb{Q}$) to provide a unique solution to the perturbed equation obtained by replacing n_2 by $n_{2,j}$ subject to the unmodified initial conditions. Then we proceed as in case 2 (but applying Lemma 8.6 instead of Lemma 8.5) to show that the corresponding sequence of solutions converges to the unique solution of the original problem. $\qquad \square$

Later in this section, we will give an alternative proof of this result, cf. Remark 8.5.

We can now use the Gronwall Lemmas to prove the structural stability of the initial value problem (8.1) even under small perturbations in the orders of the derivatives.

Theorem 8.8. *Let y be the solution of*

$$D_{*0}^{n_k} y(x) = f(x, y(x), D_{*0}^{n_1} y(x), D_{*0}^{n_2} y(x), \ldots, D_{*0}^{n_{k-1}} y(x))$$

with initial conditions

$$y^{(j)}(0) = y_0^{(j)}, \qquad j = 0, 1, \ldots, \lceil n_k \rceil - 1$$

and let z be the solution of

$$D_{*0}^{\tilde{n}_k} z(x) = f(x, z(x), D_{*0}^{\tilde{n}_1} z(x), D_{*0}^{\tilde{n}_2} z(x), \ldots, D_{*0}^{\tilde{n}_{k-1}} z(x))$$

with initial conditions

$$z^{(j)}(0) = y_0^{(j)}, \qquad j = 0, 1, \ldots, \lceil n_k \rceil - 1$$

where $|n_j - \tilde{n}_j| < \varepsilon$ for all $j = 1, 2, \ldots, k$. Then there exists some $h > 0$ such that both equations have a unique continuous solution on the interval $[0, h]$, and

$$\|y - z\|_{L_\infty[0,h]} = O(\varepsilon), \qquad \varepsilon \to 0.$$

Proof. The theorem follows from the observation that the difference $y - z$ is a Lipschitz function of $n_j - \tilde{n}_j$, $j = 1, 2, \ldots, k$, because of the Gronwall-type Lemmas 8.5 and 8.6. □

Remark 8.4. It follows, by the application of Theorem 8.8, that the solution of any non-commensurate multi-order fractional differential equation may be arbitrarily closely approximated over a finite interval $[0, h]$ by solutions of equations of rational order (which may in turn be solved by conversion to a system of equations of low order).

Up to this point, our main approach for handling multi-term equations was based on rewriting them in the form of a single-term equation for a vector-valued function. This latter equation has a formally very simple and appealing structure, and the entire process is quite natural, but the approach has two disadvantages. Firstly, it only works exactly in the case of commensurate equations (if $n_k < 1$), or, even more restrictively, in the case of equations with rational orders (if $n_k \geq 1$). In all other cases we cannot replace the given multi-term equation by a corresponding single-order system exactly; we must be content with an approximation. The second potential problem is not a major obstacle from the analytical point of view but it can have very unpleasant effects when one tries to employ it for the numerical solution of multi-term equations as discussed, e.g., in [35,36]: Depending on the exact values of the orders n_j of the differential operators, the parameter γ in Theorem 8.1 or 8.2 may be very small. This means that the dimension of the resulting system, which is proportional to $1/\gamma$, may be extremely large. In view of these potential difficulties we shall now look at a different approach for handling multi-term equations that will also provide us with an alternative proof of the general existence and unique result stated in Theorem 8.7 above.

This second approach (which is described, e.g., in [39]) is based on an alternative way of setting up a mathematical model involving more than one fractional derivative. Specifically, we may use a system of fractional differential equations where each equation has an order that may or may not coincide with the orders of the other equations. To put it more formally, this leads to a model of the type

$$D_*^{n_1} y_1(x) = f_1(x, y_1(x), \ldots, y_k(x)),$$

$$\vdots \quad \vdots \tag{8.13a}$$

$$D_*^{n_k} y_k(x) = f_k(x, y_1(x), \ldots, y_k(x)).$$

As we shall see it is sufficient for our purposes to assume that $0 < n_j \leq 1$ for all j. This implies that the initial conditions for the differential equation system (8.13a) have the form

$$y_j(0) = y_{j,0} \qquad (j = 1, 2, \ldots, k). \tag{8.13b}$$

A system of this class will be called a *multi-order fractional differential system*. Multi-order fractional differential systems seem to be investigated less frequently than multi-term equations, but we will now describe some close connections between the two concepts, and thus the former deserve some attention at least in view of the fact that they can be very useful tools for the analytical and numerical treatment of the latter.

Once again, we want to rewrite the given multi-term fractional differential equation in the form of a system of single-term equations. In contrast to the method described above this system will now be a multi-order system. To this end it will be useful to assume that all the integers that are contained in the interval $(0, n_k)$ are also members of the finite sequence $(n_j)_{j=1}^k$. In other words, it is impossible for two consecutive elements of the finite sequence (n_j) to lie on opposite sides of an integer number. It is obvious that such an assumption does not lead to any loss of generality. Then, given the equation (8.1a), we may write $\beta_1 := n_1$, $\beta_j := n_j - n_{j-1}$ $(j = 2, 3, \ldots, k)$, $y_1 := y$ and $y_j := D_*^{n_{j-1}} y$, $j = 2, 3, \ldots, k$. Note that under our assumptions on the n_j it is clear that $0 < \beta_j \leq 1$ for all j. Then we can conclude the following equivalence result:

Theorem 8.9. *Subject to the conditions above, the multi-term equation (8.1a) with initial conditions (8.1b) is equivalent to the system*

$$D_*^{\beta_1} y_1(x) = y_2(x),$$
$$D_*^{\beta_2} y_2(x) = y_3(x),$$
$$\vdots \quad \vdots$$
$$D_*^{\beta_{k-1}} y_{k-1}(x) = y_k(x),$$
$$D_*^{\beta_k} y_k(x) = f(x, y_1(x), y_2(x), \ldots, y_k(x)) \tag{8.14a}$$

with the initial conditions

$$y_j(0) = \begin{cases} y_0^{(0)} & \text{if } j = 1, \\ y_0^{(\ell)} & \text{if } n_{j-1} = \ell \in \mathbb{N}, \\ 0 & \text{else} \end{cases} \tag{8.14b}$$

in the following sense:

1. Whenever the function $y \in C^{\lceil n_k \rceil}[0,X]$ is a solution of the multi-term equation
   (8.1a) with initial conditions (8.1b), the vector-valued function $Y := (y_1, \ldots, y_k)^{\text{T}}$
   with

   $$y_j(x) := \begin{cases} y(x) & \text{if } j = 1, \\ D_*^{n_{j-1}} y(x) & \text{if } j \geq 2, \end{cases} \tag{8.15}$$

   is a solution of the multi-order fractional differential system (8.14a) with initial
   conditions (8.14b).
2. Whenever the vector-valued function $Y := (y_1, \ldots, y_k)^{\text{T}}$ is a solution of the multi-
   order fractional differential system (8.14a) with initial conditions (8.14b), the
   function $y := y_1$ is a solution of the multi-term equation (8.1a) with initial conditions (8.1b).

Proof. The proof proceeds along exactly the same steps as the proof of
Theorem 8.1. □

Example 8.1. Rewrite the initial value problem

$$D_*^{3.3} y(x) = f(x, y(x), D_*^{0.1} y(x), D^1 y(x), D_*^{1.2} y(x), D_*^{1.5} y(x),$$
$$D_*^{1.7} y(x), D^2 y(x), D_*^{2.2} y(x), D_*^{2.6} y(x), D^3 y(x)), \tag{8.16}$$
$$y^{(j)}(0) = y_0^{(j)}, \qquad j = 0, 1, 2, 3,$$

in the form indicated in Theorem 8.9.

In this case, the approach of Theorem 8.9 gives us the system

$$D_*^{0.1} y_1(x) = y_2(x), \qquad D_*^{0.9} y_2(x) = y_3(x),$$
$$D_*^{0.2} y_3(x) = y_4(x), \qquad D_*^{0.3} y_4(x) = y_5(x),$$
$$D_*^{0.2} y_5(x) = y_6(x), \qquad D_*^{0.3} y_6(x) = y_7(x),$$
$$D_*^{0.2} y_7(x) = y_8(x), \qquad D_*^{0.4} y_8(x) = y_9(x),$$
$$D_*^{0.4} y_9(x) = y_{10}(x), \qquad D_*^{0.3} y_{10}(x) = f(x, y_1(x), y_2(x), \ldots, y_{10}(x))$$

with initial conditions

$$y_1(0) = y_0^{(0)}, \qquad y_3(0) = y_0^{(1)},$$
$$y_7(0) = y_0^{(2)}, \qquad y_{10}(0) = y_0^{(3)},$$
$$y_j(0) = 0 \qquad \text{for } j \in \{2,4,5,6,8,9\}.$$

We additionally note the exact correspondence between the component functions y_j of the solution of the multi-order system on the one hand and the fractional derivatives of the solution y of the multi-term equation on the other hand, viz.

$$\begin{aligned}
y_1 &= y, & y_2 &= D_*^{0.1}y, \ y_3 = D^1 y, \\
y_4 &= D_*^{1.2}y, & y_5 &= D_*^{1.5}y, \ y_6 = D_*^{1.7}y, \\
y_7 &= D^2 y, & y_8 &= D_*^{2.2}y, \ y_9 = D_*^{2.6}y, \\
y_{10} &= D^3 y.
\end{aligned} \qquad (8.17)$$

Edwards et al. [59] have developed an alternative approach for the conversion of multi-term equations to multi-order systems. To describe this method we assume, for the sake of simplicity, that the highest order differential operator is not an integer-order derivative. Otherwise some small formal modifications in the notation are necessary, but the basic concept and the main results remain unchanged. The fundamental idea is best explained by looking at the problem from Example 8.1 again. We have two goals in mind. The first one is to retain the structure of the system (8.17) (in particular, the dimension of the system and the structure of the initial conditions). On the other hand we want to combine the components y_j in a different way such that the number of non-integer order derivatives is minimized. This leads us to the system

$$\begin{aligned}
D_*^{0.1}y_1(x) &= y_2(x), & D^1 y_1(x) &= y_3(x), \\
D_*^{0.2}y_3(x) &= y_4(x), & D_*^{0.5}y_3(x) &= y_5(x), \\
D_*^{0.7}y_3(x) &= y_6(x), & D^1 y_3(x) &= y_7(x), \\
D_*^{0.2}y_7(x) &= y_8(x), & D_*^{0.6}y_7(x) &= y_9(x), \\
D^1 y_7(x) &= y_{10}(x), & D_*^{0.3}y_{10}(x) &= f(x, y_1(x), y_2(x), \ldots, y_{10}(x)).
\end{aligned}$$

The approach via Theorem 8.9 created ten differential equations of strictly fractional orders whereas we now have seven strictly fractional and three first-order equations. This feature seems to be attractive from the point of view of computation cost when such systems are solved numerically: A first-order equation is cheaper to solve than a fractional-order equation because the operators involved in the latter are non-local. However, as indicated by Ford and Connolly [66], the structure of the system may be so inconvenient that the advantage gained by this locality can be lost completely.

For a general formal description of this method it is advantageous to express the multi-term equation in the form

$$D_*^{k+\delta_{k,\ell_k}} y(x) = f(x, D^0 y(x), D_*^{\delta_{0,1}} y(x), \ldots, D_*^{\delta_{0,\ell_0}} y(x),$$
$$D^1 y(x), D_*^{1+\delta_{1,1}} y(x), \ldots, D_*^{1+\delta_{1,\ell_1}} y(x), \ldots, \quad (8.18a)$$
$$D^k y(x) \ldots, D_*^{k+\delta_{k,\ell_k}-1} y(x)),$$

where $0 < \delta_{j,1} < \delta_{j,2} < \cdots < \delta_{j,\ell_j} < 1$ for all j. The corresponding initial conditions are then

$$y_j(0) = y_0^{(j)}, \qquad j = 0, 1, \ldots, k. \qquad (8.18b)$$

In order to achieve our goal, we define

$$s(\mu, \sigma) := \sigma + \mu + 1 + \sum_{j=0}^{\mu-1} \ell_j \quad \text{and} \quad N := s(k, \ell_k) - 1 = k + \sum_{j=0}^{k} \ell_j.$$

Notice that N is the total number of differential operators of strictly positive order in (8.18a). Thus, in our terminology, (8.18a) is an N-term equation. Using this notation, we come to the following statement.

Theorem 8.10. *The multi-term initial value problem (8.18) is equivalent to the N-dimensional system*

$$D_*^{\delta_{\mu,\sigma}} y_{s(\mu,0)}(x) = y_{s(\mu,\sigma)}(x), \quad \mu = 0, 1, \ldots, k, \quad \sigma = 1, 2, \ldots, \widehat{\sigma}_\mu,$$
$$D^1 y_{s(\mu,0)}(x) = y_{s(\mu+1,0)}(x), \quad \mu = 0, 1, \ldots, k-1, \qquad (8.19a)$$
$$D_*^{\delta_{k,n_k}} y_{s(k,0)}(x) = f(x, y_1(x), y_2(x), \ldots, y_N(x))$$

where $\widehat{\sigma}_\mu := \ell_\mu$ if $0 \le \mu < k$ and $\widehat{\sigma}_k := \ell_k - 1$, with the initial conditions

$$y_j(0) = \begin{cases} y_0^{(k)} & \text{if there exists } k \text{ such that } j = s(k,0), \\ 0 & \text{else} \end{cases} \qquad (8.19b)$$

in the following sense:

1. *Whenever the function $y \in C^{k+1}[0, X]$ is a solution of the multi-term equation (8.18a) with initial conditions (8.18b), the vector-valued function $Y := (y_1, \ldots, y_N)^{\mathrm{T}}$ with*

$$y_{s(\mu,\sigma)}(x) := \begin{cases} D^\mu y(x) & \text{for } \sigma = 0, \\ D_*^{\mu+\delta_{\mu,\sigma}} y(x) & \text{for } \sigma = 1, 2, \ldots, \widehat{\sigma}_\mu, \end{cases} \quad \mu = 0, 1, \ldots, k, \quad (8.20)$$

is a solution of the multi-order system (8.19a) with initial conditions (8.19b).

2. *Whenever the vector-valued function $Y := (y_1, \ldots, y_N)^{\mathrm{T}}$ is a solution of the multi-order system (8.19a) with initial conditions (8.19b), the function $y := y_1$ is a solution of the multi-term equation (8.18a) with initial conditions (8.18b).*

This result has been stated without proof for a subset of the class of equations described in (8.18a) in [59]. It is evident that the method used in the proof of Theorem 8.1 above, i.e. a repeated application of our Lemma 3.13, can once again be used to give a formal proof of Theorem 8.10. We leave the details to the reader.

A useful application of the concepts developed above can be found, as noted by Ford et al. [67], in the context of single-order fractional differential equations whose order is greater than 1. Indeed, given an initial value problem of the form

$$D^n_{*0}y(x) = f(x, y(x)), \qquad y^{(k)}(0) = y_0^{(k)} \quad (k = 0, 1, \dots, \lceil n \rceil - 1), \qquad (8.21)$$

with some non-integer $n > 1$, we may interpret the right-hand side of the differential equation formally as a function of x and $D^k y(x)$, $k = 0, 1, \dots, \lceil n \rceil - 1$, that actually does not depend on $D^k y(x)$, $k = 1, 2, \dots, \lceil n \rceil - 1$. Then, using our techniques developed in Theorem 8.9 we may rewrite the given problem in the equivalent form

$$\begin{aligned}
D^1 y_1(x) &= y_2(x), \\
D^1 y_2(x) &= y_3(x), \\
&\vdots \quad \vdots \\
D^1 y_{\lceil n \rceil - 1}(x) &= y_{\lceil n \rceil}(x), \\
D^{n - \lfloor n \rfloor}_{*0} y_{\lceil n \rceil}(x) &= f(x, y_1(x))
\end{aligned} \qquad (8.22a)$$

with initial conditions

$$y_k(0) = y_0^{(k-1)} \qquad (k = 1, 2, \dots, \lceil n \rceil), \qquad (8.22b)$$

i.e. as multi-order system with orders less than or equal to 1. Incidentally we could have applied Theorem 8.10 instead of Theorem 8.9 as well; in this special case both approaches lead to the same system. The system (8.22) is a somewhat easier object for numerical work as some algorithms tend to behave much worse when applied to equations of higher order. In addition, this concept may be considered an extension of the well known classical technique for the numerical solution of initial value problems of higher integer order which consists of rewriting the problem in the form of a first-order system and solving this system numerically with the help of an algorithm for first-order initial value problems.

We have not addressed the questions of existence and uniqueness of solutions for general multi-order systems (8.13a) with the corresponding initial conditions (8.13b) yet. For systems of the special form (8.14a) or (8.19a) we may use a reduction to multi-term equations by means of Theorem 8.9 or Theorem 8.10, respectively. However neither of these approaches covers the general case (8.13a). In order to prove an existence and uniqueness result for the latter we propose to use a completely different method of proof. Doing so, it is indeed possible to show the following result that actually covers a class of equations that is more general than that given in (8.13).

Theorem 8.11. *Let $n_j > 0$ for $j = 1, 2, \ldots, k$ and consider the initial value problem given by the multi-order fractional differential system*

$$D_*^{n_j} y_j(x) = f_j(x, y_1(x), \ldots, y_k(x)), \qquad j = 1, 2, \ldots, k, \qquad (8.23a)$$

with initial conditions

$$y_j^{(\ell)}(0) = y_{j,0}^{(\ell)}, \quad \ell = 0, 1, \ldots, \lceil n_j \rceil - 1, \quad j = 1, 2, \ldots, k. \qquad (8.23b)$$

Assume that the functions $f_j : [0, X] \times \mathbb{R}^k \to \mathbb{R}$, $j = 1, 2, \ldots, k$, are continuous and satisfy Lipschitz conditions with respect to all their arguments except for the first. Then the initial value problem has a uniquely determined continuous solution.

Proof. A componentwise application of Lemma 6.2 shows that the initial value problem (8.23) is equivalent to the Volterra equation system

$$y_j(x) = \sum_{\ell=0}^{\lceil n_j \rceil - 1} y_{j,0}^{(\ell)} \frac{x^\ell}{\ell!} + \frac{1}{\Gamma(n_j)} \int_0^x (x-t)^{n_j-1} f_j(t, y_1(t), \ldots, y_k(t)) \, dt, \qquad (8.24)$$

$j = 1, 2, \ldots, k$. We then define $n := \min_j n_j$ and rewrite (8.24) in the form

$$y_j(x) = \sum_{\ell=0}^{\lceil n_j \rceil - 1} y_{j,0}^{(\ell)} \frac{x^\ell}{\ell!} + \frac{1}{\Gamma(n)} \int_0^x (x-t)^{n-1} \widetilde{f}_j(t, y_1(t), \ldots, y_k(t)) \, dt$$

for $j = 1, 2, \ldots, k$ where

$$\widetilde{f}_j(t, y_1, \ldots, y_k) := \frac{\Gamma(n)}{\Gamma(n_j)} (x-t)^{n_j-n} f_j(t, y_1, \ldots, y_k).$$

Introducing the vector notation $Y := (y_1, \ldots, y_k)^T$, $\widetilde{F} := (\widetilde{f}_1, \ldots, \widetilde{f}_k)^T$, and a corresponding expression for the initial values, we can combine these k scalar equations into one vector equation, viz.

$$Y(x) = \sum_{\ell=0}^{\max_j \lceil n_j \rceil - 1} Y_0^{(\ell)} \frac{x^\ell}{\ell!} + \frac{1}{\Gamma(n)} \int_0^x (x-t)^{n-1} \widetilde{F}(t, Y(t)) \, dt.$$

In view of our assumptions on the f_j and the fact that $n_j \geq n$ for all j, we conclude that all the functions $\widetilde{f}_j$ are continuous and satisfy Lipschitz conditions with respect to $y_1, \ldots, y_k$. Thus, $\widetilde{F}$ is continuous and satisfies a Lipschitz condition with respect to Y, and in view of some standard results from the theory of Volterra integral equations [115, Theorem 4.8] (see also [115, Section 3.5]) we can conclude the existence and uniqueness of a continuous solution $Y = (y_1, \ldots, y_k)^T$. □

Remark 8.5. Evidently, the combination of Theorems 8.11 and 8.9 gives us a second method to prove the general existence and uniqueness result for multi-term equations, viz. Theorem 8.7.

We have thus presented three different ways to construct a system of single-term fractional differential equations from a given multi-term equation. The first approach, described in Theorem 8.1, leads to a system whose constituent equations all have the same order. While this feature may look appealing, its disadvantages are that the dimension of the system may be very large and that it is applicable only if the orders of the differential operators satisfy certain number-theoretic conditions. The second approach is contained in Theorem 8.9. In general, it will lead to a multi-order system of differential equations, i.e. a system whose equations may contain differential operators of different orders, but typically none of these orders will be integers. This concept is always applicable, and it tends to produce much smaller systems than the first approach. Finally, the third approach (see Theorem 8.10) is very similar to the second one, but it replaces fractional-order differential equations by first-order equations wherever possible. It is always applicable too, and it produces systems whose dimensions are the same as those created by the second approach.

A detailed comparison of these three methods has been provided by Ford and Connolly [66] whose main goal was to find out which of these schemes was most suitable for being combined with numerical algorithms for (single- or multi-order) fractional differential systems in an attempt to create an efficient strategy for the numerical solution of multi-term equations. Their main result was that the reformulation of the given multi-term equation in the form indicated in Theorem 8.10 usually turned out to give the weakest performance, whereas the two other approaches were much more efficient. Whether the application of Theorem 8.1 is to be preferred over Theorem 8.9 or vice versa depends on the precise nature of the given multi-term equation, particularly on the distribution of the orders of the differential operators involved. We refer to Appendix C.2 for a further discussion of this topic.

For linear equations, it is possible to derive explicit expressions for the solution. The two-parameter Mittag-Leffler functions turn out to be very useful tools in this context. We shall require the following fundamental result concerning the Laplace transform of certain functions closely related to these Mittag-Leffler functions:

Lemma 8.12. *Consider the two-parameter Mittag-Leffler function E_{n_1,n_2} for some $n_1, n_2 > 0$. Let $j \in \mathbb{N}_0$, $a \in \mathbb{R}$ and*

$$z_{\pm}(x) := x^{jn_1+n_2-1} E_{n_1,n_2}^{(j)}(\pm ax^{n_1}).$$

Then, for $s > |a|$,

$$\mathscr{L}z_{\pm}(s) = \frac{j! s^{n_1-n_2}}{(s^{n_1} \mp a)^{j+1}}.$$

We leave the proof as an exercise to the reader and proceed by using these results in order to look at one particularly important case explicitly.

Example 8.2. Solve the Bagley–Torvik equation

$$D_{*0}^2 y(x) + 2D_{*0}^{3/2} y(x) + 2y(x) = \sin x$$

with initial conditions $y(0) = y'(0) = 0$.

As a first approach, we recall that, according to Theorem 8.1, the equation may be transformed into a four-dimensional system of equations of order $1/2$. The precise form of the system is

$$\begin{aligned}
D_{*0}^{1/2} y_0(x) &= y_1(x), \\
D_{*0}^{1/2} y_1(x) &= y_2(x), \\
D_{*0}^{1/2} y_2(x) &= y_3(x), \\
D_{*0}^{1/2} y_3(x) &= -2y_0(x) - 2y_3(x) + \sin x,
\end{aligned} \tag{8.25}$$

combined with the initial conditions $y_j(0) = 0$ for $j = 0, 1, 2, 3$. We may then adapt the statement of Theorem 7.2 to this multidimensional setting, i.e. we have to take into consideration that the parameter λ appearing there is now a matrix,

$$\lambda = \begin{pmatrix} 0 & 1 & 0 & 0 \\ 0 & 0 & 1 & 0 \\ 0 & 0 & 0 & 1 \\ -2 & 0 & 0 & -2 \end{pmatrix}; \quad \text{hence} \quad \lambda^{-1} = \begin{pmatrix} 0 & 0 & -1 & -1/2 \\ 1 & 0 & 0 & 0 \\ 0 & 1 & 0 & 0 \\ 0 & 0 & 1 & 0 \end{pmatrix}.$$

Since we have homogeneous initial conditions we thus derive that the solution of the system is

$$Y(x) = \begin{pmatrix} y_0(x) \\ y_1(x) \\ y_2(x) \\ y_3(x) \end{pmatrix} = \lambda^{-1} \int_0^x u(t) \begin{pmatrix} 0 \\ 0 \\ 0 \\ \sin(x-t) \end{pmatrix} dt$$

where

$$u(x) = \frac{d}{dx} E_{1/2}(\lambda x^{1/2}) = \frac{d}{dx} \sum_{j=0}^\infty \frac{\lambda^j x^{j/2}}{\Gamma(1+j/2)}$$

is a matrix-valued function. By construction, λ^{-1} commutes with $u(t)$ (for any t), and thus

$$Y(x) = \int_0^x u(t) \lambda^{-1} \begin{pmatrix} 0 \\ 0 \\ 0 \\ \sin(x-t) \end{pmatrix} dt = -\frac{1}{2} \int_0^x u(t) \begin{pmatrix} \sin(x-t) \\ 0 \\ 0 \\ 0 \end{pmatrix} dt.$$

Because we were originally interested in the solution of the given Bagley–Torvik equation, we only need to look at the first component (with index 0) of the vector Y; we find

$$y(x) = -\frac{1}{2} \int_0^x u_{11}(t) \sin(x-t)\,dt$$

where $u_{11}(t)$ is the top left component of the matrix $u(t)$.

Since the equation is a linear equation with constant coefficients, we may alternatively use the Laplace transform method. Applying Laplace transforms to both sides of the equation we find

$$s^2 \mathscr{L}y(s) + 2s^{3/2}\mathscr{L}y(s) + 2\mathscr{L}y(s) = \mathscr{L}\sin(s) = \frac{1}{s^2+1}.$$

Thus,

$$\mathscr{L}y(s) = \frac{1}{(s^2+1)(s^2+2s^{3/2}+2)},$$

and we find

$$y(x) = \int_0^x z(x-t)\sin t\,dt$$

where

$$\mathscr{L}z(s) = \frac{1}{s^2+2s^{3/2}+2} = \frac{1}{2}\frac{2s^{-3/2}}{s^{1/2}+2}\frac{1}{1+\frac{2s^{-3/2}}{s^{1/2}+2}} = \frac{1}{2}\gamma\frac{1}{1+\gamma}$$

with $\gamma := 2s^{-3/2}/(s^{1/2}+2)$. For sufficiently large s (and for the purpose of Laplace transform theory it is sufficient to consider these values only) we have $|\gamma| < 1$, and therefore we may expand $\mathscr{L}z(s)$ using the well known geometric series. This yields

$$\mathscr{L}z(s) = \frac{1}{2}\gamma\sum_{k=0}^{\infty}(-1)^k\gamma^k = \frac{1}{2}\sum_{k=0}^{\infty}(-1)^k 2^{k+1}\frac{s^{-3(k+1)/2}}{(s^{1/2}+2)^{k+1}}$$

$$= \sum_{k=0}^{\infty}(-1)^k 2^k \frac{s^{-3(k+1)/2}}{(s^{1/2}+2)^{k+1}}.$$

A term-by-term inverse Laplace transform is possible here and gives, in view of Lemma 8.12,

$$z(x) = \sum_{k=0}^{\infty}(-1)^k \frac{2^k}{k!} x^{2k+1} E_{1/2,2+3k/2}^{(k)}(-2x^{1/2}).$$

Combining these equations we find the final result,

$$y(x) = \sum_{k=0}^{\infty}(-1)^k \frac{2^k}{k!} \int_0^x (x-t)^{2k+1} E_{1/2,2+3k/2}^{(k)}(-2(x-t)^{1/2}) \sin t\,dt.$$

The reader is encouraged to give an explicit proof that this solution is identical to the solution obtained by the first approach.

Yet another possibility to find the solution of the Bagley–Torvik equation is provided by an application of Theorem 7.13 to the homogeneous equation associated to (8.25), followed by the use of the variation-of-constants method as indicated in Remark 7.1 to find a particular solution of the inhomogeneous problem. We leave the details as an exercise.

Additional special cases of linear multi-term equations are discussed by Nkamnang [140, §3.5]; the resulting formulas are typically highly complicated and we do not repeat them here. More results may be found, e.g., in the paper of Luchko and Gorenflo [123], and in Podlubny's book [153, Chapters 4 and 5].

Let us close this chapter with a brief outlook: Looking at a general linear multi-term equation,

$$\sum_{j=0}^{k} \alpha_j(x) D_{*0}^{n_j} y(x) = g(x)$$

with some given functions $\alpha_0, \alpha_1, \ldots, \alpha_k$ and g and $0 = n_0 < n_1 < \ldots < n_k$, we see that we can rewrite such an equation as

$$\int_0^{n_k} \alpha_j(x) D_{*0}^{n_j} y(x) \, d\mu(j) = g(x)$$

where μ is a step function with jumps of unit height at the points n_j. If we allow μ to be a more general measure, then this immediately leads us to the so-called *distributed order differential equations*. First steps on this topic, concerning both analytic solution methods and the mathematical modelling of physical problems with those equations, have been described by Bagley and Torvik [10, 11]; a framework for a numerical approach can be found in the papers [42, 46]. However, the discussion of this topic is beyond the scope of this book.

Exercises

Exercise 8.1. Give a proof of Lemma 8.6.

Exercise 8.2. Give a proof of Theorem 8.10.

Exercise 8.3. Give a proof of Lemma 8.12.

Exercise 8.4. Show that the two solutions given for Example 8.2 are identical.

Exercise 8.5. Compute the solution of the system (8.25) with homogeneous initial conditions using the method of Theorem 7.13 and the variation-of-constants method and verify that its first component is the solution of the initial value problem of Example 8.2.

Exercise 8.6. Determine the solution of the Bagley–Torvik initial value problem

$$D_{*0}^2 y(x) + D_{*0}^{3/2} y(x) + 2y(x) = f(x), \qquad y(0) = y'(0) = 0,$$

with

(a) $f(x) = \begin{cases} 8 & \text{for } x > 1, \\ 0 & \text{else}; \end{cases}$

(b) $f(x) = 1$.

Exercise 8.7. Consider the initial value problem

$$D_{*0}^{1.455}y(x) = -x^{0.1}\frac{E_{1.545}(-x)}{E_{1.445}(-x)}\exp(x)y(x)D_{*0}^{0.555}y(x)+\exp(-2x)-[D_*^1y(x)]^2$$

for $0 \leq x \leq 1$, equipped with the initial conditions $y(0) = 1$ and $y'(0) = -1$.

(a) Prove that this problem has a unique solution.
(b) Verify that the solution is given by $y(x) = \exp(-x)$.
(c) Rewrite the equation in the form of an equivalent system according to Theorem 8.1. Choose the parameters such that the dimension of the system is as small as possible.
(d) Calculate the first four elements of the Picard iteration sequence for this system and compare them with the exact solution.

Appendix

Appendix A
List of Symbols

In this appendix, we collect a list of all the symbols that have been used in this book. If necessary, cross-references are also given.

Functions

$\|\cdot\|_\infty$ Chebyshev norm; $\|f\|_\infty = \sup_{a\le x\le b}|f(x)|$

$\|\cdot\|_p$ L_p norm ($1 \le p < \infty$); $\|f\|_p = (\int_a^b |f(x)|^p\,\mathrm{d}x)^{1/p}$

$\lfloor\cdot\rfloor$ Floor function, $\lfloor x\rfloor = \max\{z \in \mathbb{Z} : z \le x\}$

$\lceil\cdot\rceil$ Ceiling function, $\lceil x\rceil = \min\{z \in \mathbb{Z} : z \ge x\}$

$\binom{n}{k}$ Binomial coefficient, $\binom{n}{k} = n(n-1)(n-2)\cdots(n-k+1)/k!$ for $n \in \mathbb{R}$ and $k \in \mathbb{N}_0$

$B_N[f]$ Nth Bernstein polynomial for the function f, $B_N[f](t) = \sum_{k=0}^N \binom{N}{k}t^k(1-t)^{N-k}f(k/N)$ (see Appendix D.5)

B Euler's Beta function, $B(x,y) = \Gamma(x)\Gamma(y)/\Gamma(x+y)$ (cf. Appendix D.1)

Γ Euler's Gamma function, $\Gamma(x) = \int_0^\infty t^{x-1}\mathrm{e}^{-t}\,\mathrm{d}t$ (cf. Definition 1.2)

ψ Digamma function, $\psi(x) = \Gamma'(x)/\Gamma(x)$

E_n Mittag-Leffler function of order n, $E_n(x) = \sum_{j=0}^\infty x^j/\Gamma(jn+1)$ (cf. Definition 4.1)

E_{n_1,n_2} two-parameter Mittag-Leffler function, $E_{n_1,n_2}(x) = \sum_{j=0}^\infty x^j/\Gamma(jn_1+n_2)$ (cf. Definition 4.2)

${}_1F_1$ Kummer's confluent hypergeometric function [2, Chapter 13], $${}_1F_1(a;b;z) = \frac{\Gamma(b)}{\Gamma(a)}\sum_{k=0}^\infty \frac{\Gamma(a+k)}{\Gamma(b+k)k!}z^k \quad (a \in \mathbb{R}, -b \notin \mathbb{N}_0)$$

189

$_2F_1$ Gauss' hypergeometric function [2, Chapter 15],
$$_2F_1(a,b;c;z) = \frac{\Gamma(c)}{\Gamma(a)\Gamma(b)} \sum_{k=0}^{\infty} \frac{\Gamma(a+k)\Gamma(b+k)}{\Gamma(c+k)k!} z^k \quad (a,b \in \mathbb{R}, -c \notin \mathbb{N}_0)$$

o, O Landau symbols

$T_j[f;a]$ Taylor polynomial of degree j for the function f
           centered at the point a

Sets

$A^n, A^n[a,b]$ Set of functions with absolutely continuous derivative of
                 order $n-1$ (cf. Definition 1.5)
$C, C[a,b]$ Set of continuous functions (cf. Definition 1.3)
$C^k, C^k[a,b]$ Set of functions with continuous kth derivative
                 (cf. Definition 1.3)
$H^*, H^*[a,b]$ Cf. Definition 1.4
$H_\mu, H_\mu[a,b]$ Hölder space (cf. Definition 1.3)
$L_p, L_p[a,b]$ Lebesgue space (cf. Definition 1.3)
$\mathbb{N}$ $= \{1,2,3,\ldots\}$, the set of natural numbers
$\mathbb{N}_0$ $= \mathbb{N} \cup \{0\}$
$\mathbb{R}$ The set of real numbers
$\mathbb{R}_+$ $= \{x \in \mathbb{R} : x > 0\}$, the set of strictly positive real numbers
$\mathbb{Z}$ $= \{0, \pm 1, \pm 2, \pm 3 \ldots\}$, the set of integer numbers

Operators

Δ_h^n Finite difference of order n; cf. (2.11)
D Differential operator, $Df(x) = f'(x)$ (cf. Definition 1.1)
D^n $n \in \mathbb{N}$: n-fold iterate of the differential operator D (cf. Definition 1.1)
D_a^n $n \in \mathbb{R}_+$: Riemann–Liouville fractional differential operator (cf. Definition 2.2)
$\widetilde{D}_a^n$ $n \in \mathbb{R}_+$: Grünwald–Letnikov fractional differential operator (cf. Definition 2.3)
$\widehat{D}_a^n$ Cf. Definition 3.1
D_{*a}^n $n \in \mathbb{R}_+$: Caputo fractional differential operator (cf. Definition 3.2)
$\mathscr{D}^n$ $n \in \mathbb{R}_+$: Gel'fond–Leont'ev operator (cf. Definition 3.3)
I Identity operator
J_a Integral operator, $J_a f(x) = \int_a^x f(t)\,dt$ (cf. Definition 1.1)

J_a^n $n \in \mathbb{N}$: n-fold iterate of the integral operator J_a (cf. Definition 1.1)

   $n \in \mathbb{R}_+ \setminus \mathbb{N}$: Riemann–Liouville fractional integral operator (cf. Definition 2.1)

   $n = 0$: identity operator (cf. Definition 2.1)

$\tilde{J}_a^n$ $n \in \mathbb{R}_+$: Grünwald–Letnikov fractional integral operator

   (cf. Definition 2.4)

$\mathscr{L}$ Laplace transform operator (cf. Appendix D.3)

ω Modulus of continuity of the function $g : [a,b] \to \mathbb{R}$,

$$\omega(g;h) := \sup\{|g(y_1) - g(y_2)| : y_1, y_2 \in [a,b], |y_1 - y_2| \le h\}$$

Other Symbols

$$\sim \qquad a_j \sim b_j \Leftrightarrow \exists A, B > 0 \, \exists j_0 \in \mathbb{N} \, \forall j \ge j_0 : A \le |a_j/b_j| \le B$$

Remarks

1. The power series for both types of Mittag-Leffler functions converge in the entire complex plane (cf. Theorem 4.1).
2. The power series for Kummer's confluent hypergeometric function converges in the entire complex plane.
3. For the Gauss hypergeometric function, the power series converges for all complex z with $|z| < 1$ and may be extended analytically into the entire complex plane with a branch cut along the positive real axis from $+1$ to $+\infty$. (In the formulas in Appendix B, we need to evaluate this function for $z < 0$, so the branch cut for $z \ge 1$ gives no problems.)

Appendix B
A Table of Caputo Derivatives

For the convenience of the reader, we provide this appendix where we give some Caputo-type derivatives of certain important functions. We do not strive for completeness in any sense, but we do want to give at least the derivatives of the classical examples.

Throughout this appendix, n will always denote the order of the Caputo-type differential operator under consideration. We shall only consider the case $n > 0$ and $n \notin \mathbb{N}$, and we use the notation $m := \lceil n \rceil$ to denote the smallest integer greater than (or equal to) n. Recall that for $n \in \mathbb{N}$, the Caputo differential operator coincides with the usual differential operator of integer order, and for $n < 0$, the Caputo differential operator of negative order can be interpreted as the Riemann–Liouville differential operator of the same order. Tables of the latter are given in various places in the literature (cf., e.g., Podlubny [153] or Samko et al. [167]); we are not going to repeat those results here.

Various special functions will arise in this connection; for the precise definitions we refer to Appendix A. By $i = \sqrt{-1}$ we denote the imaginary unit.

1. Let $f(x) = x^j$. Here we have to distinguish some cases:

$$(D_{*0}^n f)(x) = \begin{cases} 0 & \text{if } j \in \mathbb{N}_0 \text{ and } j < m, \\ \dfrac{\Gamma(j+1)}{\Gamma(j+1-n)} x^{j-n} & \text{if } j \in \mathbb{N}_0 \text{ and } j \geq m \\ & \text{or } j \notin \mathbb{N} \text{ and } j > m-1. \end{cases}$$

2. Let $f(x) = (x+c)^j$ for arbitrary $c > 0$ and $j \in \mathbb{R}$. Then

$$(D_{*0}^n f)(x) = \frac{\Gamma(j+1)}{\Gamma(j+1-m)} \frac{c^{j-m-1} x^{m-n}}{\Gamma(m-n+1)} \, {}_2F_1(1, m-j; m-n+1; -x/c).$$

3. Let $f(x) = \exp(jx)$ for some $j \in \mathbb{R}$. Then

$$(D_{*0}^n f)(x) = j^m x^{m-n} E_{1,m-n+1}(jx).$$

4. Let $f(x) = x^j \ln x$ for some $j > m - 1$. Then

$$(D_{*0}^n f)(x) = x^{j-n} \sum_{k=0}^{m-1} (-1)^{m-k+1} \binom{j}{k} \frac{m!}{m-k} \frac{\Gamma(j-m+1)}{\Gamma(j-n+1)}$$

$$+ \frac{\Gamma(j+1)}{\Gamma(j-n+1)} x^{j-n} (\psi(j-m+1) - \psi(j-n+1) + \ln x).$$

5. Let $f(x) = \sin jx$ for some $j \in \mathbb{R}$. Here again we have two cases:

$$(D_{*0}^n f)(x) = \begin{cases} \dfrac{j^m i (-1)^{m/2} x^{m-n}}{2\Gamma(m-n+1)} [-{}_1F_1(1; m-n+1; ijx) \\ \qquad\qquad + {}_1F_1(1; m-n+1; -ijx)] & (m \text{ even}), \\ \dfrac{j^m (-1)^{(m-1)/2} x^{m-n}}{2\Gamma(m-n+1)} [{}_1F_1(1; m-n+1; ijx) \\ \qquad\qquad + {}_1F_1(1; m-n+1; -ijx)] & (m \text{ odd}). \end{cases}$$

6. Finally we consider $f(x) = \cos jx$ with some $j \in \mathbb{R}$. As in the previous example, we obtain two cases:

$$(D_{*0}^n f)(x) = \begin{cases} \dfrac{j^m (-1)^{m/2} x^{m-n}}{2\Gamma(m-n+1)} [{}_1F_1(1; m-n+1; ijx) \\ \qquad\qquad + {}_1F_1(1; m-n+1; -ijx)] & (m \text{ even}), \\ \dfrac{j^m i (-1)^{(m-1)/2} x^{m-n}}{2\Gamma(m-n+1)} [{}_1F_1(1; m-n+1; ijx) \\ \qquad\qquad - {}_1F_1(1; m-n+1; -ijx)] & (m \text{ odd}). \end{cases}$$

Appendix C
Numerical Solution of Fractional Differential Equations

For most fractional differential equations we cannot provide methods to compute the exact solutions analytically. Therefore it is necessary to revert to numerical methods. In order to give the reader a tool that can be applied to a very wide class of equations, we now present a method that is well understood and that has proven to be efficient in many practical applications [50, 51, 65, 182, 183]. We begin by discussing this problem for single-term equations and later extend our idea to multi-term problems.

C.1 An Algorithm for Single-Term Equations

The method can be called *indirect* because, rather than discretizing the differential equation

$$D_{*0}^n y(x) = f(x, y(x))$$

with appropriate initial conditions

$$D^k y(0) = y_0^{(k)}, \qquad k = 0, 1, \ldots, \lceil n \rceil - 1,$$

directly, it requires some preliminary analytical manipulation, namely an application of Lemma 6.2 in order to convert the initial value problem for the differential equation into an equivalent Volterra integral equation,

$$y(x) = \sum_{k=0}^{m-1} \frac{x^k}{k!} D^k y(0) + \frac{1}{\Gamma(n)} \int_0^x (x-t)^{n-1} f(t, y(t)) \, dt \qquad (C.1)$$

where $m = \lceil n \rceil$. We shall therefore now look at a method for the numerical solution of (C.1).

The algorithm that we shall consider can be interpreted as a fractional variant of the classical second-order Adams–Bashforth–Moulton method. It has been introduced and briefly discussed in [50]; more information is given in [51]. Some additional results for a specific initial value problem are contained in [44], a detailed mathematical analysis is provided in [49], and additional practical remarks can be found in [48]. Numerical experiments and comparisons with other methods are reported in [52, 65, 182, 183]. Here we shall give an even more detailed analysis under quite general assumptions.

Remark C.1. Before starting the investigations, we need to give a note of caution. It is common to construct methods for fractional differential equations by taking methods for classical (typically first-order) equations and then generalizing the concepts in an appropriate way. The resulting formulas are then usually given the same name as the underlying classical algorithm, possibly extended by the adjective "fractional". However, many classical numerical schemes can be extended in more than one way. This may lead to the problem that, in two different items of literature, two different algorithms are denoted in identical ways. Of course, this is a potential source for confusion, and the reader must be very careful in this respect. For example, the fractional Adams–Moulton rules of Galeone and Garrappa [70] do not coincide with the methods of the same name that we shall develop below.

Classical Formulation

In order to motivate the construction of the method, we shall first briefly recall the idea behind the classical Adams–Bashforth–Moulton algorithm for first-order equations. So, for a start, we focus our attention on the well-known initial-value problem for the first-order differential equation

$$Dy(x) = f(x, y(x)), \tag{C.2a}$$

$$y(0) = y_0. \tag{C.2b}$$

We assume the function f to be such that a unique solution exists on some interval $[0, T]$, say. Following [88, §III.1], we suggest to use the predictor-corrector technique of Adams where, for the sake of simplicity, we assume that we are working on a uniform grid $\{t_j = jh : j = 0, 1, \ldots, N\}$ with some integer N and $h = T/N$. In some applications it may be more efficient to use a non-uniform grid, and we will develop the numerical approximation formulas in this generalized sense. However, for the subsequent analysis of the properties of the scheme we will then restrict ourselves to the equispaced case.

The basic idea is, assuming that we have already calculated approximations $y_j \approx y(t_j)$ $(j = 1, 2, \ldots, k)$, that we try to obtain the approximation y_{k+1} by means of the equation

$$y(t_{k+1}) = y(t_k) + \int_{t_k}^{t_{k+1}} f(z, y(z)) \, dz. \tag{C.3}$$

This equation follows upon integration of (C.2a) on the interval $[t_k, t_{k+1}]$. Of course, we know neither of the expressions on the right-hand side of (C.3) exactly, but we do have an approximation for $y(t_k)$, namely y_k, that we can use instead. The integral is then replaced by the two-point trapezoidal quadrature formula

$$\int_a^b g(z)\,dz \approx \frac{b-a}{2}\left(g(a)+g(b)\right),\tag{C.4}$$

thus giving an equation for the unknown approximation y_{k+1}, it being

$$y_{k+1} = y_k + \frac{t_{k+1}-t_k}{2}\left(f(t_k,y(t_k))+f(t_{k+1},y(t_{k+1}))\right),\tag{C.5}$$

where again we have to replace $y(t_k)$ and $y(t_{k+1})$ by their approximations y_k and y_{k+1}, respectively. This yields the equation for the implicit one-step *Adams–Moulton method*, which is

$$y_{k+1} = y_k + \frac{t_{k+1}-t_k}{2}\left(f(t_k,y_k)+f(t_{k+1},y_{k+1})\right).\tag{C.6}$$

The problem with this equation is that the unknown quantity y_{k+1} appears on both sides, and due to the nonlinear nature of the function f, we cannot solve for y_{k+1} directly in general. Therefore, we may use (C.6) in an iterative process, inserting a preliminary approximation for y_{k+1} in the right-hand side in order to determine a better approximation that we can then use.

The preliminary approximation y_{k+1}^P, the so-called predictor, is obtained in a very similar way, only replacing the trapezoidal quadrature formula by the rectangle rule

$$\int_a^b g(z)\,dz \approx (b-a)g(a),\tag{C.7}$$

giving the explicit (forward Euler or one-step *Adams–Bashforth*) method

$$y_{k+1}^P = y_k + hf(t_k,y_k).\tag{C.8}$$

It is well known [88, p. 372] that the process defined by (C.8) and

$$y_{k+1} = y_k + \frac{h}{2}\left(f(t_k,y_k)+f(t_{k+1},y_{k+1}^P)\right),\tag{C.9}$$

known as the one-step *Adams–Bashforth–Moulton* technique, is convergent of order 2, i.e.

$$\max_{j=1,2,\dots,N}|y(t_j)-y_j| = O(h^2).\tag{C.10}$$

Moreover, this method behaves satisfactorily from the point of view of its numerical stability [89, Chap. IV]. It is said to be of the PECE (Predict, Evaluate, Correct, Evaluate) type because, in a concrete implementation, we would start by calculating the predictor in (C.8), then we evaluate $f(t_{k+1},y_{k+1}^P)$, use this to calculate the corrector in (C.9), and finally evaluate $f(t_{k+1},y_{k+1})$. This result is stored for future use in the next integration step.

Fractional Formulation

Having introduced this concept, we now try to carry over the essential ideas to the fractional-order problem with some unavoidable modifications. The key is to derive an equation similar to (C.3). Fortunately, such an equation is available, namely (C.1). This equation looks somewhat different from (C.3), because the range of integration now starts at 0 instead of t_k. This is a consequence of the non-local structure of the fractional-order differential operators. This however does not cause major problems in our attempts to generalize the Adams method. What we do is simply use the product trapezoidal quadrature formula to replace the integral, i.e. we use the nodes t_j $(j = 0, 1, \ldots, k+1)$ and interpret the function $(t_{k+1} - \cdot)^{n-1}$ as a weight function for the integral. In other words, we apply the approximation

$$\int_0^{t_{k+1}} (t_{k+1} - z)^{n-1} g(z) \, dz \approx \int_0^{t_{k+1}} (t_{k+1} - z)^{n-1} \tilde{g}_{k+1}(z) \, dz, \qquad (C.11)$$

where $\tilde{g}_{k+1}$ is the piecewise linear interpolant for g with nodes and knots chosen at the t_j, $j = 0, 1, 2, \ldots, k+1$.

It is clear by construction that the required weighted trapezoidal quadrature formula be represented as a weighted sum of function values of the integrand g, taken at the points t_j. Specifically, we find that we can write the integral on the right-hand side of (C.11) as

$$\int_0^{t_{k+1}} (t_{k+1} - z)^{n-1} \tilde{g}_{k+1}(z) \, dz = \sum_{j=0}^{k+1} a_{j,k+1} g(t_j) \qquad (C.12a)$$

where

$$a_{j,k+1} = \int_0^{t_{k+1}} (t_{k+1} - z)^{n-1} \phi_{j,k+1}(z) \, dz \qquad (C.12b)$$

and

$$\phi_{j,k+1}(z) = \begin{cases} (z - t_{j-1})/(t_j - t_{j-1}) & \text{if } t_{j-1} < z \le t_j, \\ (t_{j+1} - z)/(t_{j+1} - t_j) & \text{if } t_j < z < t_{j+1}, \\ 0 & \text{else.} \end{cases} \qquad (C.12c)$$

This is clear because the functions $\phi_{j,k+1}$ satisfy

$$\phi_{j,k+1}(t_\mu) = \begin{cases} 0 & \text{if } j \ne \mu, \\ 1 & \text{if } j = \mu, \end{cases}$$

and that they are continuous and piecewise linear with breakpoints at the nodes t_μ, so that they must be integrated exactly by our formula.

An easy explicit calculation yields that, for an arbitrary choice of the t_j, (C.12b) and (C.12c) produce

$$a_{0,k+1} = \frac{(t_{k+1} - t_1)^{n+1} + t_{k+1}^n [n t_1 + t_1 - t_{k+1}]}{t_1 n(n+1)}, \qquad (C.13a)$$

$$a_{j,k+1} = \frac{(t_{k+1}-t_{j-1})^{n+1} + (t_{k+1}-t_j)^n [n(t_{j-1}-t_j)+t_{j-1}-t_{k+1}]}{(t_j-t_{j-1})n(n+1)}$$
$$+ \frac{(t_{k+1}-t_{j+1})^{n+1} - (t_{k+1}-t_j)^n [n(t_j-t_{j+1})-t_{j+1}+t_{k+1}]}{(t_{j+1}-t_j)n(n+1)}, \quad \text{(C.13b)}$$

if $1 \le j \le k$, and

$$a_{k+1,k+1} = \frac{(t_{k+1}-t_k)^n}{n(n+1)}. \quad \text{(C.13c)}$$

In the case of equispaced nodes ($t_j = jh$ with some fixed h), these relations reduce to

$$a_{j,k+1} = \begin{cases} \dfrac{h^n}{n(n+1)} \left(k^{n+1} - (k-n)(k+1)^n\right), & \text{if } j = 0, \\[2ex] \dfrac{h^n}{n(n+1)} \left((k-j+2)^{n+1} + (k-j)^{n+1}\right) & \\ \qquad -2(k-j+1)^{n+1}) & \text{if } 1 < j \le k, \\[2ex] \dfrac{h^n}{n(n+1)} & \text{if } j = k+1. \end{cases} \quad \text{(C.14)}$$

This then gives us our corrector formula (i.e. the fractional variant of the one-step *Adams–Moulton method*), which is

$$y_{k+1} = \sum_{j=0}^{m-1} \frac{t_{k+1}^j}{j!} y_0^{(j)} + \frac{1}{\Gamma(n)} \left(\sum_{j=0}^{k} a_{j,k+1} f(t_j, y_j) + a_{k+1,k+1} f(t_{k+1}, y_{k+1}^P) \right). \quad \text{(C.15)}$$

The remaining problem is the determination of the predictor formula that we require to calculate the value y_{k+1}^P. The idea we use to generalize the one-step Adams–Bashforth method is the same as the one described above for the Adams–Moulton technique: We replace the integral on the right-hand side of (C.1) by the product rectangle rule

$$\int_0^{t_{k+1}} (t_{k+1}-z)^{n-1} g(z)\, dz \approx \sum_{j=0}^{k} b_{j,k+1} g(t_j), \quad \text{(C.16)}$$

where now

$$b_{j,k+1} = \int_{t_j}^{t_{j+1}} (t_{k+1}-z)^{n-1}\, dz = \frac{(t_{k+1}-t_j)^n - (t_{k+1}-t_{j+1})^n}{n}. \quad \text{(C.17)}$$

This expression for weights can be derived in a way similar to the method used in the derivation of (C.13). However, here we are dealing with a piecewise constant approximation and not a piecewise linear one, and hence we have to replace the "hat-shaped" functions ϕ_{kj} by functions being of constant value 1 on $[t_j, t_{j+1}]$ and 0 on the remaining parts of the interval $[0, t_{k+1}]$. Again, in the equispaced case, we have the simpler expression

$$b_{j,k+1} = \frac{h^n}{n} \left((k+1-j)^n - (k-j)^n \right). \tag{C.18}$$

Thus, the predictor y_{k+1}^{P} is determined by the fractional *Adams–Bashforth method*

$$y_{k+1}^{\mathrm{P}} = \sum_{j=0}^{m-1} \frac{t_{k+1}^j}{j!} y_0^{(j)} + \frac{1}{\Gamma(n)} \sum_{j=0}^{k} b_{j,k+1} f(t_j, y_j). \tag{C.19}$$

Our basic algorithm, the fractional *Adams–Bashforth–Moulton method*, is therefore completely described now by (C.19) and (C.15) with the weights $a_{j,k+1}$ and $b_{j,k+1}$ being defined according to (C.13) and (C.17), respectively.

Error Analysis

For the error analysis of this algorithm, we restrict our attention to the case of an equispaced grid, i.e. from now on we assume that $t_j = jh = jT/N$ with some $N \in \mathbb{N}$. Essentially we follow the structure of [49] and begin by stating some auxiliary results.

What we need for our purposes is some information on the errors of the quadrature formulas that we have used in the derivation of the predictor and the corrector, respectively. We first give a statement on the product rectangle rule that we have used for the predictor.

Theorem C.1. *(a) Let $z \in C^1[0, T]$. Then,*

$$\left| \int_0^{t_{k+1}} (t_{k+1} - t)^{n-1} z(t) \, dt - \sum_{j=0}^{k} b_{j,k+1} z(t_j) \right| \le \frac{1}{n} \|z'\|_\infty t_{k+1}^n h.$$

(b) Let $z(t) = t^p$ for some $p \in (0, 1)$. Then,

$$\left| \int_0^{t_{k+1}} (t_{k+1} - t)^{n-1} z(t) \, dt - \sum_{j=0}^{k} b_{j,k+1} z(t_j) \right| \le C_{n,p}^{\mathrm{Re}} t_{k+1}^{n+p-1} h$$

where $C_{n,p}^{\mathrm{Re}}$ is a constant that depends only on n and p.

Proof. By construction of the product rectangle formula, we find in both cases that the quadrature error has the representation

$$\int_0^{t_{k+1}} (t_{k+1} - t)^{n-1} z(t) \, dt - \sum_{j=0}^{k} b_{j,k+1} z(t_j)$$

$$= \sum_{j=0}^{k} \int_{jh}^{(j+1)h} (t_{k+1} - t)^{n-1} (z(t) - z(t_j)) \, dt. \tag{C.20}$$

To prove statement (a), we apply the Mean Value Theorem of Differential Calculus to the second factor of the integrand on the right-hand side of (C.20) and derive

$$\left| \int_0^{t_{k+1}} (t_{k+1} - t)^{n-1} z(t) \, dt - \sum_{j=0}^{k} b_{j,k+1} z(t_j) \right|$$

$$\leq \|z'\|_\infty \sum_{j=0}^{k} \int_{jh}^{(j+1)h} (t_{k+1} - t)^{n-1} (t - jh) \, dt$$

$$= \|z'\|_\infty \frac{h^{1+n}}{n} \sum_{j=0}^{k} \left(\frac{1}{1+n} [(k+1-j)^{1+n} - (k-j)^{1+n}] - (k-j)^n \right)$$

$$= \|z'\|_\infty \frac{h^{1+n}}{n} \left(\frac{(k+1)^{1+n}}{1+n} - \sum_{j=0}^{k} j^n \right)$$

$$= \|z'\|_\infty \frac{h^{1+n}}{n} \left(\int_0^{k+1} t^n \, dt - \sum_{j=0}^{k} j^n \right).$$

Here the term in parentheses is simply the remainder of the standard rectangle quadrature formula, applied to the function t^n, and taken over the interval $[0, k+1]$. Since the integrand is monotonic, we may apply some standard results from quadrature theory [19, Thm. 97] to find that this term is bounded by the total variation of the integrand, viz. the quantity $(k+1)^n$. Thus,

$$\left| \int_0^{t_{k+1}} (t_{k+1} - t)^{n-1} z(t) \, dt - \sum_{j=0}^{k} b_{j,k+1} z(t_j) \right| \leq \|z'\|_\infty \frac{h^{1+n}}{n} (k+1)^n.$$

Similarly, to prove (b), we use the monotonicity of z in (C.20) and derive

$$\left| \int_0^{t_{k+1}} (t_{k+1} - t)^{n-1} z(t) \, dt - \sum_{j=0}^{k} b_{j,k+1} z(t_j) \right|$$

$$\leq \sum_{j=0}^{k} |z(t_{j+1}) - z(t_j)| \int_{jh}^{(j+1)h} (t_{k+1} - t)^{n-1} \, dt$$

$$= \frac{h^{n+p}}{n} \sum_{j=0}^{k} ((j+1)^p - j^p)((k+1-j)^n - (k-j)^n)$$

$$\leq \frac{h^{n+p}}{n} \left((k+1)^n - k^n + (k+1)^p - k^p + pn \sum_{j=1}^{k-1} j^{p-1}(k-j+q)^{n-1} \right)$$

$$\leq \frac{h^{n+p}}{n} \left(n(k+q)^{n-1} + pk^{p-1} + pn \sum_{j=1}^{k-1} j^{p-1}(k-j+q)^{n-1} \right)$$

by additional applications of the Mean Value Theorem. Here $q = 0$ if $n \leq 1$, and $q = 1$ otherwise. In either case a brief asymptotic analysis using the Euler–MacLaurin formula [188, Thm. 3.7] yields that the term in parentheses is bounded from above by $C_{n,p}^{Re}(k+1)^{p+n-1}$ where $C_{n,p}^{Re}$ is a constant depending on n and p but not on k. $\square$

Next we come to a corresponding result for the product trapezoidal formula that we have used for the corrector. The proof of this theorem is very similar to the proof of Theorem C.1; we therefore omit the details.

Theorem C.2. *(a) If $z \in C^2[0,T]$ then there is a constant C_n^{Tr} depending only on n such that*

$$\left| \int_0^{t_{k+1}} (t_{k+1} - t)^{n-1} z(t)\,dt - \sum_{j=0}^{k+1} a_{j,k+1} z(t_j) \right| \leq C_n^{Tr} \|z''\|_\infty t_{k+1}^n h^2.$$

(b) Let $z \in C^1[0,T]$ and assume that z' fulfils a Lipschitz condition of order μ for some $\mu \in (0,1)$. Then, there exist positive constants $B_{n,\mu}^{Tr}$ (depending only on n and μ) and $M(z,\mu)$ (depending only on z and μ) such that

$$\left| \int_0^{t_{k+1}} (t_{k+1} - t)^{n-1} z(t)\,dt - \sum_{j=0}^{k+1} a_{j,k+1} z(t_j) \right| \leq B_{n,\mu}^{Tr} M(z,\mu) t_{k+1}^n h^{1+\mu}.$$

(c) Let $z(t) = t^p$ for some $p \in (0,2)$ and $\rho := \min(2, p+1)$. Then,

$$\left| \int_0^{t_{k+1}} (t_{k+1} - t)^{n-1} z(t)\,dt - \sum_{j=0}^{k+1} a_{j,k+1} z(t_j) \right| \leq C_{n,p}^{Tr} t_{k+1}^{n+p-\rho} h^\rho$$

where $C_{n,p}^{Tr}$ is a constant that depends only on n and p.

Remark C.2. Notice that in part (c) of Theorem C.2 it may happen that $n < 1$ and $p < 1$. This implies $\rho = p+1$. Thus, the exponent of t_{k+1} on the right-hand side of the inequality is equal to $n - 1$ which is negative. At first sight this may seem counter-intuitive because it means that the overall integration error becomes larger if the size of the interval of integration becomes smaller. The explanation for this phenomenon is that by making t_{k+1} smaller we do not only shorten the length of the integration interval (which should lead to a smaller error) but we also change the weight function in a way that makes the integral more difficult, and this second feature leads to an increase in the error.

A similar observation can be made in Theorem C.1 (b).

We now present the main results concerning the error of our Adams scheme. It is useful to distinguish a number of cases. Specifically, we shall see that the precise behaviour of the error differs depending on whether $n < 1$ or $n > 1$. Moreover, the smoothness properties of the given function f and the unknown solution y play an important role. In view of the results of Sect. 6.4, we find that smoothness of one of these functions will imply non-smoothness of the other unless some special conditions are fulfilled. Therefore we shall also investigate the error under those two different smoothness assumptions.

Based on the error estimates above we shall first present a general convergence result for the Adams–Bashforth–Moulton method. In the theorems below we shall specialize this result to particularly important special cases.

Lemma C.3. *Assume that the solution y of the initial value problem is such that*

$$\left| \int_0^{t_{k+1}} (t_{k+1} - t)^{n-1} D_{*0}^n y(t)\, dt - \sum_{j=0}^{k} b_{j,k+1} D_{*0}^n y(t_j) \right| \le C_1 t_{k+1}^{\gamma_1} h^{\delta_1}$$

and

$$\left| \int_0^{t_{k+1}} (t_{k+1} - t)^{n-1} D_{*0}^n y(t)\, dt - \sum_{j=0}^{k+1} a_{j,k+1} D_{*0}^n y(t_j) \right| \le C_2 t_{k+1}^{\gamma_2} h^{\delta_2}$$

with some $\gamma_1, \gamma_2 \ge 0$ and $\delta_1, \delta_2 > 0$. Then, for some suitably chosen $T > 0$, we have

$$\max_{0 \le j \le N} |y(t_j) - y_j| = O(h^q)$$

where $q = \min\{\delta_1 + n, \delta_2\}$ and $N = \lfloor T/h \rfloor$.

Proof. We will show that, for sufficiently small h,

$$|y(t_j) - y_j| \le C h^q \qquad (C.21)$$

for all $j \in \{0, 1, \ldots, N\}$, where C is a suitable constant. The proof will be based on mathematical induction. In view of the given initial condition, the induction basis ($j = 0$) is presupposed. Now assume that (C.21) is true for $j = 0, 1, \ldots, k$ for some $k \le N - 1$. We must then prove that the inequality also holds for $j = k + 1$. To do this, we first look at the error of the predictor y_{k+1}^P. By construction of the predictor we find that

$$|y(t_{k+1}) - y_{k+1}^P| = \frac{1}{\Gamma(n)} \left| \int_0^{t_{k+1}} (t_{k+1} - t)^{n-1} f(t, y(t))\, dt - \sum_{j=0}^{k} b_{j,k+1} f(t_j, y_j) \right|$$

$$\le \frac{1}{\Gamma(n)} \left| \int_0^{t_{k+1}} (t_{k+1} - t)^{n-1} D_{*0}^n y(t)\, dt - \sum_{j=0}^{k} b_{j,k+1} D_{*0}^n y(t_j) \right|$$

$$+ \frac{1}{\Gamma(n)} \sum_{j=0}^{k} b_{j,k+1} |f(t_j, y(t_j)) - f(t_j, y_j)|$$

$$\leq \frac{C_1 t_{k+1}^{\gamma_1}}{\Gamma(n)} h^{\delta_1} + \frac{1}{\Gamma(n)} \sum_{j=0}^{k} b_{j,k+1} LC h^q$$

$$\leq \frac{C_1 T^{\gamma_1}}{\Gamma(n)} h^{\delta_1} + \frac{CLT^n}{\Gamma(n+1)} h^q. \tag{C.22}$$

Here we have used the Lipschitz property of f, the assumption on the error of the rectangle formula, and the facts that, by construction of the quadrature formula underlying the predictor, $b_{j,k+1} > 0$ for all j and k and

$$\sum_{j=0}^{k} b_{j,k+1} = \int_0^{t_{k+1}} (t_{k+1} - t)^{n-1} \, dt = \frac{1}{n} t_{k+1}^n \leq \frac{1}{n} T^n.$$

On the basis of the bound (C.22) for the predictor error we begin the analysis of the corrector error. We recall the relation (C.14) which we shall use in particular for $j = k+1$ and find, arguing in a similar way to above, that

$$|y(t_{k+1}) - y_{k+1}|$$

$$= \frac{1}{\Gamma(n)} \left| \int_0^{t_{k+i}} (t_{k+1} - t)^{n-1} f(t, y(t)) \, dt \right.$$

$$\left. - \sum_{j=0}^{k} a_{j,k+1} f(t_j, y_j) - a_{k+1,k+1} f(t_{k+1}, y_{k+1}^{\mathrm{P}}) \right|$$

$$\leq \frac{1}{\Gamma(n)} \left| \int_0^{t_{k+1}} (t_{k+1} - t)^{n-1} D_{*0}^n y(t) \, dt - \sum_{j=0}^{k+1} a_{j,k+1} D_{*0}^n y(t_j) \right|$$

$$+ \frac{1}{\Gamma(n)} \sum_{j=0}^{k} a_{j,k+1} |f(t_j, y(t_j)) - f(t_j, y_j)|$$

$$+ \frac{1}{\Gamma(n)} a_{k+1,k+1} |f(t_{k+1}, y(t_{k+1})) - f(t_{k+1}, y_{k+1}^{\mathrm{P}})|$$

$$\leq \frac{C_2 t_{k+1}^{\gamma_2}}{\Gamma(n)} h^{\delta_2} + \frac{CL}{\Gamma(n)} h^q \sum_{j=0}^{k} a_{j,k+1} + a_{k+1,k+1} \frac{L}{\Gamma(n)} \left(\frac{C_1 T^{\gamma_1}}{\Gamma(n)} h^{\delta_1} + \frac{CLT^n}{\Gamma(n+1)} h^q \right)$$

$$\leq \left(\frac{C_2 T^{\gamma_2}}{\Gamma(n)} + \frac{CLT^n}{\Gamma(n+1)} + \frac{C_1 LT^{\gamma_1}}{\Gamma(n)\Gamma(n+2)} + \frac{CL^2 T^n}{\Gamma(n+1)\Gamma(n+2)} h^n \right) h^q$$

in view of the nonnegativity of γ_1 and γ_2 and the relations $\delta_2 \leq q$ and $\delta_1 + n \leq q$. By choosing T sufficiently small, we can make sure that the second summand in the parentheses is bounded by $C/2$. Having fixed this value for T, we can then make the sum of the remaining expressions in the parentheses smaller than $C/2$ too (for sufficiently small h) simply by choosing C sufficiently large. It is then obvious that the entire upper bound does not exceed Ch^q. $\qquad\square$

As a first application of this Lemma we assume that the given data is such that the solution y itself is sufficiently differentiable. As mentioned above, the result depends on whether $n > 1$ or $n < 1$.

Theorem C.4. *Let $0 < n$ and assume $D_{*0}^n y \in C^2[0,T]$ for some suitable T. Then,*

$$\max_{0 \le j \le N} |y(t_j) - y_j| = \begin{cases} O(h^2) & \text{if } n \ge 1, \\ O(h^{1+n}) & \text{if } n < 1. \end{cases}$$

Before we come to the proof, we note one particular point: The order of convergence depends on n, and it is a non-decreasing function of n. This is due to the fact that we discretize the integral operator in (C.1) which behaves more smoothly (and hence can be approximated with a higher accuracy) as n increases. In contrast, so-called *direct methods* like the backward differentiation method of [34] use a different approach; as the name suggests they directly discretize the differential operator in the given initial value problem. The smoothness properties of such operators (and thus the ease with which they may be approximated) deteriorate as n increases, and so we find that the convergence order of the method from [34] is a non-increasing function of n; in particular no convergence is achieved there for $n \ge 2$. It is a distinctive advantage of the Adams scheme presented here that it converges for all $n > 0$.

Remark C.3. We formally recover the error bound (C.10) if we set $n = 1$.

Proof (of Theorem C.4). In view of Theorems C.1 and C.2, we may apply Lemma C.3 with $\gamma_1 = \gamma_2 = n > 0$, $\delta_1 = 1$ and $\delta_2 = 2$. Thus, defining

$$q = \min\{1+n, 2\} = \begin{cases} 2 & \text{if } n \ge 1, \\ 1+n & \text{if } n < 1, \end{cases}$$

we find an $O(h^q)$ error bound. $\square$

Note that in a certain sense the theorem above deals with the "optimal" situation: The function that we approximate in our process is $f(\cdot, y(\cdot)) = D_{*0}^n y$. In order to obtain very good error bounds, we need to make sure that the quadrature errors for this function are (asymptotically) as small as possible. A sufficient condition for this to hold is, as is well known from quadrature theory [19], that this function is in C^2 on the interval of integration. This is precisely the setting discussed in Theorem C.4. So this theorem shows us what kind of performance the Adams method can give under optimal circumstances, and it also states sufficient conditions for such results to hold.

An objection against the use of this Adams–Bashforth–Moulton scheme may be the very slow rate of convergence if n is close to 0. However, a careful inspection of the proof of the error bound reveals a fact that is well known in the error analysis for methods of this structure for first-order equations (cf., e.g., [179]): The application of the corrector formula improves the accuracy of its input (the predictor) by a factor of h^n until an order of $O(h^2)$ (i.e. a saturation) is reached. Thus we may replace the plain PECE structure by a $P(EC)^\mu E$ method, i.e. by introducing additional corrector iterations.

Remark C.4. An interesting observation here is that by choosing a larger number of corrector iterations, we essentially leave the computational complexity unchanged: A corrector iteration is of the form

$$y_{j+1}^{[\ell]} = \sum_{r=0}^{\lceil n \rceil - 1} \frac{t_{j+1}^r}{r!} y_0^{(r)} + \frac{h^n}{\Gamma(n+2)} f(t_{j+1}, y_{j+1}^{[\ell-1]})$$

$$+ \frac{h^n}{\Gamma(n+2)} \sum_{r=0}^{j} a_{r,j+1} f(t_r, y_r),$$

cf. (C.15). Here $y_{j+1}^{[\ell]}$ denotes the approximation after ℓ corrector steps, $y_{j+1}^{[0]} = y_{j+1}^{P}$ is the predictor, and $y_{j+1} := y_{j+1}^{[\mu]}$ is the final approximation after μ corrector steps that we actually use. We can rewrite this as

$$y_{j+1}^{[\ell]} = \beta_{j+1} + \frac{h^n}{\Gamma(n+2)} f(t_{j+1}, y_{j+1}^{[\ell-1]})$$

where

$$\beta_{j+1} = \sum_{r=0}^{\lceil n \rceil - 1} \frac{t_{j+1}^r}{r!} y_0^{(r)} + \frac{h^n}{\Gamma(n+2)} \sum_{r=0}^{j} a_{r,j+1} f(t_r, y_r)$$

is independent of ℓ. Thus the total arithmetic complexity of the corrector part of the $(j+1)$st step (taking us from t_j to t_{j+1}) is $O(j)$ for the calculation of β_{j+1} plus $O(\mu)$ for the μ corrector steps, which (since μ is constant) is asymptotically the same as the complexity in the case $\mu = 1$.

For the error of the scheme outlined in Remark C.4 we find, as indicated above, by a repeated application of the considerations of the proof of Theorem C.4 (see [36] for details):

Theorem C.5. *Under the assumptions of Theorem C.4, the approximation obtained by the $P(EC)^\mu E$ method described above satisfies*

$$\max_{0 \le j \le N} |y(t_j) - y_j| = O(h^q)$$

where $q = \min\{2, 1 + \mu n\}$.

For the moment we leave the topic of general $P(EC)^\mu E$ methods in favour of a more detailed investigation of its special case $\mu = 1$, i.e. the original Adams–Bashforth–Moulton method introduced in (C.19) and (C.15). In this context we note an apparent disadvantage in the formulation of the hypotheses of the theorems above: They are stated in terms of the solution y (or, more precisely, its Caputo derivative of order n), which is unknown in general. Even though it is sometimes possible to determine the smoothness properties of $D_{*0}^n y$ from the given data, there still is some need for a corresponding error theory for the Adams method under assumptions formulated directly in terms of the given data, i.e. in terms of the function f. Such results will be the derived later in this section.

Before we come to those results however, we want to give some more information under assumptions similar to those of Theorem C.4. Specifically we want to state the conjecture that the error of our scheme, taken at a fixed abscissa, possesses an asymptotic expansion in powers of the step size h under additional smoothness conditions on $D_{*0}^n y$. If this were true (as, e.g., numerical results in Example C.1 below indicate), we could construct a Richardson extrapolation algorithm based on the Adams method in a way similar to the construction in [53] which is based on the scheme of [34]. The use of this extrapolation procedure then would permit us to obtain more accurate numerical approximations for the desired solution.

Conjecture C.1. *Let $n > 0$ and assume that $D_{*0}^n y \in C^k[0,T]$ for some $k \geq 3$ and some suitable T. Then,*

$$y(T) - y_{T/h} = \sum_{j=1}^{k_1} c_j h^{2j} + \sum_{j=1}^{k_2} d_j h^{j+n} + O(h^{k_3})$$

where k_1, k_2 and k_3 are certain constants depending only on k and satisfying $k_3 > \max(2k_1, k_2 + n)$.

Notice that the asymptotic expansion begins with an h^2 term and continues with h^{1+n} for $1 < n < 3$, whereas it begins with h^{1+n}, followed by h^2, for $0 < n < 1$.

Example C.1. Consider the equation

$$D_{*0}^n y(x) = \frac{40320}{\Gamma(9-n)} x^{8-n} - 3 \frac{\Gamma(5+n/2)}{\Gamma(5-n/2)} x^{4-n/2} + \frac{9}{4} \Gamma(n+1)$$
$$+ \left(\frac{3}{2} x^{n/2} - x^4 \right)^3 - [y(x)]^{3/2}$$

for $x \in [0,1]$ with homogeneous initial conditions $(y(0) = 0, y'(0) = 0$; the latter only in the case $n > 1)$.

The exact solution of this initial value problem is

$$y(x) = x^8 - 3x^{4+n/2} + \frac{9}{4} x^n,$$

and hence

$$D_{*0}^n y(x) = \frac{40320}{\Gamma(9-n)} x^{8-n} - 3 \frac{\Gamma(5+n/2)}{\Gamma(5-n/2)} x^{4-n/2} + \frac{9}{4} \Gamma(n+1),$$

i.e. $D_{*0}^n y \in C^2[0,1]$ if $n \leq 4$, and thus the conditions of Theorem C.4 are fulfilled. Moreover, assuming that Conjecture C.1 holds, the application of Richardson extrapolation is also justified. We display some of the results in Tables C.1 and C.2 where, e.g., the notation $-5.53(-3)$ stands for $-5.53 \cdot 10^{-3}$. In each case, the leftmost column shows the step size used, the following column gives the error of our

Table C.1 Errors for Example C.1 with $n = 1.25$, taken at $x = 1$

Step size	Error of Adams scheme	Extrapolated values			
1/10	−5.53(−3)				
1/20	−1.59(−3)	−2.80(−4)			
1/40	−4.33(−4)	−4.60(−5)	1.63(−5)		
1/80	−1.14(−4)	−8.17(−6)	1.90(−6)	2.13(−7)	
1/160	−2.97(−5)	−1.54(−6)	2.24(−7)	2.71(−8)	1.47(−8)
1/320	−7.66(−6)	−3.04(−7)	2.56(−8)	2.28(−9)	6.24(−10)
1/640	−1.96(−6)	−6.16(−8)	2.85(−9)	1.73(−10)	3.25(−11)
EOC	1.97	2.30	3.17	3.72	4.26

Table C.2 Errors for Example C.1 with $n = 0.25$, taken at $x = 1$

Step size	Error of Adams scheme	Extrapolated values			
1/10	2.50(−1)				
1/20	1.81(−2)	−1.50(−1)			
1/40	3.61(−3)	−6.91(−3)	4.09(−2)		
1/80	1.45(−3)	−1.10(−4)	2.16(−3)	−8.15(−3)	
1/160	6.58(−4)	8.19(−5)	1.46(−4)	−3.89(−4)	1.28(−4)
1/320	2.97(−4)	3.49(−5)	1.92(−5)	−1.45(−5)	1.05(−5)
1/640	1.31(−4)	1.12(−5)	3.37(−6)	−8.50(−7)	6.01(−8)
EOC	1.18	1.63	2.51	4.09	7.44

scheme at $x = 1$, and the columns after that give the extrapolated values. The bottom line (marked "EOC") states the experimentally determined order of convergence for each of the columns on the right of the table. According to our theoretical considerations, these values should be $1 + n, 2, 2 + n, 3 + n, 4, 4 + n, \ldots$ in the case $0 < n < 1$ and $2, 1 + n, 2 + n, 4, 3 + n, 4 + n, \ldots$ for $1 < n < 2$. The numerical data in the following tables show that these values are reproduced approximately at least for $n > 1$ (see Table C.1). In the case $0 < n < 1$, displayed in Table C.2, the situation seems to be less obvious. Apparently, we need to use much smaller values for h than in the case $n > 1$ before we can see that the asymptotic behaviour really sets in. This would normally correspond to the situation that the coefficients of the leading terms are small in magnitude compared to the coefficients of the higher-order terms.

Our belief in the truth of Conjecture C.1 is not only supported by the numerical results but also by the results of de Hoog and Weiss [32, §5] who show that asymptotic expansions of this form hold if we use the fractional Adams–Moulton method (i.e. if we solve the corrector equation exactly) and that a similar expansion can be derived for the fractional Adams–Bashforth method (using the predictor as the final approximation rather than correcting once with the Adams–Moulton formula). For the moment however, we leave the question of the influence of the corrector step (that combines the two approaches) on this expansion open.

Rather, we turn our attention to another related problem. In the previous theorems we had formulated our hypotheses in the form of smoothness assumptions on $D_{*0}^n y$. Now we want to replace this by similar assumptions on y itself. In view of Theorem 3.15 we must be aware of the fact that smoothness of y in general implies non-smoothness of $D_{*0}^n y$ (the function that we have to approximate), so some difficulties are likely. Fortunately Theorem 3.15 also informs us about the precise nature of the singularities in the derivatives of $D_{*0}^n y$. We can exploit this information to obtain the following results.

Theorem C.6. *Let $n > 1$ and assume that $y \in C^{1+\lceil n \rceil}[0,T]$ for some suitable T. Then,*

$$\max_{0 \leq j \leq N} |y(t_j) - y_j| = O(h^{1+\lceil n \rceil - n}).$$

Proof. By Theorem 3.15 we find that $D_{*0}^n y(x) = cx^{\lceil n \rceil - n} + g(x)$ where $g \in C^1[0,T]$ and g' fulfils a Lipschitz condition of order $\lceil n \rceil - n$. Thus, according to Theorems C.1 and C.2 we can apply Lemma C.3 with $\gamma_1 = 0$, $\gamma_2 = n - 1 > 0$, $\delta_1 = 1$ and $\delta_2 = 1 + \lceil n \rceil - n$. Because of $n > 1$ we then find that $\delta_1 + n = 1 + n > 2 > \delta_2$, and hence $\min\{\delta_1 + n, \delta_2\} = \delta_2$. So the overall error bound is $O(h^{\delta_2})$.   □

Notice that a reformulation of Theorem C.6 yields that, if $1 < n = k_1 + k_2$ with $k_1 \in \mathbb{N}$ and $0 < k_2 < 1$, then the error is $O(h^{2-k_2})$. Thus the fractional part of n plays the decisive role for the order of the error. In particular, we find slow convergence if the fractional part of n is large. Consequently, under these assumptions we cannot expect the convergence order to be a monotone function of n any more. Nevertheless we can prove that the method converges for all $n > 0$:

Theorem C.7. *Let $0 < n < 1$ and assume that $y \in C^2[0,T]$ for some suitable T. Then, for $1 \leq j \leq N$ we have*

$$|y(t_j) - y_j| \leq Ct_j^{n-1} \times \begin{cases} h^{1+n} & \text{if } 0 < n < 1/2, \\ h^{2-n} & \text{if } 1/2 \leq n < 1, \end{cases} \tag{C.23}$$

where C is a constant independent of j and h.

We obtain two immediate consequences.

Corollary C.8. *Under the assumptions of Theorem C.7, we have*

$$\max_{0 \leq j \leq N} |y(t_j) - y_j| = \begin{cases} O(h^{2n}) & \text{if } 0 < n < 1/2, \\ O(h) & \text{if } 1/2 \leq n < 1. \end{cases}$$

Moreover, for every $\varepsilon \in (0,T)$ we have

$$\max_{t_j \in [\varepsilon,T]} |y(t_j) - y_j| = \begin{cases} O(h^{1+n}) & \text{if } 0 < n < 1/2, \\ O(h^{2-n}) & \text{if } 1/2 \leq n < 1. \end{cases}$$

Proof (of Theorem C.7). The first steps of the proof are as in the proof of Theorem C.6. The key difference is that now $\gamma_2 < 0$ (note that we still have $\gamma_2 = n - 1$, but now $n < 1$). Thus we cannot apply Lemma C.3. Instead we modify its proof so that it fits to our requirements: We keep the inductive structure and remember that our claim is now (C.23) rather than (C.21). With this change in the induction hypothesis we proceed much as in the proof of Lemma C.3. However, because of this new hypothesis, we now have to estimate terms of the form $\sum_{j=1}^{k-1} b_{j,k+1} t_j^{\gamma_2}$ and $\sum_{j=1}^{k-1} a_{j,k+1} t_j^{\gamma_2}$. By the Mean Value Theorem we have $0 \leq b_{j,k+1} \leq h^n (k-j)^{n-1}$ and $0 \leq a_{j,k+1} \leq c h^n (k-j)^{n-1}$ for $1 \leq j \leq k-1$ (where the constant c is independent of j and k), respectively, so that the problem reduces to finding a bound for $S_k := \sum_{j=1}^{k-1} j^{\gamma_2} (k-j)^{n-1}$. Under our assumptions, both the exponents γ_2 and $n-1$ are in the interval $(-1, 0)$, and then it is easily seen that $S_k = O(k^{\gamma_2 + n})$. Using this relation we can complete the proof of Theorem C.7 by following along the lines of the rest of the proof of Lemma C.3. $\square$

We conclude the discussion of error bounds with a result where we formulate the hypotheses in terms of the given data and not in terms of the unknown solution. We give a result in the case $n > 1$ and later discuss properties of the numerical scheme when $n < 1$.

Theorem C.9. *Let $n > 1$. Then, if $f \in C^2(G)$,*

$$\max_{0 \leq j \leq N} |y(t_j) - y_j| = O(h^2).$$

Proof (of Theorem C.9). We begin by discussing the case $n \geq 2$. Then, according to Theorem 6.25, we find that $y \in C^2[0, T]$. Thus, in view of the smoothness assumption on f and the chain rule, $D_{*0}^n y := f(\cdot, y(\cdot)) \in C^2[0, T]$ too, and the claim follows by virtue of Theorem C.4.

For the case $1 < n < 2$, we want to apply Lemma C.3 and hence we have to determine the constants $\gamma_1, \gamma_2, \delta_1$ and δ_2 in its hypotheses. In order to do so we need more precise information about the behaviour of y. This information can be found in Theorem 6.38 which asserts that $y(x) = cx^n + \psi(x)$ with some $c \in \mathbb{R}$ and some $\psi \in C^2[0, T]$. This implies, in particular, that $y \in C^1[0, T]$. As in the case $n > 2$ above we can then deduce $D_{*0}^n y \in C^1[0, T]$ too, and by Theorem C.1(a), we find that we may choose $\gamma_1 = n$ and $\delta_1 = 1$. Moreover, using again the fact that $y(x) = cx^n + \psi(x)$ with some $c \in \mathbb{R}$ and some $\psi \in C^2[0, T]$ and applying Theorem C.2(a) and (c), we determine the correct values for the remaining quantities as $\gamma_2 = \min\{n, 2n-2\} = 2n - 2 \geq 0$ and $\delta_2 = 2$. The claim then follows from Lemma C.3. $\square$

In the case $n < 1$ the situation seems to be less clear. According to the theorems presented at the end of Sect. 6.4, smoothness conditions on f imply that the exact solution is of the form

$$y(t) = \psi(t) + \sum_{v=1}^{\hat{v}} c_v t^{vn} + \sum_{v=1}^{\tilde{v}} d_v t^{1+vn}$$

where ψ is twice differentiable. The first sum consists of terms which are not differentiable, and the second sum is of terms that are differentiable once but not twice. As remarked by Lubich [120] it seems unlikely that numerical schemes will be rapidly convergent over any interval that contains the origin. Indeed we can prove that the error $y(t_1) - y_1$ of the approximation after just one step behaves as $O(h^{2n})$ if $f \in C^2(G)$. Simple numerical experiments indicate that this result cannot be improved. However this error introduced in the initial phase is transient and from numerical results reported in [49, Table 4.5] we believe the following conjecture to be true.

Conjecture C.2. *Let $0 < n < 1$. Then, if $f \in C^3(G)$, for every $\varepsilon > 0$ we have*

$$\max_{t_j \in [\varepsilon, T]} |y(t_j) - y_j| = O(h^{1+n}).$$

C.2 Numerical Schemes for Multi-Term Equations

We now come to the extension of the numerical methods discussed in the previous section to multi-term equations. The most important theoretical properties of these multi-term equations have been discussed in Chap. 8. As in that chapter we restrict our attention to equations of the form

$$D_{*0}^{n_k} y(x) = f(x, y(x), D_{*0}^{n_1} y(x), D_{*0}^{n_2} y(x), \ldots, D_{*0}^{n_{k-1}} y(x)) \qquad \text{(C.24a)}$$

(where $0 < n_1 < n_2 < \ldots < n_k$) with a suitable function f and initial conditions

$$y^{(j)}(0) = y_0^{(j)}, \qquad j = 0, 1, \ldots, \lceil n_k \rceil - 1. \qquad \text{(C.24b)}$$

For initial value problems of this type we shall discuss the various approaches introduced in Chap. 8 and find out their respective advantages and disadvantages.

Conversion to Single-Order Systems

Our first attempt consists in a direct application of the result of Theorem 8.1 or 8.2 (depending on whether $n_k > 1$ or not) to the given initial value problem. In this way we transform the given initial value problem into a system of equations of the form

$$D_{*0}^{\gamma} y_0(x) = y_1(x),$$
$$D_{*0}^{\gamma} y_1(x) = y_2(x),$$
$$\vdots \qquad\qquad\qquad\qquad\qquad\qquad\qquad \text{(C.25a)}$$
$$D_{*0}^{\gamma} y_{N-2}(x) = y_{N-1}(x),$$
$$D_{*0}^{\gamma} y_{N-1}(x) = f(x, y_0(x), y_{n_1/\gamma}(x), \dots, y_{n_{k-1}/\gamma}(x)),$$

together with the initial conditions

$$y_j(0) = \begin{cases} y_0^{(j\gamma)} & \text{if } j\gamma \in \mathbb{N}_0, \\ 0 & \text{else,} \end{cases} \qquad\qquad \text{(C.25b)}$$

with the precise choice of the new parameters γ and N being according to Theorems 8.1 or 8.2, as appropriate. We have thus formally obtained an equation of the type

$$D_{*0}^{\gamma} Y(x) = F(x, Y(x)), \qquad Y(0) = Y_0, \qquad\qquad \text{(C.26)}$$

with certain vector-valued functions F (known) and Y (unknown) and an initial condition vector Y_0, i.e. a single-term equation of order γ with vector-valued data. Thus we may apply any numerical method for such single-term equations and calculate an approximate solution for this system; for the sake of simplicity we shall restrict ourselves to the Adams–Bashforth–Moulton scheme developed above. The first component of the (numerical) solution vector (with index 0) is then the required approximate solution for the original equation. We illustrate the procedure by a simple example taken from [44].

Example C.2. Solve the Bagley–Torvik equation

$$Ay''(x) + BD_{*0}^{3/2} y(x) + Cy(x) = C(x+1)$$

(where $A \neq 0$ and $B, C \in \mathbb{R}$) with initial conditions

$$y(0) = y'(0) = 1$$

with the approach described above.

It is easily verified that the exact solution of this problem is

$$y(x) = x + 1$$

independent of the choice of the coefficients A, B, and C. Thus, the resulting system is

$$D_{*0}^{1/2} \begin{pmatrix} y_0(x) \\ y_1(x) \\ y_2(x) \\ y_3(x) \end{pmatrix} = \frac{1}{A} \begin{pmatrix} y_1(x) \\ y_2(x) \\ y_3(x) \\ -By_3(x) - Cy_0(x) + C(x+1) \end{pmatrix},$$

$$\begin{pmatrix} y_0(0) \\ y_1(0) \\ y_2(0) \\ y_3(0) \end{pmatrix} = \begin{pmatrix} 1 \\ 0 \\ 1 \\ 0 \end{pmatrix}.$$

We have solved this problem on the interval $[0,5]$ with the Adams–Bashforth–Moulton scheme. The numerical results at $x = 5$ were as shown in Table C.3.

These data indicate that the convergence behaviour is $O(h^{3/2})$. To understand how this relates to the error estimates of Sect. C.1, we must recall that we now construct an approximate solution for the entire system and not just for its first component. Thus we see that, as a by-product of the method, we do not only obtain information on y but also on its fractional derivatives of order $\gamma, 2\gamma, \ldots, (N-1)\gamma$. Depending on the task at hand this information can be anything between highly useful and absolutely unnecessary. In any case the error estimate is dominated by the worst error estimate of the four components. Since the exact solution for the system is $(x+1, x^{1/2}/\Gamma(3/2), 1, 0)^T$ we see that the second, third and fourth components of the exact solution have smooth derivatives of order $1/2$; thus they may be approximated with order $O(h^{3/2})$ according to Theorem C.4. The first component is smooth itself; it allows an application of Corollary C.8 that gives an $O(h)$ error estimate on the full interval $[0,T]$ and an $O(h^{3/2})$ estimate on each interval of the form $[\varepsilon, T]$ with $\varepsilon > 0$. Since the latter case covers the problem considered in Table C.3, we have agreement between theoretical and numerical results.

In order to explain the weaknesses of this concept, we look at a second example, already considered in [45] and [35].

Example C.3. Solve the nonlinear three-term equation

$$D_{*0}^{1.455} y(x) = -x^{0.1} \frac{E_{1.545}(-x)}{E_{1.445}(-x)} e^x y(x) D_{*0}^{0.555} y(x) + e^{-2x} - [D_{*0}^1 y(x)]^2 \qquad (C.27)$$

for $0 \le x \le 1$, equipped with the initial conditions $y(0) = 1$ and $y'(0) = -1$, with the same algorithm.

The exact solution of this problem is $y(x) = \exp(-x)$. When applying our idea to this equation, we first need to calculate the order γ of the new system as described in Theorem 8.1. In our case the result is $\gamma = 1/200$, and hence the dimension of the resulting system is $N = 1.455/\gamma = 291$ – a rather large number. In a first attempt we have tried to solve the system with the Adams–Bashforth–Moulton scheme as

Table C.3 Bagley–Torvik equation solved with Adams method

Step size	Numerical solution	Error	Estimated order of convergence
0.5	6.15131473519232	−0.15131473519232	
0.25	6.04684102179946	−0.04684102179946	1.69
0.125	6.01602947553912	−0.01602947553912	1.55
0.0625	6.00562770408881	−0.00562770408881	1.51

Table C.4 Numerical results for Example C.3 (system with $N = 291$, $\gamma = 0.005$) using the standard PECE-type Adams algorithm

Step size	Maximal error	Run time
1/200	0.3904	101.2 s
1/400	0.2193	368.4 s
1/800	0.1164	1358.0 s
1/1,600	0.0600	5017.4 s

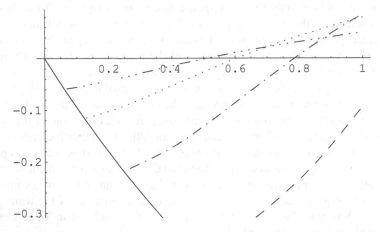

Fig. C.1 Approximation errors for Example C.3 with $h = 1/100$ (*solid line*), $h = 1/200$ (*dashed line*), $h = 1/400$ (*dot–dashed line*), $h = 1/800$ (*dotted line*), $h = 1/1,600$ (*dashed and triple-dotted line*)

above. The resulting errors are reported in Table C.4 and Fig. C.1. For the purpose of comparison with later methods we have also included information about the run time of the algorithm on a standard 500 MHz Pentium PC in double precision arithmetic.

It is clearly seen that the results for $h = 1/100$ and $h = 1/200$ are totally unacceptable. There is a simple explanation for this phenomenon which becomes evident when one takes a look at the numerical solution of the initial value problem and not at the approximation error: In each case the first 99 steps of the algorithm do not change the first component of the approximate solution. In other words we get stuck at the initial value instead of following the exact solution. The reason for this behaviour can be found in the structure of the function F on the right-hand side of the system (C.26) and of its initial condition: The first component (index 0) is y_0, the component with index 200 ($= 1/\gamma$) is y_0' ($\neq 0$), and all the components in between vanish. The interaction of the Adams–Bashforth–Moulton method with the function F now implies that the non-zero component is propagated by one row in each predictor step and another row in each corrector step, so in the first 99 steps only a total of 198 zeros are added to the initial value of the leading component of the numerical solution. The last (199th) zero is then used in the predictor of the 100th step, and the corrector of step 100 is actually the first operation where a non-zero entry reaches the first component of the solution. Thus we always have an initial interval of $2/\gamma - 1$ steps where the numerical solution is constant before it can start to make progress towards the exact solution. Notice that in our Bagley–Torvik example above we had $\gamma = 1/2$, and so (since here $2/\gamma - 1 = 3$) the effect is negligible. This means that, if one wants to keep the structure of the algorithm and the uniform step size, then the only way to reduce this effect is a drastic reduction of the step size h, essentially by the rule $h = O(\gamma)$. A look at Fig. C.1 confirms this observation.

Let us briefly summarize what we have achieved so far: The approach described above has the advantage of producing a system of equations with a very simple structure. As a consequence of this structure, numerical schemes for this system can be implemented on a parallel computer architecture in a rather efficient way. However we have also seen some disadvantages:

(a)   The method only works in the case of commensurate multi-term equations (if $n_k \leq 1$) or under the even more restrictive assumption that all the n_j are rational (if $n_k > 1$)

(b)   The dimension of the system can be very large (depending on the precise values of the parameters of the original equation); this can lead to very long run times on sequential machines

(c)   The structure of the initial conditions of the new system can be problematic for some types of numerical algorithms; in particular we may be forced to use excessively small step sizes

In order to overcome these problems (at least partially), we propose two possible strategies. The first one, taken from [45], is a slight refinement of the idea used so far that works in two stages. We recall that the problem is mainly due to the large dimension of the system, and this in turn is a consequence of the size of the greatest common divisor of the orders of the differential operators in the given system. Therefore we introduce a stage of preliminary manipulations before actually starting the numerical algorithm.

This first stage consists of replacing the given initial value problem (C.24) by a new differential equation

$$D_{*0}^{\tilde{n}_k}\tilde{y}(x) = f(x, \tilde{y}(x), D_{*0}^{\tilde{n}_1}\tilde{y}(x), D_{*0}^{\tilde{n}_2}\tilde{y}(x), \ldots, D_{*0}^{\tilde{n}_{k-1}}\tilde{y}(x)) \qquad \text{(C.28)}$$

with identical initial conditions (C.24b). We thus perturb the orders of the differential operators, but all other parameters of the given problem (the function f on the right-hand side and the initial conditions) remain unchanged.

The essence of this idea is that, according to Theorem 8.8, the exact solution $\tilde{y}$ of this new initial value problem and the exact solution y of the original problem differ only by

$$\|y - \tilde{y}\|_\infty = O\left(\max_{j=1,2,\ldots,k} |n_j - \tilde{n}_j| \right). \qquad \text{(C.29)}$$

Here by $\|\cdot\|_\infty$ we denote the Chebyshev norm taken over a suitable finite interval $[0, T]$, say, where both problems have a solution.

In order to exploit the capabilities of this approach, we need to choose the new parameters $\tilde{n}_1, \ldots, \tilde{n}_k$ in such a way that they have the following three properties:

(a) $\tilde{n}_1, \ldots, \tilde{n}_k \in \mathbb{Q}$
(b)   The least common multiple of the denominators of $\tilde{n}_1, \ldots, \tilde{n}_k$ is small
(c) $\max_j |n_j - \tilde{n}_j|$ is small

Here condition (a) asserts that a conversion of (C.28) to a single-term system (as described in Chap. 8) is possible. Specifically, since only the new values $\tilde{n}_j$ enter the later stages of the scheme, such a conversion is always possible, without any restrictions on the original values n_j. The purpose of condition (b) is to keep the dimension N of this system small (remember that in Theorem 8.1 we had seen that essentially $N = Mn_k$ where M is the least common multiple mentioned in condition (b)), which – according to our initial idea – was the main point of the concept. Condition (c) finally makes sure that the error introduced by this perturbation remains small, cf. (C.29).

It must be noted of course that there is a conflict between conditions (b) and (c): In many cases it will be possible to improve the approximation required in (c) at the price of increasing the least common multiple mentioned in (b). A proper compromise must be found in this case. It seems to be impossible however to state a generally valid strategy for the solution of this conflict; a good compromise will likely depend on the specific parameters of the equation under consideration.

This completes the first stage of the algorithm. At the end of this stage we have found a new initial value problem that consists of the perturbed differential equation (C.28) together with the original (unperturbed) initial conditions (C.24b).

The second stage of the algorithm is then the stage where the initial value problem that was constructed in stage 1 will be solved numerically. In practice we will first use the approach described at the beginning of this section: We convert the new initial value problem into a single-term system, and then we will solve this system numerically (for example by means of the Adams method).

Example C.4. Construct an approximate solution for the problem from Example C.3 by the two-stage strategy outlined above.

As a first attempt to solve the problem with our refined method, we approximate (C.27) by

$$D_{*0}^{1.5}\tilde{y}(x) = -x^{0.1}\frac{E_{1.545}(-x)}{E_{1.445}(-x)}e^x\tilde{y}(x)D_{*0}^{0.5}\tilde{y}(x) + e^{-2x} - [D_{*0}^1\tilde{y}(x)]^2, \qquad \text{(C.30)}$$

convert (C.30) to a three-dimensional system of order $\gamma = 0.5$, and solve this system numerically with the Adams method in its standard form using various step sizes. The results are described in Table C.5.

Table C.5 Numerical results of first approximation ($N = 3$, $\gamma = 0.5$)

Step size	Maximal error	Run time
1/10	0.136	0.07 s
1/20	0.124	0.18 s
1/40	0.118	0.56 s

Fig. C.2 Exact solution and
first approximation ($N = 3$,
$\gamma = 0.5$, step size $h = 0.1$)

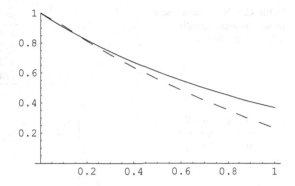

We can see that there is almost no improvement when we change the step size
from $1/20$ to $1/40$. This indicates that the error of the Adams scheme (i.e. the error
introduced in the second stage) is already very small compared to the error of the
first stage (i.e. the error introduced by perturbing the differential equation). Therefore there is no need to look for an improved scheme for the solution of this simple
system. Note in particular (see Fig. C.2) that even the crudest of these three approximations (the dashed line) gives a *qualitatively* correct picture of the exact solution
(the solid line). Certainly this cannot be said to be true for our plain and simple first
approach discussed above.

In order to obtain a better approximation with our method we must now reduce
the error of stage 1, i.e. we need to introduce smaller perturbations in the orders of
the differential operators. We thus try to approximate the given equation (C.27) not
by (C.30) but by

$$D_{*0}^{1.45}\tilde{y}(x) = -x^{0.1}\frac{E_{1.545}(-x)}{E_{1.445}(-x)}e^x\tilde{y}(x)D_{*0}^{0.55}\tilde{y}(x) + e^{-2x} - [D_{*0}^{1}\tilde{y}(x)]^2 \qquad \text{(C.31)}$$

and proceed as above. Consequently we find that we have to solve a 29-dimensional
system of order 0.05 numerically. This task is (in particular due to the nature of
the initial conditions) much more difficult than the previous one, and therefore we
need to put more effort into the numerical scheme. For the moment we interpret this
requirement as a demand for a smaller step size; an alternative will be considered
later. The results are given in Table C.6.

For the purpose of comparison with the previous example we have included the
case of a step size of $1/40$. As can be seen by comparing Tables C.5 and C.6, the
error is much larger now than it was before. The reason is the problem that we
mentioned above: Since the dimension of the system has been increased, the numerical solution needs more time to get away from the initial value. An even more
obvious picture of the situation appears when we look at the graphical data provided
in Fig. C.3. Here again the solid line is the exact solution, the other lines correspond
to the numerical solutions (dashed line: $h = 1/40$; dash-dotted line: $h = 1/100$;

Table C.6 Numerical results
of second approximation
($N = 29$, $\gamma = 0.05$)

Step size	Maximal error	Run time
1/40	0.2015	0.9 s
1/100	0.0861	5.5 s
1/200	0.0440	21.5 s
1/400	0.0222	82.4 s

Fig. C.3 Exact solution
and second approximation
($N = 29$, $\gamma = 0.05$, various
step sizes)

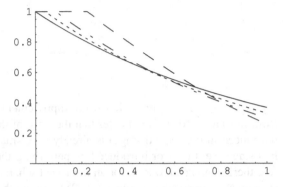

dotted line: $h = 1/200$). We thus have to say that the graph for $h = 1/40$ does *not* give a qualitatively correct picture of the true solution.

As pointed out above, we will sometimes be forced to choose the parameters in such a way that the dimension of the system is larger than desirable. In this context the present 29-dimensional system may be considered to be such a case. That means that also the number of zeros in the initial condition of the resulting system is larger than one would like it to be, which forces us to use a very small step size.

This is where our second strategy mentioned above comes into play as a possible alternative. Specifically, it may be useful to replace the plain PECE structure by a $P(EC)^\mu E$ method (i.e. by introducing additional corrector iterations) as described in Theorem C.5. This allows for a quicker propagation of the non-zero elements, and it may be possible to avoid the use of excessively small step sizes. We shall provide a numerical example now. This flexibility in the number of corrector steps is actually one of the main reasons why we suggest the Adams scheme and not, e.g., the method of [34]. Recall that, as derived in Remark C.4, by (for example) doubling the number of corrector iterations, we essentially leave the computational complexity unchanged.

If we would use the other option and reduce the step size by a factor of two, then the run time would increase by a factor of four because the complexity of the algorithm is $O(h^{-2})$. Both approaches would reduce the size of the initial interval where the numerical solution gets stuck at the initial value by a factor of $1/2$.

The data obtained by our $P(EC)^\mu E$ approach are given in Table C.7. Note that the data of Table C.6 correspond to this method with $\mu = 1$.

It is clearly seen that there is a significant advantage in this approach: By choosing $\mu = 10$ and $h = 1/40$ for example, we obtain an absolute error that is about 25% smaller than in the case $\mu = 1$ and $h = 1/200$, and at the same time the run time is

Table C.7 Numerical results of second approximation ($N = 29$, $\gamma = 0.05$) as in (C.31) with $P(EC)^\mu E$ algorithm

Number μ of corrector iterations	Step size	Maximal error	Run time
10	1/40	0.03175	4.9 s
10	1/100	0.01174	28.6 s
20	1/40	0.00989	9.3 s
20	1/100	0.00379	55.0 s

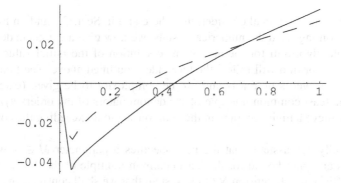

Fig. C.4 Errors for second approximation ($N = 29$, $\gamma = 0.05$) as in (C.31) with various combinations of step size and number of corrector steps

75% shorter. The reason is the following. In the case $\mu = 1$ the numerical solution gets stuck at the initial value for a rather long interval. At the end of this interval the true solution has moved away significantly from the initial value, and here the error attains its maximum. Over the remainder of the interval $[0, 1]$ the numerical solution then has to creep towards the exact solution, and the error gets smaller. If we choose a larger value for μ, we make the problematic initial interval smaller, and therefore we also diminish the error attained over this interval. This is apparent from Fig. C.4 where we have compared the absolute errors for $\mu = 1$, $h = 1/200$ (solid line) and $\mu = 10$, $h = 1/40$ (dashed line).

In this example one can of course now apply the idea of using many corrector steps also to the given equation (C.27) itself. This is equivalent to skipping stage 1 of our two-stage process. It is clear that the run times of the plain PECE scheme (see Table C.4) are not competitive. Therefore we once again revert to the $P(EC)^\mu E$ structure with larger values for μ and larger step sizes as before. Some results are stated in Table C.8.

Comparing Tables C.4 and C.8 we once again find a significant run time advantage in the $P(EC)^\mu E$ method as compared to the PECE method without losing accuracy, but even the approximations obtained by the faster $P(EC)^\mu E$ approach are less accurate and more time consuming than the results presented in Table C.7 where we had used a simpler differential equation system.

Table C.8 Numerical results for unperturbed equation
($N = 291$, $\gamma = 0.005$), with $P(EC)^\mu E$ algorithm

Number μ of corrector iterations	Step size	Maximal error	Run time
10	1/100	0.16473	117.3 s
10	1/200	0.08607	446.8 s
20	1/100	0.08607	222.8 s
20	1/200	0.04400	811.0 s

Based on our theoretical considerations here and in Sect. C.1 and on heuristical arguments coming from the numerical results, we now give a complete description of a possible algorithm for the approximate solution of the initial value problem (C.24). The algorithm will follow the basic ideas outlined above. The fundamental concept is that we assume a bound on the complexity to be given (expressed in terms of the least common multiple of the denominators of the orders $\tilde{n}_j$) and that we try to achieve a high accuracy in the solution without exceeding the complexity limit.

Specifically, we assume that the user specifies a parameter $M \in \mathbb{N}$ which we interpret as an upper bound for the least common multiple of the denominators of $\tilde{n}_1, \ldots, \tilde{n}_k$. Since the dimension N of the system that we shall construct in stage 2 of the algorithm is given by $N = M\tilde{n}_k \approx Mn_k$, this data gives us an upper bound on the dimension and hence an upper bound on the arithmetic complexity.

We begin by constructing the perturbations required for the first stage. This is very simple; for $j = 1, 2, \ldots, k$ we only have to set $\tilde{n}_j := \alpha_j/M$ where $\alpha_j \in \mathbb{N}$ is chosen to be the natural number closest to Mn_j (i.e. $\alpha_j = \lfloor Mn_j + 0.5 \rfloor$). In this way we make sure that, for every single j, the quantity $|\tilde{n}_j - n_j|$ is minimized under the condition that the least common multiple of the denominators of $\tilde{n}_1, \ldots, \tilde{n}_k$ is bounded by M. This essentially completes the first stage.

The second stage begins by rewriting the perturbed equations as a system of order $\gamma = 1/M$ and dimension N as described in Theorem 8.1. This system is solved by the $P(EC)^\mu E$ scheme indicated above. To avoid the problems caused by the large number of zeros in the new initial condition, we choose the parameter μ in a way that depends on the number of zeros (i.e. on M); specifically we set $\mu := M$ as suggested in [36]. Note that it follows from our considerations that it is neither necessary nor helpful to introduce additional flexibility by choosing different values for the parameter μ in each step. The choice that we propose here is sufficient to avoid the problems caused by the (possibly) large number of zeros in the initial condition. Choosing μ larger than this would not give a better order of accuracy, so there is no point in doing that (cf. the considerations on (C.32) below). Choosing μ smaller (permanently or temporarily) would mean that the problem cannot be avoided totally, so one would have to assume a deterioration of the approximation quality, but on the other hand it would not lead to a significantly faster algorithm because the arithmetic complexity of the entire scheme is (asymptotically) independent of μ.

Another advantage of the $P(EC)^\mu E$ scheme with the choice of μ indicated above can be explained by a look at Theorem C.5: The algorithm converges to the true solution of the perturbed equation with an error of

$$\max_{j=0,1,\ldots,N} |y(t_j) - y_h(t_j)| = O(h^p) \quad \text{where} \quad p = \min(2, 1 + \gamma\mu). \tag{C.32}$$

Since here we have $\mu = M = 1/\gamma$ by construction, this means that in the proposed scheme we actually have $p = 2$ in every case, so we find slightly better convergence behaviour than in the simple PECE approach; indeed this is the maximum order than one can possibly obtain by an algorithm that uses the approximation method underlying our scheme.

This approach is particularly useful when one is looking for a computationally inexpensive but still reasonably accurate approximation. In many applications this will be what is desired because often one needs to solve a great number of such initial value problems whose solutions are then required as input data for other problems. Additionally, high accuracy is frequently impossible to obtain anyway because the given data (in particular the orders n_j of the differential operators) are something like material constants known only up to a certain (usually moderate) precision.

Conversion to Multi-Order Systems

An alternative approach has been suggested in [59]. Two variants are possible; we begin with the one that is simpler to describe and discuss the other one later.

The basic idea is to use a completely different transformation of the given initial value problem into the form of a system of fractional differential equations. Essentially this amounts to replacing the path that uses Theorem 8.1 or 8.2 that we had used above by the transformation to a multi-order system according to Theorem 8.9 or Theorem 8.10. We begin with the former and, as above, assume that the original initial value problem is given in the form

$$D_{*0}^{n_k} y(x) = f(x, y(x), D_{*0}^{n_1} y(x), D_{*0}^{n_2} y(x), \ldots, D_{*0}^{n_{k-1}} y(x)) \tag{C.33a}$$

(where $0 < n_1 < n_2 < \ldots < n_k$) with a suitable function f and initial conditions

$$y^{(j)}(0) = y_0^{(j)}, \qquad j = 0, 1, \ldots, \lceil n_k \rceil - 1. \tag{C.33b}$$

However we now assume (without loss of generality) that additionally we have

$$\{1, 2, \ldots, \lceil n_k \rceil - 1\} \subset \{n_1, n_2, \ldots, n_k\}. \tag{C.34}$$

This implies $n_j - n_{j-1} \leq 1$ for all j. Consider now the differences $d_j := n_j - n_{j-1}$ for $j = 1, 2, 3, \ldots, k$, where we have defined $n_0 := 0$. We then introduce the new functions

$$y_0 := y, \qquad y_j := D_{*0}^{d_j} y_{j-1} \quad (j = 1, 2, \ldots, k),$$

such that

$$y_1 = D_{*0}^{d_1} y_0 = D_{*0}^{n_1 - n_0} y_0 = D_{*0}^{n_1} y,$$

$$y_2 = D_{*0}^{d_2} y_1 = D_{*0}^{n_2 - n_1} D_{*0}^{n_1} y = D_{*0}^{n_2} y,$$

$$\vdots$$

$$y_k = D_{*0}^{d_k} y_{k-1} = D_{*0}^{n_k - n_{k-1}} D_{*0}^{n_{k-1}} y = D_{*0}^{n_k} y.$$

For the derivation of these identities one can proceed as in the proof of Lemma 3.13; the key point is that – due to our assumption (C.34) – we never jump across an integer when moving from n_{j-1} to n_j. Thus we may rewrite (C.33a) in the form

$$D_{*0}^{d_1} y_0 = y_1,$$

$$D_{*0}^{d_2} y_1 = y_2,$$

$$\vdots$$

$$D_{*0}^{d_{k-1}} y_{k-2} = y_{k-1},$$

$$D_{*0}^{d_k} y_{k-1} = f(x, y_0(x), y_1(x), y_2(x), \ldots, y_{k-1}(x)). \tag{C.35a}$$

We have found the required system of differential equations; the corresponding initial values obviously have to be

$$y_j(0) = \begin{cases} y_0^{(n_j)} & \text{if } n_j \in \mathbb{N}_0, \\ 0 & \text{else} \end{cases} \tag{C.35b}$$

in view of the fact that $y_j = D_{*0}^{n_j} y$, Lemma 3.11 and the given initial values (C.33b). A comparison of this multidimensional initial value problem with its counterpart (C.25) constructed above reveals a number of substantial differences even though formally they are equivalent in the sense that the first components of the solutions of the two problems coincide:

(a) The dimension of the new system is k (a small number in typical applications), independent of the values of the n_j; we had seen above (see, e.g., Example C.3) that the other approach could give rise to systems of very large dimension even if k was small.

(b) The number of zeros in the initial condition (C.33b) relates to the number of zeros in its counterpart (C.25b) in the same way as the dimensions.

(c) The structure of the left-hand side of the new system (C.35a) is much more complicated than it was in the old system (C.25a).

(d) The formal structures of the right-hand sides of the two system do not differ from each other at all.

It turns out that, in view of the considerations with respect to the approach using single-order systems, the first two points mentioned above indicate the capability of the new approach to avoid the potentially serious problems encountered in the first approach. On the other hand, the third item reveals that we have to pay a certain price for this improvement: Instead of requiring an approximation for only one differential operator we now need to work with operators of order $d_1, d_2, \ldots, d_k$. These orders may or may not coincide with each other.

Example C.5. We rewrite the equations from Examples C.2 and C.3 as systems of equations according to the ideas outlined above.

For the Bagley–Torvik equation

$$Ay''(x) = -BD_{*0}^{3/2}y(x) - Cy(x) + C(x+1), \qquad y(0) = y'(0) = 1,$$

from Example C.2, we have

$$n_1 = 1, \quad n_2 = \frac{3}{2} \quad \text{and} \quad n_3 = 2,$$

such that

$$d_1 = 1, \quad d_2 = \frac{1}{2} \quad \text{and} \quad d_3 = \frac{1}{2}.$$

The resulting system thus is three-dimensional and reads

$$D_{*0}^1 y_0(x) = y_1(x),$$
$$D_{*0}^{1/2} y_1(x) = y_2(x),$$
$$D_{*0}^{1/2} y_2(x) = -\frac{B}{A}y_2(x) - \frac{C}{A}y_0(x) + \frac{C}{A}(x+1)$$

with initial conditions

$$y_0(0) = y_1(0) = 1 \quad \text{and} \quad y_2(0) = 0.$$

A comparison with Example C.2 shows that the differences for this simple example are small: The dimension is reduced by one, and two different fractional derivatives appear on the left-hand side of the system.

In the other example, the equation was

$$D_{*0}^{1.455}y(x) = -x^{0.1}\frac{E_{1.545}(-x)}{E_{1.445}(-x)}\exp(x)y(x)D_{*0}^{0.555}y(x)$$
$$+\exp(-2x) - [D_{*0}^1 y(x)]^2,$$

with initial conditions $y(0) = 1$ and $y'(0) = -1$. Here the new approach uses the parameters

$$n_1 = 0.555, \quad n_2 = 1 \quad \text{and} \quad n_3 = 1.455,$$

such that
$$d_1 = 0.555, \quad d_2 = 0.445 \quad \text{and} \quad d_3 = 0.455.$$

Once again we obtain a three-dimensional system; this time it has the form

$$D_{*0}^{0.555}y_0(x) = y_1(x),$$
$$D_{*0}^{0.445}y_1(x) = y_2(x),$$
$$D_{*0}^{0.455}y_2(x) = -x^{0.1}\frac{E_{1.545}(-x)}{E_{1.445}(-x)}\exp(x)y_0(x)y_1(x) + \exp(-2x) - [y_2(x)]^2$$

with initial conditions

$$y_0(0) = 1, \quad y_1(0) = 0 \quad \text{and} \quad y_2(0) = -1.$$

Now the difference to the single-order system approach is enormous: The dimension of the system is reduced from 291 to 3, but of course we now have to work with three different differential operators.

Before we come to the question for a suitable numerical scheme for systems of this structure, let us briefly introduce a small modification of the idea presented so far that may lead to a slightly more efficient scheme. This is motivated by the observation from the two examples above that integer order differential operators that are local by nature are decomposed into two (or more) non-local fractional differential operators: In the Bagley–Torvik example we have $y'' = D_{*0}^{1/2}y_2 = D_{*0}^{1/2}D_{*0}^{1/2}y_1 = D_{*0}^{1/2}D_{*0}^{1/2}D_{*0}^1 y_0$, and so in any approximation method the possibility to save time by making use of the locality is lost. A similar decomposition $y_2 = y' = D_{*0}^{0.445}y_1 = D_{*0}^{0.445}D_{*0}^{0.555}y_0$ is used in the other example. It would thus be preferable to use the alternative systems

$$D_{*0}^1 y_0(x) = y_1(x),$$
$$D_{*0}^{1/2}y_1(x) = y_2(x),$$
$$D_{*0}^1 y_1(x) = -\frac{B}{A}y_2(x) - \frac{C}{A}y_0(x) + \frac{C}{A}(x+1)$$

with initial conditions

$$y_0(0) = y_1(0) = 1 \quad \text{and} \quad y_2(0) = 0$$

for the Bagley–Torvik problem and

$$D_{*0}^{0.555}y_0(x) = y_1(x),$$
$$D_{*0}^1 y_0(x) = y_2(x),$$
$$D_{*0}^{0.455}y_2(x) = -x^{0.1}\frac{E_{1.545}(-x)}{E_{1.445}(-x)}\exp(x)y_0(x)y_1(x) + \exp(-2x) - [y_2(x)]^2$$

with initial conditions

$$y_0(0) = 1, \quad y_1(0) = 0 \quad \text{and} \quad y_2(0) = -1$$

for the other example. In this way we simplify the structure somewhat because some of the fractional differential operators on the left-hand side can be replaced by integer-order operators. This modification is equivalent to using Theorem 8.10 instead of Theorem 8.9 in our multi-order system approach.

For the numerical solution of these systems of equations one can then use a numerical scheme for scalar fractional differential equations for each component separately. Of course one needs to take into account that the individual equations now will typically not be of the same order, so numerical methods for different orders must be used. In general, there do not seem to be any advantages in using differently constructed methods for the individual equations; rather one would usually prefer to use just one class of numerical schemes and merely change the orders of the algorithms as prescribed by the orders of the differential operators on the left-hand side of the system.

In either of the two multi-order system approaches, Edwards et al. [59] have investigated the use of the formula of [34] for the numerical solution of the resulting system of equations; later results [65] indicate that our Adams–Bashforth–Moulton method is likely to be more efficient. As far as the error is concerned, it turns out that the behaviour of the entire scheme is dominated by the component with the worst behaviour.

A comparison with the single-order system approach shows that the multi-order system approach is always applicable (there are no number-theoretic restrictions on the orders $n_1, \ldots, n_k$), and it will in many cases lead to a system with a considerably smaller dimension. However the structure of the left-hand side becomes more complicated.

The numerical experiments of Ford and Connolly [66] indicate that the conversion to a multi-order system via Theorem 8.10 tends to be the computationally least efficient of the approaches presented here. They found the conversion to a multi-order system via Theorem 8.9 and the approach using the transformation to single-order systems by means of Theorem 8.1 or 8.2 to be usually preferable. Which of these ideas works best seems to depend on the particular problem under consideration, so a generally valid advice cannot be given.

Exercise

Exercise C.1. Give an explicit proof of Theorem C.2.

Appendix D
Useful Results from Analysis

In this chapter we collect some information on some concepts from Analysis that is useful in the remainder of the text.

D.1 Euler's Gamma Function

We begin with the Gamma function.

We recall the definition

$$\Gamma(x) = \int_0^\infty t^{x-1} e^{-t} \, dt$$

for $x > 0$. Elementary considerations from the theory of improper integrals reveal that the integral exists. Moreover, upon setting $x = 1$ we easily see

$$\Gamma(1) = \int_0^\infty e^{-t} \, dt = \lim_{z \to \infty} \int_0^z e^{-t} \, dt = \lim_{z \to \infty} [-e^{-t}]_0^z = \lim_{z \to \infty} (1 - e^{-z}) = 1. \quad \text{(D.1)}$$

Additionally we may, for arbitrary $x > 0$, manipulate the integral in the definition of the Gamma function by means of a partial integration. This yields

$$\Gamma(x+1) = \int_0^\infty t^x e^{-t} \, dt = \lim_{z \to \infty, y \to 0+} \int_y^z t^x e^{-t} \, dt$$

$$= \lim_{z \to \infty, y \to 0+} \left([-e^{-t} t^x]_{t=y}^{t=z} + x \int_y^z t^{x-1} e^{-t} \, dt \right)$$

$$= x \int_0^\infty t^{x-1} e^{-t} \, dt = x \Gamma(x).$$

We have thus shown

Theorem D.1 (Functional Equation for Γ). *If $x > 0$ then $x\Gamma(x) = \Gamma(x+1)$.*

Now we may prove the all important relation between the Gamma function and the factorial:

Proof (of Theorem 1.3). The proof uses mathematical induction. The induction basis $(n = 1)$ reads $\Gamma(1) = 0! = 1$ which is true in view of (D.1). For the induction step, we use the functional equation and the induction hypothesis:

$$\Gamma(n+1) = n\Gamma(n) = n(n-1)! = n!$$

as desired.   □

There is one other important application of the functional equation of the Gamma function. We solve it for $\Gamma(x)$; it then reads

$$\Gamma(x) = \frac{\Gamma(x+1)}{x} \tag{D.2}$$

if $x > 0$. Now the expression on the right-hand side is meaningful not only if $x > 0$ but also in the case $-1 < x < 0$. Therefore we may use it as a definition for the left-hand side, i.e. for $\Gamma(x)$, in that case (which is not covered by the original definition because the defining integral is divergent for $x < 0$). Having done this extension, the Gamma function is also defined for $-1 < x < 0$, and we may return to (D.2) with x in that range. This allows us to extend the definition to $-2 < x < -1$. Proceeding in this manner, we find a definition for the Gamma function that can be applied for all $x \in \mathbb{R}$ with the exception of those for which $-x \in \mathbb{N}_0$.

As a consequence of these considerations, we find another important identity involving the Gamma function:

Theorem D.2. *Let $n \notin \mathbb{Z}$ and $k \in \mathbb{N}_0$. Then,*

$$(-1)^{k+1}\Gamma(n-k)\Gamma(k+1-n) = \Gamma(-n)\Gamma(n+1).$$

Another useful identity in this context is

Theorem D.3 (Reflection Formula for Γ). *Let $0 < x < 1$. Then,*

$$\Gamma(x)\Gamma(1-x) = \frac{\pi}{\sin \pi x}.$$

It is also possible to find an alternative representation, due to Gauss, for the Gamma function. This representation actually holds for the extension indicated above. However, in practical calculations one frequently observes that the integral representation is easier to handle.

Theorem D.4 (Gauss' Product Formula for Γ). *Let $x \in \mathbb{R}$, $-x \notin \mathbb{N}_0$. Then,*

$$\Gamma(x) = \lim_{n \to \infty} \frac{n!n^x}{x(x+1)(x+2)\cdots(x+n)}.$$

The asymptotic behaviour of $\Gamma(x)$ as $x \to \infty$ is sometimes important; it can be described by the following result [2, Chapter 6].

Theorem D.5 (Stirling's Formula). *For $x \to \infty$ we have*

$$\Gamma(x+1) = \left(\frac{x}{e}\right)^x \sqrt{2\pi x}(1+o(1)).$$

One last result that we shall mention explicitly is the following integral identity. We leave the proof as an exercise.

Theorem D.6. *Let $\alpha, \beta \in \mathbb{R}_+$. Then*

$$\int_0^1 t^{\alpha-1}(1-t)^{\beta-1}\,dt = \frac{\Gamma(\alpha)\Gamma(\beta)}{\Gamma(\alpha+\beta)},$$

and hence

$$\int_0^x t^{\alpha-1}(x-t)^{\beta-1}\,dt = x^{\alpha+\beta-1}\frac{\Gamma(\alpha)\Gamma(\beta)}{\Gamma(\alpha+\beta)}.$$

The integral in the first equation of Theorem D.6 is known as *Euler's integral of the first kind* or *Euler's Beta function $B(\alpha, \beta)$*.

More information on the Gamma function may be found, e.g., in the classical work of Artin [6] or in the usual reference works on special functions like [2, Chapter 6] or [62, Chapter I].

D.2   Fixed Point Theorems

The proofs of various existence and uniqueness theorems throughout this text have been based on classical theorems asserting existence or uniqueness of fixed points of certain operators.

The first of these theorems is the following generalization of Banach's fixed point theorem that we take from [189].

Theorem D.7 (Weissinger's Fixed Point Theorem). *Assume (U,d) to be a non-empty complete metric space, and let $\alpha_j \geq 0$ for every $j \in \mathbb{N}_0$ and such that $\sum_{j=0}^{\infty} \alpha_j$ converges. Furthermore, let the mapping $A : U \to U$ satisfy the inequality*

$$d(A^j u, A^j v) \leq \alpha_j d(u,v) \tag{D.3}$$

for every $j \in \mathbb{N}$ and every $u, v \in U$. Then, A has a uniquely determined fixed point u^*. Moreover, for any $u_0 \in U$, the sequence $(A^j u_0)_{j=1}^{\infty}$ converges to this fixed point u^*.

An immediate consequence is

Corollary D.8 (Banach's Fixed Point Theorem). *Assume (U,d) to be a non-empty complete metric space, let $0 \leq \alpha < 1$, and let the mapping $A : U \to U$ satisfy the inequality*

$$d(Au, Av) \leq \alpha d(u, v) \tag{D.4}$$

for every $u, v \in U$. Then, A has a uniquely determined fixed point u^*. Furthermore, for any $u_0 \in U$, the sequence $(A^j u_0)_{j=1}^{\infty}$ converges to this fixed point u^*.

Moreover we also used a slightly different result that asserts only the existence but not the uniqueness of a fixed point. Here we may work with weaker assumptions on the operator in question. A proof may be found, e.g., in [29].

Theorem D.9 (Schauder's Fixed Point Theorem). *Let (E, d) be a complete metric space, let U be a closed convex subset of E, and let $A : U \to U$ be a mapping such that the set $\{Au : u \in U\}$ is relatively compact in E. Then A has at least one fixed point.*

In this context we recall a definition:

Definition D.1. Let (E, d) be a metric space and $F \subseteq E$. The set F is called *relatively compact in E* if the closure of F is a compact subset of E.

A helpful classical result from Analysis in connection with such sets is as follows. The proof can be found in many standard textbooks, e.g. in [30, p. 30].

Theorem D.10 (Arzelà–Ascoli). *Let $F \subseteq C[a, b]$ for some $a < b$, and assume the sets to be equipped with the Chebyshev norm. Then, F is relatively compact in $C[a, b]$ if F is equicontinuous (i.e. for every $\varepsilon > 0$ there exists some $\delta > 0$ such that for all $f \in F$ and all $x, x^* \in [a, b]$ with $|x - x^*| < \delta$ we have $|f(x) - f(x^*)| < \varepsilon$) and uniformly bounded (i.e. there exists a constant $C > 0$ such that $\|f\|_{\infty} \leq C$ for every $f \in F$).*

D.3 The Laplace Transform

The Laplace transform method is an extremely useful tool for the analysis of linear (fractional or classical) initial value problems. In particular, it allows us to replace a differential equation by an algebraic equation. We take the fundamental definition from the classical book of Doetsch [56] where the interested reader may find a comprehensive treatment of the Laplace transform.

Definition D.2. Let $f : [0, \infty) \to \mathbb{R}$ be given. The function F defined by

$$F(s) := \mathscr{L}f(s) := \int_0^{\infty} f(x) e^{-sx} \, dx$$

is called the *Laplace transform* of f whenever the integral exists.

It is rather simple to calculate the Laplace transform of some elementary functions.

Example D.1. (a) For $f(x) = \exp(ax)$ with $a \in \mathbb{R}$ we have $\mathscr{L}f(s) = 1/(s-a)$
whenever $s > a$.
(b) For $f(x) = x^k$ with $k > -1$ we find $\mathscr{L}f(s) = \Gamma(k+1)/s^{k+1}$ whenever $s > 0$.
(c) For $f(x) = \sin \omega x$ with $\omega > 0$ we have $\mathscr{L}f(s) = \omega/(s^2 + \omega^2)$, again for $s > 0$.

We cite the most important rules for Laplace transforms.

Theorem D.11. *Assume the functions f_1, f_2 and f_3 to be given on $[0, \infty)$ and to be such that their Laplace transforms exist for all $s \geq s_0$ with some suitable $s_0 \in \mathbb{R}$. Then we have the following rules.*

(a) *If $f_3 = a_1 f_1 + a_2 f_2$ with arbitrary real constants a_1 and a_2 then*

$$\mathscr{L}f_3(s) = a_1 \mathscr{L}f_1(s) + a_2 \mathscr{L}f_2(s)$$

(linearity of the Laplace transform).
(b) *If f_3 is the convolution of f_1 and f_2, i.e. if*

$$f_3(x) = \int_0^x f_1(x-t) f_2(t)\,dt,$$

then

$$\mathscr{L}f_3(s) = \mathscr{L}f_1(s) \cdot \mathscr{L}f_2(s)$$

(the convolution theorem). In other words: The convolution in the original domain corresponds to the usual product in the Laplace domain.
(c) *If $f_3(x) = \int_0^x f_1(t)\,dt$ then we have for $s > \max\{0, s_0\}$*

$$\mathscr{L}f_3(s) = \frac{1}{s}\mathscr{L}f_1(s)$$

(the integration theorem).
(d) *Let $m \in \mathbb{N}$. If $f_3 = D^m f_1$ is the mth derivative of f_1 then*

$$\mathscr{L}f_3(s) = s^m \mathscr{L}f_1(s) - \sum_{k=1}^{m} s^{m-k} f_1^{(k-1)}(0)$$

(the differentiation theorem).
(e) *Let $a > 0$ and $f_3(x) = f_1(ax)$. Then*

$$\mathscr{L}f_3(s) = \frac{1}{a}\mathscr{L}f_1(s/a).$$

(f) *Let $a \in \mathbb{R}$ and $f_3(x) = e^{-ax} f_1(x)$. Then*

$$\mathscr{L}f_3(s) = \mathscr{L}f_1(s+a).$$

(g)   Let $m \in \mathbb{N}$ and $f_3(x) = x^m f_1(x)$. Then

$$\mathscr{L} f_3(s) = (-1)^m \frac{d^m}{ds^m} \mathscr{L} f_1(s).$$

(h)   Let $f_3(x) = f_1(x)/x$. Then

$$\mathscr{L} f_3(s) = \int_s^\infty \mathscr{L} f_1(\sigma) \, d\sigma.$$

(i)   Let $a \in \mathbb{R}$ and

$$f_3(x) = \begin{cases} 0 & \text{for } x < a, \\ f_1(x-a) & \text{for } x \geq a. \end{cases}$$

Then

$$\mathscr{L} f_3(s) = e^{-as} \mathscr{L} f_1(s).$$

Part (d), the differentiation theorem, is of particular interest to us. Specifically this result needed to be generalized to $m \notin \mathbb{N}$ with a suitable definition of the differential operator. We have dealt with this question in Theorem 7.1.

Of course it is not sufficient to have the Laplace transform; for practical work the inverse transform is required too. There are various ways to express this inverse; one possibility is contained in the following result. We refer to the standard books on Laplace transforms for details on the "suitable assumptions".

Theorem D.12. *Under suitable assumptions on f we have*

$$f(x) = \frac{1}{2\pi i} \int_{c-i\infty}^{c+i\infty} \exp(sx) \mathscr{L} f(s) \, ds.$$

Under certain conditions, the long-term behaviour of functions may also be expressed with the help of Laplace transforms, see [27, 76, 158] and the references cited therein:

Theorem D.13 (Final Value Theorem). *Assume that $\mathscr{L} f$ does not have any singularities in the closed right half-plane $\{s \in \mathbb{C} : \text{Re}\, s \geq 0\}$, except for possibly a simple pole at the origin. Then,*

$$\lim_{x \to \infty} f(x) = \lim_{s \to 0+} s \mathscr{L} f(s).$$

Remark D.1. The condition on the singularities of $\mathscr{L} f$ is essential here: If $\mathscr{L} f$ has a pole with positive real part, then $f(x)$ is unbounded as $x \to \infty$, and if $\mathscr{L} f$ has a pole on the imaginary axis (but not at the origin) then f has persistent oscillations, so $\lim_{x \to \infty} f(x)$ does not exist either.

D.4 Hadamard's Finite-Part Integral

The integral $\int_a^b (x-a)^{-\mu} f(x) \, dx$ is divergent for $\mu \geq 1$ whenever $f(a) \neq 0$. Nevertheless it is sometimes useful to assign a finite value to such integrals. This has been observed by Hadamard [86] in connection with solution methods for certain partial differential equations, and he introduced the following idea for the solution of this problem, known as the *finite-part integral*. We shall mainly require this concept for $\mu \notin \mathbb{N}$, and therefore we will restrict our attention to these values of μ. The consideration of integer values requires some small modifications.

The Hadamard finite-part of the integral (that we will, for the sake of simplicity, denote by the same symbol as the standard integral) is, roughly speaking, defined by a Taylor expansion of f at $x = a$ where the resulting singular integrals are defined by

$$\int_a^b (x-a)^{-\mu} \, dx = \frac{1}{1-\mu} (b-a)^{1-\mu} \qquad (\mu > 1). \tag{D.5}$$

Essentially this means that we first replace the integral $\int_a^b (x-a)^{-\mu} \, dx$ by the expression $\int_{a+\varepsilon}^b (x-a)^{-\mu} \, dx$ for $\varepsilon > 0$. This is a convergent integral; its value is simply $(1-\mu)^{-1}[(b-a)^{1-\mu} - \varepsilon^{1-\mu}]$. Then we let $\varepsilon \to 0$. Of course the limit does not exist for $\mu \geq 1$, and so Hadamard suggested simply to ignore the unbounded contribution $\lim_{\varepsilon \to 0} \varepsilon^{1-\mu}/(1-\mu)$ and to assign the value of the remaining (finite) expression $(1-\mu)^{-1}(b-a)^{1-\mu}$ – hence the name "finite-part integral".

A precise way to define the finite-part integral is (for $\mu \notin \mathbb{N}$)

$$\int_a^b (x-a)^{-\mu} f(x) \, dx := \sum_{k=0}^{\lfloor \mu \rfloor -1} \frac{f^{(k)}(a)(b-a)^{k+1-\mu}}{(k+1-\mu)k!} \tag{D.6a}$$

$$+ \int_a^b (x-a)^{-\mu} R_{\lfloor \mu \rfloor -1}(x,a) \, dx.$$

Here,

$$R_p(x,a) := \frac{1}{p!} \int_a^x (x-y)^p f^{(p+1)}(y) \, dy \tag{D.6b}$$

is the remainder of the pth degree Taylor polynomial of f with expansion point a. It is well known that a sufficient condition for the existence of the integral (D.6a) is that $f \in C^s[a,b]$ with $\mu - 1 < s \in \mathbb{N}$. This is due to the fact that then the remainder term of the Taylor expansion has a zero at a whose order is so high that the singularity in the other factor in the last integral in (D.6a) is almost being cancelled; the remaining singularity is weak and integrable in the improper sense.

An alternative representation that is helpful for us can be taken from [61, eq. (A17)]:

Theorem D.14. *Let $\mu > 1$ but $\mu \notin \mathbb{N}$ and $m := \lceil \mu - 1 \rceil$. For $f \in C^m[a,b]$ we have*

$$\frac{1}{\Gamma(1-\mu)} \int_a^b (b-x)^{-\mu} f(x) \, dx = \sum_{k=0}^{m-1} \frac{(b-a)^{k-\mu+1}}{\Gamma(k-\mu+2)} f^{(k)}(a) + J_a^{m-\mu+1} f^{(m)}(b).$$

We mention here the most important properties of the finite-part integral:

- In contrast to the classical Riemann or Lebesgue integral, the finite-part integral is not a positive functional, i.e. the inequality

$$\left| \int_a^b (x-a)^{-\mu} f(x) \, dx \right| \leq \int_a^b (x-a)^{-\mu} |f(x)| \, dx$$

  is not true in general.
- The finite-part integral is a consistent extension of the concept of regular integrals, i.e. whenever the integral $\int_a^b (x-a)^{-\mu} f(x) \, dx$ exists in the classical sense, then it also exists in the finite-part sense, and the two integrals have the same value.
- The finite-part integral is additive with respect to the union of integration intervals and invariant with respect to translation.
- The finite-part integral is linear.
- The usual change-of-variables rule remains valid if $\mu \notin \mathbb{N}$.

A very useful result on these integrals is as follows.

Theorem D.15. *Let $f \in C^k[a,b]$ for some $k \in \mathbb{N}_0$, and let $p < k$. Then, for $a < x < b$,*

$$\frac{d}{dx} \int_a^x f(t)(x-t)^{-p} \, dt = -p \int_a^x f(t)(x-t)^{-p-1} \, dt.$$

We leave the proof as an exercise to the reader.

D.5 Approximation Theory

A well-known concept from approximation theory that we had to use in the proof of Theorem 2.25 was the *Bernstein polynomial*. A classical reference is the book of Lorentz [117]. The definition is

$$B_N[f](t) := \sum_{k=0}^{N} \binom{N}{k} t^k (1-t)^{N-k} f\left(\frac{k}{N}\right)$$

where $f : [0,1] \to \mathbb{R}$. The fundamental result that we require is a convergence theorem:

Theorem D.16. *Let $f \in C^\ell[0,1]$ for some $\ell \in \mathbb{N}_0$. Then, for all $\mu \in \{0,1,2,\ldots,\ell\}$, the sequence $(D^\mu B_N[f])_{N=1}^\infty$ converges uniformly towards $D^\mu f$.*

A proof may be found in [117, §1.8].

Exercises

Exercise D.1. Give a proof for Theorem D.6.

Exercise D.2. Give a proof for Theorem D.2.

Exercise D.3. Prove the relations stated in Example D.1.

Exercise D.4. Evaluate the finite-part integral $\int_0^1 x^{-\mu}\,dx$ for $\mu > 1$.

Exercise D.5. Give a proof for Theorem D.15.

References

1. Abel, N.H.: Auflösung einer mechanischen Aufgabe. J. Reine Angew. Math. **1**, 153–157 (1826)
2. Abramowitz, M., Stegun, I.A.: Handbook of Mathematical Functions, 2nd printing with corrections. National Bureau of Standards, Washington (1964); republished by Dover, New York (1965)
3. Adomian, G.: Solving Frontier Problems of Physics: the Decomposition Method. Kluwer, Dordrecht (1994)
4. Agarwal, R.P., Benchohra, M., Hamani, S.: A survey on existence results for boundary value problems of nonlinear fractional differential equations and inclusions. Acta Appl. Math. **109**, 973–1033 (2010)
5. Ahmad, W.M., El-Khazali, R.: Fractional-order dynamical models of love. Chaos Solitons Fractals **33**, 1367–1375 (2007)
6. Artin, E.: Einführung in die Theorie der Gammafunktion. Teubner, Leipzig (1931); English translation: The Gamma Function. Holt, Rinehart and Winston, New York (1964)
7. Audounet, J., Roquejoffre, J.-M.: An asymptotic fractional differential model of spherical flame. In: Matignon, D., Montseny, G. (eds.) Fractional Differential Systems: Models, Methods and Applications, pp. 15–27. SMAI, Paris (1998)
8. Audounet, J., Giovangigli, V., Roquejoffre, J.-M.: A threshold phenomenon in the propagation of a point source initiated flame. Physica D **121**, 295–316 (1998)
9. Bagley, R.L., Calico, R.A.: Fractional order state equations for the control of viscoelastically damped structures. J. Guid. Contr. Dynam. **14**, 304–311 (1991)
10. Bagley, R.L., Torvik, P.J.: On the existence of the order domain and the solution of distributed order equations – Part I. Int. J. Appl. Math. **2**, 865–882 (2000)
11. Bagley, R.L., Torvik, P.J.: On the existence of the order domain and the solution of distributed order equations – Part II. Int. J. Appl. Math. **2**, 965–987 (2000)
12. Bai, J., Feng, X.-C.: Fractional-order anisotropic diffusion for image denoising. IEEE Trans. Image Process. **16**, 2492–2502 (2007)
13. Baleanu, D., Diethelm, K., Scalas, E., Trujillo, J.J.: Fractional Calculus: Models and Numerical Methods. World Scientific, Singapore (2012)
14. Basset, A.B.: On the motion of a sphere in a viscous liquid. Philos. Trans. R. Soc. A **179**, 43–63 (1888)
15. Basset, A.B.: On the descent of a sphere in a viscous liquid. Q. J. Pure Appl. Math. **41**, 369–381 (1910)
16. Benchohra, M., Hamani, S., Ntouyas, S.K.: Boundary value problems for differential equations with fractional order. Surv. Math. Appl. **3**, 1–12 (2008)
17. Benson, D.A.: The fractional advection–dispersion equation: development and application. Ph.D. thesis, University of Nevada at Reno (1998)
18. Bonilla, B., Rivero, M., Trujillo, J.J.: On systems of linear fractional differential equations with constant coefficients. Appl. Math. Comput. **187**, 68–78 (2007)
19. Braß, H.: Quadraturverfahren. Vandenhoeck & Ruprecht, Göttingen (1977)

20. Brunner, H., van der Houwen, P.J.: The Numerical Solution of Volterra Equations. North-Holland, Amsterdam (1986)

21. Brunner, H., Pedas, A., Vainikko, G.: The piecewise polynomial collocation method for non-linear weakly singular Volterra equations. Math. Comput. **68**, 1079–1095 (1999)

22. Caponetto, R., Dongola, G., Fortuna, L., Petráš, I.: Fractional Order Systems: Modeling and Control Applications. World Scientific, River Edge, NJ (2010)

23. Caputo, M.: Linear models of dissipation whose Q is almost frequency independent – II. Geophys. J. Roy. Astron. Soc. **13**, 529–539 (1967); reprinted in Fract. Calc. Appl. Anal. **11**, 4–14 (2008)

24. Caputo, M., Mainardi, F.: A new dissipation model based on memory mechanism. Pure Appl. Geophys. **91**, 134–147 (1971); reprinted in Fract. Calc. Appl. Anal. **10**, 310–323 (2007)

25. Caputo, M., Mainardi, F.: Linear models of dissipation in anelastic solids. Rivista del Nuovo Cimento **1**, 161–198 (1971)

26. Chatterjee, A.: Statistical origins of fractional derivatives in viscoelasticity. J. Sound Vib. **284**, 1239–1245 (2005)

27. Chen, J., Lundberg, K.H., Davison, D.E., Bernstein, D.S.: The final value theorem revisited – infinite limits and irrational functions. IEEE Contr. Syst. Mag. **27** (3), 97–99 (2007)

28. Chern, J.-T.: Finite element modeling of viscoelastic materials on the theory of fractional calculus. Ph.D. thesis, Pennsylvania State University (1993)

29. Collatz, L.: Funktionalanalysis und numerische Mathematik. Springer, Berlin (1968)

30. Corduneanu, C.: Principles of Differential and Integral Equations, 2nd edn. Chelsea Publ. Comp., New York (1977)

31. Cuesta, E., Finat Codes, J.: Image processing by means of a linear integro-differential equation. In: Hamza, M.H. (ed.) Visualization, Imaging, and Image Processing 2003, Paper 91. ACTA Press, Calgary (2003)

32. de Hoog, F., Weiss, R.: Asymptotic expansions for product integration. Math. Comput. **27**, 295–306 (1973)

33. Deng, W.: Short memory principle and a predictor-corrector approach for fractional differential equations. J. Comput. Appl. Math. **206**, 174–188 (2007)

34. Diethelm, K.: An algorithm for the numerical solution of differential equations of fractional order. Electron. Trans. Numer. Anal. **5**, 1–6 (1997)

35. Diethelm, K.: Predictor-corrector strategies for single- and multi-term fractional differential equations. In: Lipitakis, E.A. (ed.) Proceedings of the 5th Hellenic-European Conference on Computer Mathematics and Its Applications, pp. 117–122. LEA Press, Athens (2002)

36. Diethelm, K.: Efficient solution of multi-term fractional differential equations using P(EC)^{m}E methods. Computing **71**, 305–319 (2003)

37. Diethelm, K.: Smoothness properties of solutions of Caputo-type fractional differential equations. Fract. Calc. Appl. Anal. **10**, 151–160 (2007)

38. Diethelm, K.: An investigation of some nonclassical methods for the numerical approximation of Caputo-type fractional derivatives. Numer. Algorithms **47**, 361–390 (2008)

39. Diethelm, K.: Multi-term fractional differential equations, multi-order fractional differential systems and their numerical solution. J. Eur. Syst. Autom. **42**, 665–676 (2008)

40. Diethelm, K.: On the separation of solutions of fractional differential equations. Fract. Calc. Appl. Anal. **11**, 259–268 (2008)

41. Diethelm, K.: An improvement of a nonclassical numerical method for the computation of fractional derivatives. J. Vib. Acoust. **131**, 014502 (2009)

42. Diethelm, K., Ford, N.J.: Numerical solution methods for distributed order differential equations. Fract. Calc. Appl. Anal. **4**, 531–542 (2001)

43. Diethelm, K., Ford, N.J.: Analysis of fractional differential equations. J. Math. Anal. Appl. **265**, 229–248 (2002)

44. Diethelm, K., Ford, N.J.: Numerical solution of the Bagley–Torvik equation. BIT **42**, 490–507 (2002)

45. Diethelm, K., Ford, N.J.: Multi-order fractional differential equations and their numerical solution. Appl. Math. Comput. **154**, 621–640 (2004)

46. Diethelm, K., Ford, N.J.: Numerical analysis for distributed order differential equations. J. Comput. Appl. Math. **225**, 96–104 (2009)
47. Diethelm, K., Ford, N.J.: Volterra integral equations and fractional calculus: do neighbouring solutions intersect? J. Integr. Equ. Appl. (in press)
48. Diethelm, K., Ford, N.J., Freed, A.D.: A predictor-corrector approach for the numerical solution of fractional differential equations. Nonlinear Dynam. **29**, 3–22 (2002)
49. Diethelm, K., Ford, N.J., Freed A.D.: Detailed error analysis for a fractional Adams method. Numer. Algorithms **36**, 31–52 (2004)
50. Diethelm, K., Freed, A.D.: On the solution of nonlinear fractional differential equations used in the modeling of viscoplasticity. In: Keil, F., Mackens, W., Voß, H., Werther, J. (eds.) Scientific Computing in Chemical Engineering II: Computational Fluid Dynamics, Reaction Engineering, and Molecular Properties, pp. 217–224. Springer, Heidelberg (1999)
51. Diethelm, K., Freed, A.D.: The FracPECE subroutine for the numerical solution of differential equations of fractional order. In: Heinzel, S., Plesser, T. (eds.) Forschung und wissenschaftliches Rechnen: Beiträge zum Heinz-Billing-Preis 1998, pp. 57–71. Gesellschaft für wissenschaftliche Datenverarbeitung, Göttingen (1999)
52. Diethelm, K., Luchko, Y.: Numerical solution of linear multi-term initial value problems of fractional order. J. Comput. Anal. Appl. **6**, 243–263 (2004)
53. Diethelm, K., Walz, G.: Numerical solution of fractional order differential equations by extrapolation. Numer. Algorithms **16**, 231–253 (1997)
54. Diethelm, K., Weilbeer, M.: A numerical approach for Joulin's model of a point source initiated flame. Fract. Calc. Appl. Anal. **7**, 191–212 (2004)
55. Dixon, J., McKee, S.: Weakly singular discrete Gronwall inequalities. Z. Angew. Math. Mech. **66**, 535–544 (1986)
56. Doetsch, G.: Anleitung zum praktischen Gebrauch der Laplace-Transformation und der Z-Transformation, 6th edn. Oldenbourg, München (1989)
57. Dokoumetzidis, A., Magin, R., Macheras, P.: A commentary on fractionalization of multi-compartmental models. J. Pharmacokinet. Pharmacodyn. **37**, 203–207 (2010)
58. Dzherbashyan, M.M., Nersesian, A.B.: Fractional derivatives and the Cauchy problem for differential equations of fractional order. Izv. Akad. Nauk Arm. SSR, Mat. **3**, 3–29 (1968) (in Russian)
59. Edwards, J.T., Ford, N.J., Simpson, A.C.: The numerical solution of linear multi-term fractional differential equations: systems of equations. J. Comput. Appl. Math. **148**, 401–418 (2002)
60. Edwards, J.T., Roberts, J.A., Ford, N.J.: A comparison of Adomian's decomposition method and Runge–Kutta methods for approximate solution of some predator prey model equations. Numerical Analysis Report 309, Manchester Centre for Computational Mathematics (1997)
61. Elliott, D.: An asymptotic analysis of two algorithms for certain Hadamard finite-part integrals. IMA J. Numer. Anal. **13**, 445–462 (1993)
62. Erdelyi, A., Magnus, W., Oberhettinger, F., Tricomi, F.G. (eds.): Higher Transcendental Functions, vol. I. McGraw-Hill, New York (1953)
63. Erdelyi, A., Magnus, W., Oberhettinger, F., Tricomi, F.G. (eds.): Higher Transcendental Functions, vol. III. McGraw-Hill, New York (1955)
64. Fitzgerald, W.J., Leung, C.M.: A critical review of the Adomian method. Tech. Report CUED/F-INFENG/TR. 277. University of Cambridge (1996)
65. Ford, N.J., Connolly, J.A.: Comparison of numerical methods for fractional differential equations. Commun. Pure Appl. Anal. **5**, 289–307 (2006)
66. Ford, N.J., Connolly, J.A.: Systems-based decomposition schemes for the approximate solution of multi-term fractional differential equations. J. Comput. Appl. Math. **229**, 382–391 (2009)
67. Ford, N.J., Simpson, A.C.: The approximate solution of fractional differential equations of order greater than 1. In: Deville, M., Owens, R. (eds.) Proceedings of the 16th IMACS World Congress on Scientific Computation, Applied Mathematics, and Simulation, Paper 213-1. IMACS, New Brunswick (2000)

68. Freed, A.D., Diethelm, K.: Fractional calculus in biomechanics: a 3D viscoelastic model using regularized fractional-derivative kernels with application to the human calcaneal fat pad. Biomech. Model. Mechanobiol. **5**, 203–215 (2006)

69. Freed, A.D., Diethelm, K., Luchko, Y.: Fractional-order viscoelasticity (FOV): constitutive development using the fractional calculus (first annual report). Technical Memorandum 2002-211914, NASA Glenn Research Center, Cleveland (2002)

70. Galeone, L., Garrappa, R.: Fractional Adams–Moulton methods. Math. Comput. Simul. **79**, 1358–1367 (2008)

71. Gaul, L., Klein, P., Kempfle, S.: Damping description involving fractional operators. Mech. Syst. Signal Process. **5**, 81–88 (1991)

72. Gel'fand, I.M., Shilov, G.E.: Generalized Functions, vol. 1. Academic Press, New York (1964)

73. Gel'fond, A.O., Leont'ev, A.F.: On a generalization of the Fourier series. Mat. Sb. (N. S.) **29**, 477–500 (1951) (in Russian)

74. Gerasimov, A.N.: A generalization of linear laws of deformation and its application to the problems of internal friction. Prikl. Mat. Mekh. **12**, 251–260 (1948) (in Russian)

75. Glöckle, W.G., Nonnenmacher, T.F.: A fractional calculus approach to self-similar protein dynamics. Biophys. J. **68**, 46–53 (1995)

76. Gluskin, E.: Let us teach this generalization of the final-value theorem. Eur. J. Phys. **24**, 591–597 (2003)

77. Gorenflo, R.: Fractional calculus: some numerical methods. In: Carpinteri, A., Mainardi, F. (eds.) Fractals and Fractional Calculus in Continuum Mechanics, pp. 277–290. Springer, Wien (1997)

78. Gorenflo, R., De Fabritiis, G., Mainardi, F.: Discrete random walk models for symmetric Lévy-Feller diffusion processes. Physica A **269**, 79–89 (1999)

79. Gorenflo, R., Loutchko, J., Luchko, Y.: Computation of the Mittag-Leffler function $E_{\alpha,\beta}(z)$ and its derivative. Fract. Calc. Appl. Anal. **5**, 491–518 (2002); Corrections: Fract. Calc. Appl. Anal. **6**, 111–112 (2003)

80. Gorenflo, R., Mainardi, F.: Fractional oscillations and Mittag-Leffler functions. In: Proceedings of the International Workshop on the Recent Advances in Applied Mathematics (RAAM '96), pp. 193–208. Kuwait University, Department of Mathematics and Computer Science, Kuwait (1996)

81. Gorenflo, R., Mainardi, F.: Fractional calculus: integral and differential equations of fractional order. In: Carpinteri, A., Mainardi, F. (eds.) Fractals and Fractional Calculus in Continuum Mechanics, pp. 223–276. Springer, Wien (1997)

82. Gorenflo, R., Rutman, R.: On ultraslow and intermediate processes. In: Rusev, P., Dimovski, I., Kiryakova, V. (eds.) Transform Methods and Special Functions, pp. 61–81. Science Culture Technology, Singapore (1995)

83. Gorenflo, R., Vivoli, A.: Fully discrete random walks for space-time fractional diffusion equations. Signal Process. **83**, 2411–2420 (2003)

84. Gross, B.: On creep and relaxation. J. Appl. Phys. **18**, 212–221 (1947)

85. Grünwald, A.K.: Über "begrenzte" Derivationen und deren Anwendung. Z. Angew. Math. Phys. **12**, 441–480 (1867)

86. Hadamard, J.: Lectures on Cauchy's Problem in Linear Partial Differential Equations. Yale Univ. Press, New Haven (1923); reprinted by Dover, New York (1952)

87. Hadid, S.B., Ta'ani, A.A., Momani, S.M.: Some existence theorems on differential equations of generalized order through a fixed-point theorem. J. Fract. Calc. **9**, 45–49 (1996)

88. Hairer, E., Nørsett, S.P., Wanner, G.: Solving Ordinary Differential Equations I: Nonstiff Problems, 2nd edn. Springer, Berlin (1993)

89. Hairer, E., Wanner, G.: Solving Ordinary Differential Equations II: Stiff and Differential-Algebraic Problems. Springer, Berlin (1991)

90. Haubold, H.J., Mathai, A.M., Saxena, R.K.: Mittag-Leffler functions and their applications. arXiv:0909.0230v2 [math.CA] (2009)

91. He, J.H.: Approximate solution of non linear differential equations with convolution product nonlinearities. Comput. Methods Appl. Mech. Eng. **167**, 69–73 (1998)

92. He, J.H.: Variational iteration method – some recent results and new interpretations. J. Comput. Appl. Math. **207**, 3–17 (2007)
93. Heaviside, O.: Electromagnetic Theory, vol. II, 2nd reprint. Benn, London (1925)
94. Hilfer, R. (ed.): Applications of Fractional Calculus in Physics. World Scientific, Singapore (2000)
95. Hille, E.: Lectures on Ordinary Differential Equations. Addison-Wesley, Reading (1969)
96. Hille, E.: Ordinary Differential Equations in the Complex Domain. Dover, New York (1997)
97. Jacob, N., Krägeloh, A.M.: The Caputo fractional derivative, Feller semigroups, and the fractional power of the first order derivative on $C_\infty(\mathbb{R}_0^+)$. Fract. Calc. Appl. Anal. **5**, 395–410 (2002)
98. Johnson, W.P.: The curious history of Faà di Bruno's formula. Am. Math. Mon. **109**, 217–234 (2002)
99. Joulin, G.: Point-source initiation of lean spherical flames of light reactants: an asymptotic theory. Combust. Sci. Technol. **43**, 99–113 (1985)
100. Kilbas, A.A., Srivastava, H.M., Trujillo, J.J.: Theory and Applications of Fractional Differential Equations. Elsevier, Amsterdam (2006)
101. Kilbas, A.A., Trujillo, J.J.: Differential equations of fractional order. Methods, results and problems. I. Appl. Anal. **78**, 153–192 (2001)
102. Kilbas, A.A., Trujillo, J.J.: Differential equations of fractional order. Methods, results and problems. II. Appl. Anal. **81**, 435–493 (2002)
103. Kiryakova, V.: The multi-index Mittag-Leffler functions as an important class of special functions of fractional calculus. Comput. Math. Appl. **59**, 1885–1895 (2010)
104. Kiryakova, V.: The special functions of fractional calculus as generalized fractional calculus operators of some basic functions. Comput. Math. Appl. **59**, 1128–1141 (2010)
105. Klages, R., Radons, G., Sokolov, I.M. (eds.): Anomalous Transport: Foundations and Applications. Wiley-VCH, Weinheim (2008)
106. Kochubei, A.N.: Fractional differential equations: α-entire solutions, regular and irregular singularities. Fract. Calc. Appl. Anal. **12**, 135–158 (2009)
107. Koeller, R.C.: Polynomial operators, Stieltjes convolution, and fractional calculus in hereditary mechanics. Acta Mech. **58**, 251–264 (1986)
108. Krägeloh, A.M.: Feller semigroups generated by fractional derivatives and pseudo-differential operators. Ph.D. thesis, Universität Erlangen (2001)
109. Kress, R.: Linear Integral Equations, 2nd edn. Springer, New York (1999)
110. Lederman, C., Roquejoffre, J.-M., Wolanski, N.: Mathematical justification of a nonlinear integro-differential equation for the propagation of spherical flames. C. R. Math. Acad. Sci. Paris **334**, 569–574 (2002)
111. Le Mehauté, A., Tenreiro Machado, J.A., Trigeassou, J.C., Sabatier, J. (eds.): Fractional Differentiation and Its Applications. Ubooks, Neusäß (2005)
112. Letnikov, A.V.: Theory of differentiation with an arbitrary index. Mat. Sb. **3**, 1–66 (1868) (in Russian)
113. Liang, S., Jeffrey, D.J.: Comparison of homotopy analysis method and homotopy perturbation method through an evolution equation. Commun. Nonlinear Sci. Numer. Simulat. **14**, 4057–4064 (2009)
114. Liao, S.J.: Beyond Perturbation: Introduction to the Homotopy Analysis Method. Chapman & Hall/CRC Press, Boca Raton (2003)
115. Linz, P.: Analytical and Numerical Methods for Volterra Equations. SIAM, Philadelphia (1985)
116. Liouville J.: Memoire sur quelques questions de géometrie et de mécanique, et sur un nouveau gentre pour resoudre ces questions. J. Ecole Polytech. **13**, 1–69 (1832)
117. Lorentz, G.G.: Bernstein Polynomials, 2nd edn. Chelsea Publ. Comp., New York (1986)
118. Lu, J.-F., Hanyga, A.: Wave field simulation for heterogeneous porous media with singular memory drag force. J. Comput. Phys. **208**, 651–674 (2005)
119. Lu, J.G.: Chaotic dynamics of the fractional-order Lü system and its synchronization. Phys. Lett. A **354**, 305–311 (2006)

120. Lubich, C.: Runge-Kutta theory for Volterra and Abel integral equations of the second kind. Math. Comput. **41**, 87–102 (1983)

121. Lubich, C.: Discretized fractional calculus. SIAM J. Math. Anal. **17**, 704–719 (1986)

122. Lubinsky, D.S.: A survey of weighted polynomial approximation with exponential weights. Surv. Approx. Theory **3**, 1–105 (2007)

123. Luchko, Y., Gorenflo, R.: An operational method for solving fractional differential equations with the Caputo derivatives, Acta Math. Vietnamica **24**, 207–233 (1999)

124. Magin, R.: Fractional Calculus in Bioengineering. Begell House, Redding (2006)

125. Mainardi, F.: Fractional calculus: some basic problems in continuum and statistical mechanics. In: Carpinteri, A., Mainardi, F. (eds.) Fractals and Fractional Calculus in Continuum Mechanics, pp. 291–348. Springer, Wien (1997)

126. Mainardi, F.: Fractional Calculus and Waves in Linear Viscoelasticity. Imperial College Press, London (2010)

127. Mainardi, F., Gorenflo, R.: On Mittag-Leffler-type functions in fractional evolution processes. J. Comput. Appl. Math. **118**, 283–299 (2000)

128. Mainardi, F., Raberto, M., Gorenflo, R., Scalas, E.: Fractional calculus and continuous-time finance II: the waiting-time distribution. Physica A **287**, 468–481 (2000)

129. Marchaud, A.: Sur les dérivées et sur les différences des fonctions de variables réelles. J. Math. **6**, 337–425 (1927)

130. Marks, R.J. II, Hall, M.W.: Differintegral interpolation from a bandlimited signal's samples. IEEE Trans. Acoust. Speech Signal Process. **29**, 872–877 (1981)

131. Mathai, A.M., Haubold, H.J.: Special Functions for Applied Scientists. Springer, New York (2008)

132. Matignon, D.: Représentations en variables d'état de modèles de guides d'ondes avec dérivation fractionnaire. Ph.D. thesis, Université Paris XI (1994)

133. Matignon, D., Montseny, G. (eds.): Fractional Differential Systems: Models, Methods, and Applications. SMAI, Paris (1998)

134. Metzler, R., Schick, W., Kilian, H.-G., Nonnenmacher, T.F.: Relaxation in filled polymers: a fractional calculus approach. J. Chem. Phys. **103**, 7180–7186 (1995)

135. Miller, K.S., Ross, B.: An Introduction to the Fractional Calculus and Fractional Differential Equations. Wiley, New York (1993)

136. Mittag-Leffler, G.M.: Une généralisation de l'intégrale de Laplace-Abel. C. R. Acad. Sci. Paris (Ser. II) **136**, 537–539 (1903)

137. Mittag-Leffler, G.M.: Sur la nouvelle fonction $E_\alpha(x)$. C. R. Acad. Sci. Paris (Ser. II) **137**, 554–558 (1903)

138. Momani, S.M.: Local and global uniqueness theorems on differential equations of noninteger order via Bihari's and Gronwall's inequalities. Rev. Téc. Fac. Ing. Univ. Zulia **23**, 66–69 (2000)

139. Momani, S., Odibat, Z.: Homotopy perturbation method for nonlinear partial differential equations of fractional order. Phys. Lett. A **365**, 345–350 (2007)

140. Nkamnang, A.R.: Diskretisierung von mehrgliedrigen Abelschen Integralgleichungen und gewöhnlichen Differentialgleichungen gebrochener Ordnung. Ph.D. thesis, Freie Universität Berlin (1999)

141. Nonnenmacher, T.F., Metzler, R.: On the Riemann–Liouville fractional calculus and some recent applications. Fractals **3**, 557–566 (1995)

142. Nutting, P.G.: A new general law of deformation. J. Franklin Inst. **191**, 679–685 (1921)

143. Nutting, P.G.: A general stress–strain–time formula. J. Franklin Inst. **235**, 513–524 (1943)

144. Odibat, Z.M.: Analytic study on linear systems of fractional differential equations. Comput. Math. Appl. **59**, 1171–1183 (2010)

145. Odibat, Z., Momani, S., Erturk, V.S.: Generalized differential transform method: application to differential equations of fractional order. Appl. Math. Comput. **197**, 467–477 (2008)

146. Oldham, K.B., Spanier, J.: The Fractional Calculus. Academic, New York (1974)

147. Olmstead, W.E., Handelsman, R.A.: Diffusion in a semi-infinite region with nonlinear surface dissipation. SIAM Rev. **18**, 275–291 (1976)

148. Perron, O.: Über Integration von gewöhnlichen Differentialgleichungen durch Reihen. Sitz.-Ber. Heidelberger Akademie der Wissenschaften A Math.-Nat. Klasse Abhandlung 2 (1919)
149. Perron, O.: Über Integration von gewöhnlichen Differentialgleichungen durch Reihen, II. Sitz.-Ber. Heidelberger Akademie der Wissenschaften A Math.-Nat. Klasse Abhandlung 8 (1919)
150. Perron, O.: Über Integration von gewöhnlichen Differentialgleichungen durch Reihen, III. Sitz.-Ber. Heidelberger Akademie der Wissenschaften A Math.-Nat. Klasse Abhandlung 12 (1919)
151. Perron, O.: Über Integration partieller Differentialgleichungen durch Reihen. Sitz.-Ber. Heidelberger Akademie der Wissenschaften A Math.-Nat. Klasse Abhandlung 9 (1920)
152. Podlubny, I.: Fractional-order systems and fractional-order controllers. Technical report UEF-03-94, Institute for Experimental Physics, Slovak Acad. Sci. (1994)
153. Podlubny, I.: Fractional Differential Equations. Academic Press, San Diego (1999)
154. Podlubny, I.: Geometric and physical interpretation of fractional integration and fractional differentiation. Fract. Calc. Appl. Anal. 5, 367–386 (2002)
155. Podlubny, I., Dorcak, L., Misanek, J.: Application of fractional-order derivatives to calculation of heat load intensity change in blast furnace walls. Trans. Tech. Univ. Košice 5, 137–144 (1995)
156. Popović, J.K., Atanacković, M.T., Pilipović, A.S., Rapaić, M.R., Pilipović, S., Atanacković, T.M.: A new approach to the compartmental analysis in pharmacokinetics: fractional time evolution of Diclofenac. J. Pharmacokinet. Pharmacodyn. 37, 119–134 (2010)
157. Rabotnov, Yu.N.: Polzuchest Elementov Konstruktsii. Nauka, Moscow (1966); English translation: Creep Problems in Structural Members. North-Holland, Amsterdam (1969)
158. Rasof, B.: The initial- and final-value theorems in Laplace transform theory. J. Franklin Inst. 274, 165–177 (1962)
159. Répaci, A.: Nonlinear dynamical systems: on the accuracy of Adomian's decomposition method. Appl. Math. Lett. 3(4), 35–39 (1990)
160. Riesz, F., Sz.-Nagy, B.: Vorlesungen über Funktionalanalysis. Deutscher Verlag der Wissenschaften, Berlin (1956)
161. Roquejoffre, J.-M., Vázquez, J.-L.: Ignition and propagation in an integro-differential model for spherical flames. Discrete Contin. Dyn. Syst. Ser. B 2, 379–387 (2002)
162. Ross, B.: The development of fractional calculus 1695–1900. Hist. Math. 4, 75–89 (1977)
163. Rossikhin, Yu.A.: Reflections on two parallel ways in the progress of fractional calculus in mechanics of solids. Appl. Mech. Rev. 63, 010701 (2010)
164. Rossikhin, Yu.A., Shitikova, M.V.: Comparative analysis of viscoelastic models involving fractional derivatives of different orders. Fract. Calc. Appl. Anal. 10, 111–121 (2007)
165. Rudin, W.: Principles of Mathematical Analysis. McGraw-Hill, New York (1953)
166. Sabatier, J., Agrawal, O.P., Tenreiro Machado, J.A. (eds.): Advances in Fractional Calculus. Springer, Dordrecht (2007)
167. Samko, S.G., Kilbas, A.A., Marichev, O.I.: Fractional Integrals and Derivatives: Theory and Applications. Gordon and Breach, Yverdon (1993)
168. Sard, A.: Integral representation of remainders. Duke Math. J. 15, 333–345 (1948)
169. Scalas, E., Gorenflo, R., Mainardi, F.: Fractional calculus and continuous-time finance. Physica A 284, 376–384 (2000)
170. Scalas, E., Gorenflo, R., Mainardi, F.: Uncoupled continuous-time random walks: analytic solution and limiting behaviour of the master equation. Phys. Rev. E 69, 011107 (2004)
171. Schmidt, A., Gaul, L.: On a critique of a numerical scheme for the calculation of fractionally damped dynamical systems. Mech. Res. Commun. 33, 99–107 (2006)
172. Scott Blair, G.W., Reiner, M.: The rheological law underlying the Nutting equation. Appl. Sci. Res. 2, 225–234 (1951)
173. Sedletskij, A.M.: Asymptotic formulae for zeros of a function of Mittag-Leffler's type. Anal. Math. 20, 117–132 (1994) (in Russian)
174. Seybold, H.J., Hilfer, R.: Numerical results for the generalized Mittag-Leffler functions. Fract. Calc. Appl. Anal. 8, 127–139 (2005)

175. Seybold, H.J., Hilfer, R.: Numerical algorithm for calculating the generalized Mittag-Leffler function. SIAM J. Numer. Anal. **47**, 69–88 (2008)
176. Shaw, S., Warby, M.K., Whiteman, J.R.: A comparison of hereditary integral and internal variable approaches to numerical linear solid elasticity. In: Proceedings of the XIII Polish Conference on Computer Methods in Mechanics, Poznan (1997)
177. Singh, S.J., Chatterjee, A.: Galerkin projections and finite elements for fractional order derivatives. Nonlinear Dynam. **45**, 183–206 (2006)
178. Song, L., Xu, S.Y., Yang, J.Y.: Dynamical models of happiness with fractional order. Commun. Nonlinear Sci. Numer. Simulat. **15**, 616–628 (2010)
179. Strehmel, K., Weiner, R.: Numerik gewöhnlicher Differentialgleichungen. Teubner, Stuttgart (1995)
180. Taş, K., Tenreiro Machado, J.A., Baleanu, D. (eds.): Mathematical Methods in Engineering. Springer, Dordrecht (2007)
181. Tatari, M., Dehghan, M.: On the convergence of He's variational iteration method. J. Comput. Appl. Math. **207**, 121–128 (2007)
182. Tavazoei, M.S.: Comments on stability analysis of a class of nonlinear fractional-order systems. IEEE Trans. Circ. Syst. II **56**, 519–520 (2009)
183. Tavazoei, M.S., Haeri, M., Bolouki, S., Siami, M.: Stability preservation analysis for frequency-based methods in numerical simulation of fractional-order systems. SIAM J. Numer. Anal. **47**, 321–328 (2008)
184. Torvik, P.J., Bagley, R.L.: On the appearance of the fractional derivative in the behavior of real materials. J. Appl. Mech. **51**, 294–298 (1984)
185. Trinks, C., Ruge, P.: Treatment of dynamic systems with fractional derivatives without evaluating memory-integrals. Comput. Mech. **29**, 471–476 (2002)
186. Uchaikin, V.V.: Method of Fractional Derivatives. Artishok Publ. House, Ul'yanovsk (2008) (in Russian)
187. Verotta, D.: Fractional compartmental models and multi-term Mittag-Leffler response functions. J. Pharmacokinet. Pharmacodyn. **37**, 209–215 (2010)
188. Walz, G.: Asymptotics and Extrapolation. Akademie, Berlin (1996)
189. Weissinger, J.: Zur Theorie und Anwendung des Iterationsverfahrens. Math. Nachr. **8**, 193–212 (1952)
190. Williamson, J.H.: Lebesgue Integration. Holt, Rinehart and Winston, New York (1962)
191. Wiman, A.: Über den Fundamentalsatz in der Teorie der Funktionen $E_\alpha(x)$. Acta Math. **29**, 191–201 (1905)
192. Wiman, A.: Über die Nullstellen der Funktionen $E_\alpha(x)$. Acta Math. **29**, 217–234 (1905)
193. Woon, S.C.: Analytic continuation of operators. Applications: from number theory and group theory to quantum field and string theories. Rev. Math. Phys. **11**, 463–501 (1999)
194. Yuan, L., Agrawal, O.P.: A numerical scheme for dynamic systems containing fractional derivatives. J. Vib. Acoust. **124**, 321–324 (2002)

Index

A

absolute continuity, 10
Adams–Bashforth method
    classical, 197
    fractional, 200
Adams–Bashforth–Moulton method
    classical, 197
    fractional, 200
Adams–Moulton method
    classical, 197
    fractional, 199
Arzelà–Ascoli theorem, 230
Ascoli, *see* Arzelà–Ascoli theorem
asymptotic stability of solutions, 157

B

Bagley–Torvik equation, 167
Banach's fixed point theorem, 229
Bashforth, *see* Adams–Bashforth method, *see*
    Adams–Bashforth–Moulton method
Basset equation, 167
Bernstein polynomial, 44, 234
Beta function, 229
binomial coefficient, 31
boundary value problem, 127

C

Caputo fractional derivative, 50
ceiling function, 189
chain rule
    for Caputo operators, 59
    for Riemann–Liouville derivatives, 35
commensuracy, 168
compactness, relative, 230
completely monotonic, 139
continuity, absolute, 10
convolution, 231
    theorem, 231

D

derivative
    Caputo fractional, 50
    Grünwald–Letnikov fractional, 43
    Riemann–Liouville fractional, 27
differentiation theorem, 231
direct method, 205
distributed order equation, 185

E

equation
    Bagley–Torvik, 167
    Basset, 167
    distributed order, 185
    functional – of the Gamma function, 227
    oscillation, 138
        fractional, 140
    relaxation, 138
        fractional, 138
    resolvent, 147
Euler's
    Beta function, 229
    Gamma function, 9, 227
    integral of the first kind, *see* Euler's Beta
        function
    integral of the second kind, *see* Euler's
        Gamma function
expansion, Taylor, 41, 54

F

Faà di Bruno's formula, 34
    for Caputo operators, 59
    for Riemann–Liouville operators, 35
final value theorem, 232
finite-part integral, 233
fixed point theorem
    Banach's, 229
    Schauder's, 230
    Weissinger's, 229

floor function, 189
formula
    Faà di Bruno's, 34
        for Caputo operators, 59
        for Riemann–Liouville operators, 35
    Gauss' product – for the Gamma function,
        228
    Leibniz', 32
        for Riemann–Liouville operators, 33
    reflection – for the Gamma function, 228
    Stirling's, 229
fractional
    derivative
        Caputo, 50
        Grünwald–Letnikov, 43
        Riemann–Liouville, 27
    integral
        Grünwald–Letnikov, 46
        Liouville, 29
        Riemann–Liouville, 13
function
    absolutely continuous, 10
    Beta, 229
    ceiling, 189
    floor, 189
    Gamma, 9, 227
    hypergeometric
        Gauss', 190
        Kummer's confluent, 189
    Mittag-Leffler, 67, 189
        multi-index, 69
        two-parameter, 67, 189
functional equation of the Gamma function,
    227
fundamental theorem of calculus,
    7, 10

G
Gamma function, 9, 227
Gauss'
    hypergeometric function, 190
    product formula for the Gamma function,
        228
Gel'fond-Leont'ev operator, 55
Grünwald–Letnikov
    fractional derivative, 43
    fractional integral, 46
Gronwall inequality, 111, 172, 173

H
Hölder space, 9
Hadamard, see finite-part integral

Hooke's law, 11
hypergeometric function
    Gauss', 190
    Kummer's confluent, 189

I
indirect method, 195
inequality, Gronwall, 111, 172, 173
initial value problem
    for Caputo equations, 85
    for Riemann–Liouville equations, 77
integral
    Euler's
        of the first kind, see Euler's Beta
            function
        of the second kind, see Euler's Gamma
            function
    finite-part, 233
    Grünwald–Letnikov fractional, 46
    Liouville fractional, 29
    Riemann–Liouville fractional, 13
integration theorem, 231
inverse Laplace transform, 232
iterated kernel, 144
iteration, Picard, 82, 94

K
kernel
    iterated, 144
    resolvent, 144
Kummer's confluent hypergeometric function,
    189

L
Laplace transform, 133, 230
    inverse, 232
Lebesgue space, 9
Leibniz' formula
    classical, 32
    for Caputo operators, 59
    for Riemann–Liouville operators, 33
Leont'ev, see Gel'fond-Leont'ev operator
Letnikov, see Grünwald–Letnikov fractional
    derivative
linearity, 32, 58
Liouville, see Riemann–Liouville
    fractional integral, 29
Lipschitz space, see Hölder space
local operator, 37, 52

M
memory, 87
method
  direct, 205
  indirect, 195
Mittag-Leffler function, 67, 189
  multi-index, 69
  two-parameter, 67, 189
modulus of continuity, 25
Moulton, *see* Adams–Bashforth–Moulton
      method, *see* Adams–Moulton
      method
multi-index Mittag-Leffler function, 69
multi-order fractional differential system, 176
multi-term fractional differential equation, 167

N
Newton's law, 10
Nutting's law, 12

O
oscillation equation, 138
  fractional, 140

P
PECE method, 197
Picard iteration, 82, 94
polynomial, Bernstein, 44, 234
problem
  boundary value, 127
  initial value
      for Caputo equations, 85
      for Riemann–Liouville equations, 77
  terminal value, 108
product rule, Leibniz', *see* Leibniz' formula

R
reflection formula for the Gamma function,
      228
relatively compact, 230
relaxation equation, 138
  fractional, 138
resolvent equation, 147

resolvent kernel, 144
Riemann–Liouville
  fractional derivative, 27
  fractional integral, 13

S
Schauder's fixed point theorem, 230
semigroup, 14, 30, 56
space
  Hölder, 9
  Lebesgue, 9
  Lipschitz, *see* Hölder space
stability
  of solutions, 157
      asymptotic, 157
Stirling's formula, 229

T
Taylor expansion, 41, 54
terminal value problem, 108
theorem
  Arzelà–Ascoli, 230
  Banach's fixed point, 229
  convolution, 231
  differentiation, 231
  final value, 232
  fundamental – of calculus, 7, 10
  integration, 231
  Schauder's fixed point, 230
  Weissinger's fixed point, 229
Torvik, *see* Bagley–Torvik equation
transform, Laplace, 133, 230
  inverse, 232

U
ultraslow, 139

V
variation of constants, 136

W
Weissinger's fixed point theorem, 229

Lecture Notes in Mathematics

For information about earlier volumes
please contact your bookseller or Springer
LNM Online archive: springerlink.com

Vol. 1817: E. Koelink, W. Van Assche (Eds.), Orthogonal Polynomials and Special Functions. Leuven 2002 (2003)

Vol. 1818: M. Bildhauer, Convex Variational Problems with Linear, nearly Linear and/or Anisotropic Growth Conditions (2003)

Vol. 1819: D. Masser, Yu. V. Nesterenko, H. P. Schlickewei, W. M. Schmidt, M. Waldschmidt, Diophantine Approximation. Cetraro, Italy 2000. Editors: F. Amoroso, U. Zannier (2003)

Vol. 1820: F. Hiai, H. Kosaki, Means of Hilbert Space Operators (2003)

Vol. 1821: S. Teufel, Adiabatic Perturbation Theory in Quantum Dynamics (2003)

Vol. 1822: S.-N. Chow, R. Conti, R. Johnson, J. Mallet-Paret, R. Nussbaum, Dynamical Systems. Cetraro, Italy 2000. Editors: J. W. Macki, P. Zecca (2003)

Vol. 1823: A. M. Anile, W. Allegretto, C. Ringhofer, Mathematical Problems in Semiconductor Physics. Cetraro, Italy 1998. Editor: A. M. Anile (2003)

Vol. 1824: J. A. Navarro González, J. B. Sancho de Salas, $\mathscr{C}^\infty$ – Differentiable Spaces (2003)

Vol. 1825: J. H. Bramble, A. Cohen, W. Dahmen, Multiscale Problems and Methods in Numerical Simulations, Martina Franca, Italy 2001. Editor: C. Canuto (2003)

Vol. 1826: K. Dohmen, Improved Bonferroni Inequalities via Abstract Tubes. Inequalities and Identities of Inclusion-Exclusion Type. VIII, 113 p, 2003.

Vol. 1827: K. M. Pilgrim, Combinations of Complex Dynamical Systems. IX, 118 p, 2003.

Vol. 1828: D. J. Green, Grbner Bases and the Computation of Group Cohomology. XII, 138 p, 2003.

Vol. 1829: E. Altman, B. Gaujal, A. Hordijk, Discrete-Event Control of Stochastic Networks: Multimodularity and Regularity. XIV, 313 p, 2003.

Vol. 1830: M. I. Gil', Operator Functions and Localization of Spectra. XIV, 256 p, 2003.

Vol. 1831: A. Connes, J. Cuntz, E. Guentner, N. Higson, J. E. Kaminker, Noncommutative Geometry, Martina Franca, Italy 2002. Editors: S. Doplicher, L. Longo (2004)

Vol. 1832: J. Azéma, M. Émery, M. Ledoux, M. Yor (Eds.), Séminaire de Probabilités XXXVII (2003)

Vol. 1833: D.-Q. Jiang, M. Qian, M.-P. Qian, Mathematical Theory of Nonequilibrium Steady States. On the Frontier of Probability and Dynamical Systems. IX, 280 p, 2004.

Vol. 1834: Yo. Yomdin, G. Comte, Tame Geometry with Application in Smooth Analysis. VIII, 186 p, 2004.

Vol. 1835: O.T. Izhboldin, B. Kahn, N.A. Karpenko, A. Vishik, Geometric Methods in the Algebraic Theory of Quadratic Forms. Summer School, Lens, 2000. Editor: J.-P. Tignol (2004)

Vol. 1836: C. Năstăsescu, F. Van Oystaeyen, Methods of Graded Rings. XIII, 304 p, 2004.

Vol. 1837: S. Tavaré, O. Zeitouni, Lectures on Probability Theory and Statistics. Ecole d'Eté de Probabilités de Saint-Flour XXXI-2001. Editor: J. Picard (2004)

Vol. 1838: A.J. Ganesh, N.W. O'Connell, D.J. Wischik, Big Queues. XII, 254 p, 2004.

Vol. 1839: R. Gohm, Noncommutative Stationary Processes. VIII, 170 p, 2004.

Vol. 1840: B. Tsirelson, W. Werner, Lectures on Probability Theory and Statistics. Ecole d'Eté de Probabilités de Saint-Flour XXXII-2002. Editor: J. Picard (2004)

Vol. 1841: W. Reichel, Uniqueness Theorems for Variational Problems by the Method of Transformation Groups (2004)

Vol. 1842: T. Johnsen, A. L. Knutsen, K_3 Projective Models in Scrolls (2004)

Vol. 1843: B. Jefferies, Spectral Properties of Noncommuting Operators (2004)

Vol. 1844: K.F. Siburg, The Principle of Least Action in Geometry and Dynamics (2004)

Vol. 1845: Min Ho Lee, Mixed Automorphic Forms, Torus Bundles, and Jacobi Forms (2004)

Vol. 1846: H. Ammari, H. Kang, Reconstruction of Small Inhomogeneities from Boundary Measurements (2004)

Vol. 1847: T.R. Bielecki, T. Bjrk, M. Jeanblanc, M. Rutkowski, J.A. Scheinkman, W. Xiong, Paris-Princeton Lectures on Mathematical Finance 2003 (2004)

Vol. 1848: M. Abate, J. E. Fornaess, X. Huang, J. P. Rosay, A. Tumanov, Real Methods in Complex and CR Geometry, Martina Franca, Italy 2002. Editors: D. Zaitsev, G. Zampieri (2004)

Vol. 1849: Martin L. Brown, Heegner Modules and Elliptic Curves (2004)

Vol. 1850: V. D. Milman, G. Schechtman (Eds.), Geometric Aspects of Functional Analysis. Israel Seminar 2002-2003 (2004)

Vol. 1851: O. Catoni, Statistical Learning Theory and Stochastic Optimization (2004)

Vol. 1852: A.S. Kechris, B.D. Miller, Topics in Orbit Equivalence (2004)

Vol. 1853: Ch. Favre, M. Jonsson, The Valuative Tree (2004)

Vol. 1854: O. Saeki, Topology of Singular Fibers of Differential Maps (2004)

Vol. 1855: G. Da Prato, P.C. Kunstmann, I. Lasiecka, A. Lunardi, R. Schnaubelt, L. Weis, Functional Analytic Methods for Evolution Equations. Editors: M. Iannelli, R. Nagel, S. Piazzera (2004)

Vol. 1856: K. Back, T.R. Bielecki, C. Hipp, S. Peng, W. Schachermayer, Stochastic Methods in Finance, Bressanone/Brixen, Italy, 2003. Editors: M. Fritelli, W. Runggaldier (2004)

Vol. 1857: M. Émery, M. Ledoux, M. Yor (Eds.), Séminaire de Probabilités XXXVIII (2005)

Vol. 1858: A.S. Cherny, H.-J. Engelbert, Singular Stochastic Differential Equations (2005)

Vol. 1859: E. Letellier, Fourier Transforms of Invariant Functions on Finite Reductive Lie Algebras (2005)

Vol. 1860: A. Borisyuk, G.B. Ermentrout, A. Friedman, D. Terman, Tutorials in Mathematical Biosciences I. Mathematical Neurosciences (2005)

Vol. 1861: G. Benettin, J. Henrard, S. Kuksin, Hamiltonian Dynamics – Theory and Applications, Cetraro, Italy, 1999. Editor: A. Giorgilli (2005)

Vol. 1862: B. Helffer, F. Nier, Hypoelliptic Estimates and Spectral Theory for Fokker-Planck Operators and Witten Laplacians (2005)

Vol. 1863: H. Führ, Abstract Harmonic Analysis of Continuous Wavelet Transforms (2005)

Vol. 1864: K. Efstathiou, Metamorphoses of Hamiltonian Systems with Symmetries (2005)

Vol. 1865: D. Applebaum, B.V. R. Bhat, J. Kustermans, J. M. Lindsay, Quantum Independent Increment Processes I. From Classical Probability to Quantum Stochastic Calculus. Editors: M. Schürmann, U. Franz (2005)

Vol. 1866: O.E. Barndorff-Nielsen, U. Franz, R. Gohm, B. Kümmerer, S. Thorbjønsen, Quantum Independent Increment Processes II. Structure of Quantum Lévy Processes, Classical Probability, and Physics. Editors: M. Schürmann, U. Franz, (2005)

Vol. 1867: J. Sneyd (Ed.), Tutorials in Mathematical Biosciences II. Mathematical Modeling of Calcium Dynamics and Signal Transduction. (2005)

Vol. 1868: J. Jorgenson, S. Lang, $Pos_n(R)$ and Eisenstein Series. (2005)

Vol. 1869: A. Dembo, T. Funaki, Lectures on Probability Theory and Statistics. Ecole d'Eté de Probabilités de Saint-Flour XXXIII-2003. Editor: J. Picard (2005)

Vol. 1870: V.I. Gurariy, W. Lusky, Geometry of Mntz Spaces and Related Questions. (2005)

Vol. 1871: P. Constantin, G. Gallavotti, A.V. Kazhikhov, Y. Meyer, S. Ukai, Mathematical Foundation of Turbulent Viscous Flows, Martina Franca, Italy, 2003. Editors: M. Cannone, T. Miyakawa (2006)

Vol. 1872: A. Friedman (Ed.), Tutorials in Mathematical Biosciences III. Cell Cycle, Proliferation, and Cancer (2006)

Vol. 1873: R. Mansuy, M. Yor, Random Times and Enlargements of Filtrations in a Brownian Setting (2006)

Vol. 1874: M. Yor, M. Émery (Eds.), In Memoriam Paul-Andr Meyer - Sminaire de Probabilits XXXIX (2006)

Vol. 1875: J. Pitman, Combinatorial Stochastic Processes. Ecole d'Et de Probabilits de Saint-Flour XXXII-2002. Editor: J. Picard (2006)

Vol. 1876: H. Herrlich, Axiom of Choice (2006)

Vol. 1877: J. Steuding, Value Distributions of L-Functions (2007)

Vol. 1878: R. Cerf, The Wulff Crystal in Ising and Percolation Models, Ecole d'Et de Probabilités de Saint-Flour XXXIV-2004. Editor: Jean Picard (2006)

Vol. 1879: G. Slade, The Lace Expansion and its Applications, Ecole d'Et de Probabilits de Saint-Flour XXXIV-2004. Editor: Jean Picard (2006)

Vol. 1880: S. Attal, A. Joye, C.-A. Pillet, Open Quantum Systems I, The Hamiltonian Approach (2006)

Vol. 1881: S. Attal, A. Joye, C.-A. Pillet, Open Quantum Systems II, The Markovian Approach (2006)

Vol. 1882: S. Attal, A. Joye, C.-A. Pillet, Open Quantum Systems III, Recent Developments (2006)

Vol. 1883: W. Van Assche, F. Marcellàn (Eds.), Orthogonal Polynomials and Special Functions, Computation and Application (2006)

Vol. 1884: N. Hayashi, E.I. Kaikina, P.I. Naumkin, I.A. Shishmarev, Asymptotics for Dissipative Nonlinear Equations (2006)

Vol. 1885: A. Telcs, The Art of Random Walks (2006)

Vol. 1886: S. Takamura, Splitting Deformations of Degenerations of Complex Curves (2006)

Vol. 1887: K. Habermann, L. Habermann, Introduction to Symplectic Dirac Operators (2006)

Vol. 1888: J. van der Hoeven, Transseries and Real Differential Algebra (2006)

Vol. 1889: G. Osipenko, Dynamical Systems, Graphs, and Algorithms (2006)

Vol. 1890: M. Bunge, J. Funk, Singular Coverings of Toposes (2006)

Vol. 1891: J.B. Friedlander, D.R. Heath-Brown, H. Iwaniec, J. Kaczorowski, Analytic Number Theory, Cetraro, Italy, 2002. Editors: A. Perelli, C. Viola (2006)

Vol. 1892: A. Baddeley, I. Bárány, R. Schneider, W. Weil, Stochastic Geometry, Martina Franca, Italy, 2004. Editor: W. Weil (2007)

Vol. 1893: H. Hanßmann, Local and Semi-Local Bifurcations in Hamiltonian Dynamical Systems, Results and Examples (2007)

Vol. 1894: C.W. Groetsch, Stable Approximate Evaluation of Unbounded Operators (2007)

Vol. 1895: L. Molnár, Selected Preserver Problems on Algebraic Structures of Linear Operators and on Function Spaces (2007)

Vol. 1896: P. Massart, Concentration Inequalities and Model Selection, Ecole d'Été de Probabilités de Saint-Flour XXXIII-2003. Editor: J. Picard (2007)

Vol. 1897: R. Doney, Fluctuation Theory for Lévy Processes, Ecole d'Été de Probabilités de Saint-Flour XXXV-2005. Editor: J. Picard (2007)

Vol. 1898: H.R. Beyer, Beyond Partial Differential Equations, On linear and Quasi-Linear Abstract Hyperbolic Evolution Equations (2007)

Vol. 1899: Séminaire de Probabilités XL. Editors: C. Donati-Martin, M. Émery, A. Rouault, C. Stricker (2007)

Vol. 1900: E. Bolthausen, A. Bovier (Eds.), Spin Glasses (2007)

Vol. 1901: O. Wittenberg, Intersections de deux quadriques et pinceaux de courbes de genre 1, Intersections of Two Quadrics and Pencils of Curves of Genus 1 (2007)

Vol. 1902: A. Isaev, Lectures on the Automorphism Groups of Kobayashi-Hyperbolic Manifolds (2007)

Vol. 1903: G. Kresin, V. Maz'ya, Sharp Real-Part Theorems (2007)

Vol. 1904: P. Giesl, Construction of Global Lyapunov Functions Using Radial Basis Functions (2007)

Vol. 1905: C. Prévôt, M. Röckner, A Concise Course on Stochastic Partial Differential Equations (2007)

Vol. 1906: T. Schuster, The Method of Approximate Inverse: Theory and Applications (2007)

Vol. 1907: M. Rasmussen, Attractivity and Bifurcation for Nonautonomous Dynamical Systems (2007)

Vol. 1908: T.J. Lyons, M. Caruana, T. Lévy, Differential Equations Driven by Rough Paths, Ecole d'Été de Probabilités de Saint-Flour XXXIV-2004 (2007)

Vol. 1909: H. Akiyoshi, M. Sakuma, M. Wada, Y. Yamashita, Punctured Torus Groups and 2-Bridge Knot Groups (I) (2007)

Vol. 1910: V.D. Milman, G. Schechtman (Eds.), Geometric Aspects of Functional Analysis. Israel Seminar 2004-2005 (2007)

Vol. 1911: A. Bressan, D. Serre, M. Williams, K. Zumbrun, Hyperbolic Systems of Balance Laws. Cetraro, Italy 2003. Editor: P. Marcati (2007)

Vol. 1912: V. Berinde, Iterative Approximation of Fixed Points (2007)

Vol. 1913: J.E. Marsden, G. Misiołek, J.-P. Ortega, M. Perlmutter, T.S. Ratiu, Hamiltonian Reduction by Stages (2007)

Vol. 1914: G. Kutyniok, Affine Density in Wavelet Analysis (2007)

Vol. 1915: T. Bıyıkoğlu, J. Leydold, P.F. Stadler, Laplacian Eigenvectors of Graphs. Perron-Frobenius and Faber-Krahn Type Theorems (2007)

Vol. 1916: C. Villani, F. Rezakhanlou, Entropy Methods for the Boltzmann Equation. Editors: F. Golse, S. Olla (2008)

Vol. 1917: I. Veselić, Existence and Regularity Properties of the Integrated Density of States of Random Schrdinger (2008)

Vol. 1918: B. Roberts, R. Schmidt, Local Newforms for GSp(4) (2007)

Vol. 1919: R.A. Carmona, I. Ekeland, A. Kohatsu-Higa, J.-M. Lasry, P.-L. Lions, H. Pham, E. Taflin, Paris-Princeton Lectures on Mathematical Finance 2004. Editors: R.A. Carmona, E. inlar, I. Ekeland, E. Jouini, J.A. Scheinkman, N. Touzi (2007)

Vol. 1920: S.N. Evans, Probability and Real Trees. Ecole d'Été de Probabilités de Saint-Flour XXXV-2005 (2008)

Vol. 1921: J.P. Tian, Evolution Algebras and their Applications (2008)

Vol. 1922: A. Friedman (Ed.), Tutorials in Mathematical BioSciences IV. Evolution and Ecology (2008)

Vol. 1923: J.P.N. Bishwal, Parameter Estimation in Stochastic Differential Equations (2008)

Vol. 1924: M. Wilson, Littlewood-Paley Theory and Exponential-Square Integrability (2008)

Vol. 1925: M. du Sautoy, L. Woodward, Zeta Functions of Groups and Rings (2008)

Vol. 1926: L. Barreira, V. Claudia, Stability of Nonautonomous Differential Equations (2008)

Vol. 1927: L. Ambrosio, L. Caffarelli, M.G. Crandall, L.C. Evans, N. Fusco, Calculus of Variations and Non-Linear Partial Differential Equations. Cetraro, Italy 2005. Editors: B. Dacorogna, P. Marcellini (2008)

Vol. 1928: J. Jonsson, Simplicial Complexes of Graphs (2008)

Vol. 1929: Y. Mishura, Stochastic Calculus for Fractional Brownian Motion and Related Processes (2008)

Vol. 1930: J.M. Urbano, The Method of Intrinsic Scaling. A Systematic Approach to Regularity for Degenerate and Singular PDEs (2008)

Vol. 1931: M. Cowling, E. Frenkel, M. Kashiwara, A. Valette, D.A. Vogan, Jr., N.R. Wallach, Representation Theory and Complex Analysis. Venice, Italy 2004. Editors: E.C. Tarabusi, A. D'Agnolo, M. Picardello (2008)

Vol. 1932: A.A. Agrachev, A.S. Morse, E.D. Sontag, H.J. Sussmann, V.I. Utkin, Nonlinear and Optimal Control Theory. Cetraro, Italy 2004. Editors: P. Nistri, G. Stefani (2008)

Vol. 1933: M. Petkovic, Point Estimation of Root Finding Methods (2008)

Vol. 1934: C. Donati-Martin, M. Émery, A. Rouault, C. Stricker (Eds.), Séminaire de Probabilités XLI (2008)

Vol. 1935: A. Unterberger, Alternative Pseudodifferential Analysis (2008)

Vol. 1936: P. Magal, S. Ruan (Eds.), Structured Population Models in Biology and Epidemiology (2008)

Vol. 1937: G. Capriz, P. Giovine, P.M. Mariano (Eds.), Mathematical Models of Granular Matter (2008)

Vol. 1938: D. Auroux, F. Catanese, M. Manetti, P. Seidel, B. Siebert, I. Smith, G. Tian, Symplectic 4-Manifolds and Algebraic Surfaces. Cetraro, Italy 2003. Editors: F. Catanese, G. Tian (2008)

Vol. 1939: D. Boffi, F. Brezzi, L. Demkowicz, R.G. Durán, R.S. Falk, M. Fortin, Mixed Finite Elements, Compatibility Conditions, and Applications. Cetraro, Italy 2006. Editors: D. Boffi, L. Gastaldi (2008)

Vol. 1940: J. Banasiak, V. Capasso, M.A.J. Chaplain, M. Lachowicz, J. Miękisz, Multiscale Problems in the Life Sciences. From Microscopic to Macroscopic. Będlewo, Poland 2006. Editors: V. Capasso, M. Lachowicz (2008)

Vol. 1941: S.M.J. Haran, Arithmetical Investigations. Representation Theory, Orthogonal Polynomials, and Quantum Interpolations (2008)

Vol. 1942: S. Albeverio, F. Flandoli, Y.G. Sinai, SPDE in Hydrodynamic. Recent Progress and Prospects. Cetraro, Italy 2005. Editors: G. Da Prato, M. Rckner (2008)

Vol. 1943: L.L. Bonilla (Ed.), Inverse Problems and Imaging. Martina Franca, Italy 2002 (2008)

Vol. 1944: A. Di Bartolo, G. Falcone, P. Plaumann, K. Strambach, Algebraic Groups and Lie Groups with Few Factors (2008)

Vol. 1945: F. Brauer, P. van den Driessche, J. Wu (Eds.), Mathematical Epidemiology (2008)

Vol. 1946: G. Allaire, A. Arnold, P. Degond, T.Y. Hou, Quantum Transport. Modelling, Analysis and Asymptotics. Cetraro, Italy 2006. Editors: N.B. Abdallah, G. Frosali (2008)

Vol. 1947: D. Abramovich, M. Mariño, M. Thaddeus, R. Vakil, Enumerative Invariants in Algebraic Geometry and String Theory. Cetraro, Italy 2005. Editors: K. Behrend, M. Manetti (2008)

Vol. 1948: F. Cao, J-L. Lisani, J-M. Morel, P. Mus, F. Sur, A Theory of Shape Identification (2008)

Vol. 1949: H.G. Feichtinger, B. Helffer, M.P. Lamoureux, N. Lerner, J. Toft, Pseudo-Differential Operators. Quantization and Signals Cetraro Italy 2006 Editors: L. Rodino, M.W. Wong (2008)

Vol. 1950: M. Bramson, Stability of Queueing Networks, Ecole d'Eté de Probabilits de Saint-Flour XXXVI-2006 (2008)

Vol. 1951: A. Moltó, J. Orihuela, S. Troyanski, M. Valdivia, A Non Linear Transfer Technique for Renorming (2009)

Vol. 1952: R. Mikhailov, I.B.S. Passi, Lower Central and Dimension Series of Groups (2009)

Vol. 1953: K. Arwini, C.T.J. Dodson, Information Geometry (2008)

Vol. 1954: P. Biane, L. Bouten, F. Cipriani, N. Konno, N. Privault, Q. Xu, Quantum Potential Theory. Editors: U. Franz, M. Schuermann (2008)

Vol. 1955: M. Bernot, V. Caselles, J.-M. Morel, Optimal Transportation Networks (2008)

Vol. 1956: C.H. Chu, Matrix Convolution Operators on Groups (2008)

Vol. 1957: A. Guionnet, On Random Matrices: Macroscopic Asymptotics, Ecole d'Eté de Probabilits de Saint-Flour XXXVI-2006 (2009)

Vol. 1958: M.C. Olsson, Compactifying Moduli Spaces for Abelian Varieties (2008)

Vol. 1959: Y. Nakkajima, A. Shiho, Weight Filtrations on Log Crystalline Cohomologies of Families of Open Smooth Varieties (2008)

Vol. 1960: J. Lipman, M. Hashimoto, Foundations of Grothendieck Duality for Diagrams of Schemes (2009)

Vol. 1961: G. Buttazzo, A. Pratelli, S. Solimini, E. Stepanov, Optimal Urban Networks via Mass Transportation (2009)

Vol. 1962: R. Dalang, D. Khoshnevisan, C. Mueller, D. Nualart, Y. Xiao, A Minicourse on Stochastic Partial Differential Equations (2009)

Vol. 1963: W. Siegert, Local Lyapunov Exponents (2009)

Vol. 1964: W. Roth, Operator-valued Measures and Integrals for Cone-valued Functions and Integrals for Cone-valued Functions (2009)

Vol. 1965: C. Chidume, Geometric Properties of Banach Spaces and Nonlinear Iterations (2009)

Vol. 1966: D. Deng, Y. Han, Harmonic Analysis on Spaces of Homogeneous Type (2009)

Vol. 1967: B. Fresse, Modules over Operads and Functors (2009)

Vol. 1968: R. Weissauer, Endoscopy for GSP(4) and the Cohomology of Siegel Modular Threefolds (2009)

Vol. 1969: B. Roynette, M. Yor, Penalising Brownian Paths (2009)

Vol. 1970: M. Biskup, A. Bovier, F. den Hollander, D. Ioffe, F. Martinelli, K. Netočný, F. Toninelli, Methods of Contemporary Mathematical Statistical Physics. Editor: R. Kotecký (2009)

Vol. 1971: L. Saint-Raymond, Hydrodynamic Limits of the Boltzmann Equation (2009)

Vol. 1972: T. Mochizuki, Donaldson Type Invariants for Algebraic Surfaces (2009)

Vol. 1973: M.A. Berger, L.H. Kauffmann, B. Khesin, H.K. Moffatt, R.L. Ricca, De W. Sumners, Lectures on Topological Fluid Mechanics. Cetraro, Italy 2001. Editor: R.L. Ricca (2009)

Vol. 1974: F. den Hollander, Random Polymers: École d'Été de Probabilités de Saint-Flour XXXVII – 2007 (2009)

Vol. 1975: J.C. Rohde, Cyclic Coverings, Calabi-Yau Manifolds and Complex Multiplication (2009)

Vol. 1976: N. Ginoux, The Dirac Spectrum (2009)

Vol. 1977: M.J. Gursky, E. Lanconelli, A. Malchiodi, G. Tarantello, X.-J. Wang, P.C. Yang, Geometric Analysis and PDEs. Cetraro, Italy 2001. Editors: A. Ambrosetti, S.-Y.A. Chang, A. Malchiodi (2009)

Vol. 1978: M. Qian, J.-S. Xie, S. Zhu, Smooth Ergodic Theory for Endomorphisms (2009)

Vol. 1979: C. Donati-Martin, M. Émery, A. Rouault, C. Stricker (Eds.), Séminaire de Probablitiés XLII (2009)

Vol. 1980: P. Graczyk, A. Stos (Eds.), Potential Analysis of Stable Processes and its Extensions (2009)

Vol. 1981: M. Chlouveraki, Blocks and Families for Cyclotomic Hecke Algebras (2009)

Vol. 1982: N. Privault, Stochastic Analysis in Discrete and Continuous Settings. With Normal Martingales (2009)

Vol. 1983: H. Ammari (Ed.), Mathematical Modeling in Biomedical Imaging I. Electrical and Ultrasound Tomographies, Anomaly Detection, and Brain Imaging (2009)

Vol. 1984: V. Caselles, P. Monasse, Geometric Description of Images as Topographic Maps (2010)

Vol. 1985: T. Linß, Layer-Adapted Meshes for Reaction-Convection-Diffusion Problems (2010)

Vol. 1986: J.-P. Antoine, C. Trapani, Partial Inner Product Spaces. Theory and Applications (2009)

Vol. 1987: J.-P. Brasselet, J. Seade, T. Suwa, Vector Fields on Singular Varieties (2010)

Vol. 1988: M. Broué, Introduction to Complex Reflection Groups and Their Braid Groups (2010)

Vol. 1989: I.M. Bomze, V. Demyanov, Nonlinear Optimization. Cetraro, Italy 2007. Editors: G. di Pillo, F. Schoen (2010)

Vol. 1990: S. Bouc, Biset Functors for Finite Groups (2010)

Vol. 1991: F. Gazzola, H.-C. Grunau, G. Sweers, Polyharmonic Boundary Value Problems (2010)

Vol. 1992: A. Parmeggiani, Spectral Theory of Non-Commutative Harmonic Oscillators: An Introduction (2010)

Vol. 1993: P. Dodos, Banach Spaces and Descriptive Set Theory: Selected Topics (2010)

Vol. 1994: A. Baricz, Generalized Bessel Functions of the First Kind (2010)

Vol. 1995: A.Y. Khapalov, Controllability of Partial Differential Equations Governed by Multiplicative Controls (2010)

Vol. 1996: T. Lorenz, Mutational Analysis. A Joint Framework for Cauchy Problems *In* and *Beyond* Vector Spaces (2010)

Vol. 1997: M. Banagl, Intersection Spaces, Spatial Homology Truncation, and String Theory (2010)

Vol. 1998: M. Abate, E. Bedford, M. Brunella, T.-C. Dinh, D. Schleicher, N. Sibony, Holomorphic Dynamical Systems. Cetraro, Italy 2008. Editors: G. Gentili, J. Guenot, G. Patrizio (2010)

Vol. 1999: H. Schoutens, The Use of Ultraproducts in Commutative Algebra (2010)

Vol. 2000: H. Yserentant, Regularity and Approximability of Electronic Wave Functions (2010)

Vol. 2001: T. Duquesne, O.E. Barndorff-Nielson, O. Reichmann, J. Bertoin, K.-I. Sato, J. Jacod, C. Schwab, C. Klüppelberg, Lévy Matters I (2010)

Vol. 2002: C. Pötzsche, Geometric Theory of Discrete Nonautonomous Dynamical Systems (2010)

Vol. 2003: A. Cousin, S. Crépey, O. Guéant, D. Hobson, M. Jeanblanc, J.-M. Lasry, J.-P. Laurent, P.-L. Lions, P. Tankov, Paris-Princeton Lectures on Mathematical Finance 2010. Editors: R.A. Carmona, E. Cinlar, I. Ekeland, E. Jouini, J.A. Scheinkman, N. Touzi (2010)

Vol. 2004: K. Diethelm, The Analysis of Fractional Differential Equations (2010)

Recent Reprints and New Editions

Vol. 1702: J. Ma, J. Yong, Forward-Backward Stochastic Differential Equations and their Applications. 1999 – Corr. 3rd printing (2007)

Vol. 830: J.A. Green, Polynomial Representations of GL_n, with an Appendix on Schensted Correspondence and Littelmann Paths by K. Erdmann, J.A. Green and M. Schoker 1980 – 2nd corr. and augmented edition (2007)

Vol. 1693: S. Simons, From Hahn-Banach to Monotonicity (Minimax and Monotonicity 1998) – 2nd exp. edition (2008)

Vol. 470: R.E. Bowen, Equilibrium States and the Ergodic Theory of Anosov Diffeomorphisms. With a preface by D. Ruelle. Edited by J.-R. Chazottes. 1975 – 2nd rev. edition (2008)

Vol. 523: S.A. Albeverio, R.J. Høegh-Krohn, S. Mazzucchi, Mathematical Theory of Feynman Path Integral. 1976 – 2nd corr. and enlarged edition (2008)

Vol. 1764: A. Cannas da Silva, Lectures on Symplectic Geometry 2001 – Corr. 2nd printing (2008)

LECTURE NOTES IN MATHEMATICS   🐎 Springer

Edited by J.-M. Morel, F. Takens, B. Teissier, P.K. Maini

Editorial Policy (for the publication of monographs)

1. Lecture Notes aim to report new developments in all areas of mathematics and their applications - quickly, informally and at a high level. Mathematical texts analysing new developments in modelling and numerical simulation are welcome.

   Monograph manuscripts should be reasonably self-contained and rounded off. Thus they may, and often will, present not only results of the author but also related work by other people. They may be based on specialised lecture courses. Furthermore, the manuscripts should provide sufficient motivation, examples and applications. This clearly distinguishes Lecture Notes from journal articles or technical reports which normally are very concise. Articles intended for a journal but too long to be accepted by most journals, usually do not have this "lecture notes" character. For similar reasons it is unusual for doctoral theses to be accepted for the Lecture Notes series, though habilitation theses may be appropriate.

2. Manuscripts should be submitted either to Springer's mathematics editorial in Heidelberg, or to one of the series editors. In general, manuscripts will be sent out to 2 external referees for evaluation. If a decision cannot yet be reached on the basis of the first 2 reports, further referees may be contacted: The author will be informed of this. A final decision to publish can be made only on the basis of the complete manuscript, however a refereeing process leading to a preliminary decision can be based on a pre-final or incomplete manuscript. The strict minimum amount of material that will be considered should include a detailed outline describing the planned contents of each chapter, a bibliography and several sample chapters.

   Authors should be aware that incomplete or insufficiently close to final manuscripts almost always result in longer refereeing times and nevertheless unclear referees' recommendations, making further refereeing of a final draft necessary.

   Authors should also be aware that parallel submission of their manuscript to another publisher while under consideration for LNM will in general lead to immediate rejection.

3. Manuscripts should in general be submitted in English. Final manuscripts should contain at least 100 pages of mathematical text and should always include

   - a table of contents;
   - an informative introduction, with adequate motivation and perhaps some historical remarks: it should be accessible to a reader not intimately familiar with the topic treated;
   - a subject index: as a rule this is genuinely helpful for the reader.

   For evaluation purposes, manuscripts may be submitted in print or electronic form, in the latter case preferably as pdf- or zipped ps-files. Lecture Notes volumes are, as a rule, printed digitally from the authors' files. To ensure best results, authors are asked to use the LaTeX2e style files available from Springer's web-server at:

   ftp://ftp.springer.de/pub/tex/latex/svmonot1/ (for monographs).

Additional technical instructions, if necessary, are available on request from: lnm@springer.com.

4. Careful preparation of the manuscripts will help keep production time short besides ensuring satisfactory appearance of the finished book in print and online. After acceptance of the manuscript authors will be asked to prepare the final LaTeX source files (and also the corresponding dvi-, pdf- or zipped ps-file) together with the final printout made from these files. The LaTeX source files are essential for producing the full-text online version of the book (see www.springerlink.com/content/110312 for the existing online volumes of LNM).

   The actual production of a Lecture Notes volume takes approximately 12 weeks.

5. Authors receive a total of 50 free copies of their volume, but no royalties. They are entitled to a discount of 33.3% on the price of Springer books purchased for their personal use, if ordering directly from Springer.

6. Commitment to publish is made by letter of intent rather than by signing a formal contract. Springer-Verlag secures the copyright for each volume. Authors are free to reuse material contained in their LNM volumes in later publications: a brief written (or e-mail) request for formal permission is sufficient.

Addresses:
Professor J.-M. Morel, CMLA,
École Normale Supérieure de Cachan,
61 Avenue du Président Wilson, 94235 Cachan Cedex, France
E-mail: Jean-Michel.Morel@cmla.ens-cachan.fr

Professor F. Takens, Mathematisch Instituut,
Rijksuniversiteit Groningen, Postbus 800,
9700 AV Groningen, The Netherlands
E-mail: F.Takens@math.rug.nl

Professor B. Teissier, Institut Mathématique de Jussieu,
UMR 7586 du CNRS, Équipe "Géométrie et Dynamique",
175 rue du Chevaleret
75013 Paris, France
E-mail: teissier@math.jussieu.fr

For the "Mathematical Biosciences Subseries" of LNM:

Professor P.K. Maini, Center for Mathematical Biology,
Mathematical Institute, 24-29 St Giles,
Oxford OX1 3LP, UK
E-mail: maini@maths.ox.ac.uk

Springer, Mathematics Editorial I, Tiergartenstr. 17
69121 Heidelberg, Germany,
Tel.: +49 (6221) 487-8259
Fax: +49 (6221) 4876-8259
E-mail: lnm@springer.com